Advances in Experimental Medicine and Biology

Proteomics, Metabolomics, Interactomics and Systems Biology

Series Editor

Daniel Martins-de-Souza
University of Campinas (UNICAMP)
Institute of Biology
Laboratory of Neuroproteomics
Campinas, Brazil

This series of volumes focuses on concepts, techniques and recent advances in the field of proteomics, interactomics, metabolomics and systems biology. Recent advances in various 'omics' technologies enable quantitative monitoring of myriad various biological molecules in a high-throughput manner, and allow determination of their variation between different biological states on a genomic scale. Now that the sequencing of various genomes, from prokaryotes to humans, has provided the list and linear sequence of proteins and RNA that build living organisms, defining the complete set of interactions that sustain life constitutes one of the key challenges of the postgenomic era. This series is intended to cover experimental approaches for defining protein-protein, protein-RNA, protein-DNA and protein-lipid interactions; as well as theoretical approaches dealing with data analysis, integration and modeling and ethical issues.

More information about this series at http://www.springer.com/series/15040

Paul C. Guest
Editor

Reviews on Biomarker Studies in Psychiatric and Neurodegenerative Disorders

Springer

Editor
Paul C. Guest
Department of Biochemistry and Tissue Biology
Institute of Biology
University of Campinas (UNICAMP)
Campinas, Brazil

ISSN 0065-2598 ISSN 2214-8019 (electronic)
Advances in Experimental Medicine and Biology
ISBN 978-3-030-05541-7 ISBN 978-3-030-05542-4 (eBook)
https://doi.org/10.1007/978-3-030-05542-4

Library of Congress Control Number: 2018967961

This Springer imprint is published by the registered company Springer Nature Switzerland AG
The registered company address is: Gewerbestrasse 11, 6330 Cham, Switzerland

Preface

Although hundreds of biomarker candidates for psychiatric and neurodegenerative illnesses have been proposed over the last 20 years, only a handful of these have been approved by the regulatory authorities and put into practice in the clinic. This shortfall is most likely due to inconsistencies across laboratories during the early discovery phases and may be a result of factors, such as technical deviations within and across platforms, insufficient validation of candidates, and a widespread lack of awareness across the scientific and medical communities on the essential criteria and regulatory requirements for incorporating biomarkers into the research and drug development pipelines. Success in this area has also been complicated by the reality that psychiatric and neurodegenerative diseases are not homogeneous in nature and can consist of multiple etiologies and subtypes, rendering their complete characterization extremely difficult.

It is now recognized throughout the scientific and medical communities that robust tests incorporating biomarker readouts are urgently needed to improve diagnosis of patients suffering from conditions such as schizophrenia, depression, bipolar disorder, autism spectrum disorders, and Alzheimer's disease. The availability of such tests have many anticipated benefits stemming from the fact that an accurate and early diagnosis will lead to improved outcomes by helping to place patients on the correct treatment at the earliest possible time. This is critical as pathological effects appear to be increasingly ingrained over time. Biomarker tests can also serve as surrogate markers of response to treatment as well as for the risk of developing unwanted side effects. In a similar manner, biomarker-based tests could also be used as surrogate response agents in the development of new and better drugs for treatment of these complex disorders. This would be a significant breakthrough as only a limited number of drug entities have been developed in this arena over the last 20 years, and most of these come at the price of having significant side effects.

This book includes a series of reviews on general aspects of biomarker use in the study of psychiatric and neurodegenerative diseases and the development of novel medications in these areas. The chapters come from international experts in these fields and arise from five continents, including the countries of Brazil, China, Denmark, France, Germany, Italy, Japan, Poland, Spain, the United Kingdom, and the United States. Each chapter describes the pros and cons of the various approaches and covers the successes and failures in this research field. It is only by a thorough understanding of the shortcomings that progress can be made. The overall goal is to facilitate a better understanding for improved treatment of these disorders by providing a viable mechanism of catching up with other areas of modern medicine, such as diabetes and heart disease. Finally, it is anticipated that the development and application of valid biomarker tests will help the treatment of individuals suffering with these disorders of the mind move into the area of personalized medicine where the right patients can receive the right medication at the right time for the best possible outcome.

Campinas, SP, Brazil Paul C. Guest

Contents

Contributors

Joseph Abraham Cleveland Clinic Lerner College of Medicine of Case Western Reserve University, Cleveland, OH, USA

Antonello Bellomo Psychiatric Unit, Department of Clinical and Experimental Medicine, University of Foggia, Foggia, Italy

Deborah Benevenuto Translational Psychiatry Program, Department of Psychiatry and Behavioral Sciences, McGovern Medical School, The University of Texas Health Science Center at Houston (UTHealth), Houston, TX, USA

Pamela T. Carlson Laboratory of Neuroproteomics, Department of Biochemistry and Tissue Biology, Institute of Biology, University of Campinas (UNICAMP), Campinas, SP, Brazil

Tássia Brena Barroso Carneiro Costa Laboratory of Chemical Biology, Department of Organic Chemistry, Institute of Chemistry, University of Campinas (UNICAMP), Campinas, São Paulo, Brazil

Guilherme Cruz Laboratory of Chemical Biology, Department of Organic Chemistry, Institute of Chemistry, University of Campinas (UNICAMP), Campinas, São Paulo, Brazil

Jeffrey Cummings Cleveland Clinic Lou Ruvo Center for Brain Health, Las Vegas, NV, USA

Francesca D'Urso Psychiatric Unit, Department of Clinical and Experimental Medicine, University of Foggia, Foggia, Italy

Antonio Daniele Institute of Neurology, Catholic University of Sacred Heart, Rome, Italy

Fondazione Policlinico Universitario A. Gemelli IRCCS, Rome, Italy

Valéria de Almeida Laboratory of Neuroproteomics, Department of Biochemistry and Tissue Biology, Institute of Biology, University of Campinas (UNICAMP), Campinas, SP, Brazil

Marie Dekerle Centre Ressource de Réhabilitation Psychosociale et Remédiation Cognitive, Lyon, France

Chengyun Feng Maternal and Child Health Hospital of Baoan, Shenzhen, People's Republic of China

Nicolas Franck Centre Ressource de Réhabilitation Psychosociale et Remédiation Cognitive, UMR 5229 CNRS, Centre Hospitalier Le Vinatier, University of Lyon, Lyon, France

Gabriel R. Fries Translational Psychiatry Program, Department of Psychiatry and Behavioral Sciences, McGovern Medical School, The University of Texas Health Science Center at Houston (UTHealth), Houston, TX, USA

Yan Gao Maternal and Child Health Hospital of Baoan, Shenzhen, People's Republic of China

Laura González-Vallespí Consorci Sanitari del Maresme, Mataró, Spain

Tony Grace Department of Genomics, Central University of Kerala, Kasaragod, Kerala, India

Division of Biology, Kansas State University, Manhattan, KS, USA

Antonio Greco Geriatric Unit, Fondazione IRCCS Casa Sollievo della Sofferenza, Foggia, Italy

Paul C. Guest Laboratory of Neuroproteomics, Department of Biochemistry and Tissue Biology, Institute of Biology, University of Campinas (UNICAMP), Campinas, SP, Brazil

Francesca L. Guest Taunton and Somerset NHS Trust, Musgrove Park Hospital, Taunton, Somerset, UK

Qi Hong Maternal and Child Health Hospital of Baoan, Shenzhen, People's Republic of China

Akihiro Iguchi Laboratory of Neuropathology and Neuroscience, Graduate School of Pharmaceutical Sciences, The University of Tokyo, Tokyo, Japan

Javed Iqbal College of Life Science and Oceanography, Shenzhen University, Shenzhen, People's Republic of China

Keiko Iwata Venetian Institute of Molecular Medicine, Padova, Italy

Research Center for Child Mental Development, University of Fukui, Fukui, Japan

Maddalena La Montagna Psychiatric Unit, Department of Clinical and Experimental Medicine, University of Foggia, Foggia, Italy

Xu Kun Liu College of Life Science and Oceanography, Shenzhen University, Shenzhen, People's Republic of China

Giancarlo Logroscino Neurodegenerative Disease Unit, Department of Basic Medical Sciences, Neuroscience and Sense Organs, University of Bari Aldo Moro, Bari, Italy

Department of Clinical Research in Neurology, University of Bari Aldo Moro, Lecce, Italy

Madia Lozupone Neurodegenerative Disease Unit, Department of Basic Medical Sciences, Neuroscience and Sense Organs, University of Bari Aldo Moro, Bari, Italy

Daniel Martins-de-Souza Laboratory of Neuroproteomics, Department of Biochemistry and Tissue Biology, Institute of Biology, University of Campinas (UNICAMP), Campinas, SP, Brazil

Instituto Nacional de Biomarcadores em Neuropsiquiatria (INBION), Conselho Nacional de Desenvolvimento Científico e Tecnológico (CNPq), São Paulo, Brazil

Agnisrota Mazumder Department of Genomics, Central University of Kerala, Kasaragod, Kerala, India

Marvin R. Natowicz Cleveland Clinic Lerner College of Medicine of Case Western Reserve University, Cleveland, OH, USA

Pathology & Laboratory Medicine, Genomic Medicine, Neurology and Pediatrics Institutes, Cleveland Clinic, Cleveland, OH, USA

Francesco Panza Neurodegenerative Disease Unit, Department of Basic Medical Sciences, Neuroscience and Sense Organs, University of Bari Aldo Moro, Bari, Italy

Geriatric Unit, Fondazione IRCCS Casa Sollievo della Sofferenza, Foggia, Italy

Department of Clinical Research in Neurology, University of Bari Aldo Moro, Lecce, Italy

Dong Ik Park Danish Research Institute of Translational Neuroscience (DANDRITE), Department of Biomedicine, Aarhus University, Aarhus, Denmark

Rafael Penadés Hospital Clínic Barcelona, IDIBAPS, CIBERSAM, University of Barcelona, Barcelona, Spain

Ryszard Pluta Laboratory of Ischemic and Neurodegenerative Brain Research, Mossakowski Medical Research Centre, Polish Academy of Sciences, Warsaw, Poland

João G. M. Pontes Laboratory of Microbial Chemical Biology, Institute of Chemistry, University of Campinas (UNICAMP), Campinas, São Paulo, Brazil

Joby Pulikkan Department of Genomics, Central University of Kerala, Kasaragod, Kerala, India

Joao Quevedo Translational Psychiatry Program, Department of Psychiatry and Behavioral Sciences, McGovern Medical School, The University of Texas Health Science Center at Houston (UTHealth), Houston, TX, USA

Center of Excellence on Mood Disorders, Department of Psychiatry and Behavioral Sciences, McGovern Medical School, The University of Texas Health Science Center at Houston (UTHealth), Houston, TX, USA

Neuroscience Graduate Program, The University of Texas MD Anderson Cancer Center UTHealth Graduate School of Biomedical Sciences, Houston, TX, USA

Translational Psychiatry Laboratory, Graduate Program in Health Sciences, University of Southern Santa Catarina (UNESC), Criciúma, SC, Brazil

Melissa Quintero Laboratory of Chemical Biology, Department of Organic Chemistry, Institute of Chemistry, University of Campinas (UNICAMP), Campinas, São Paulo, Brazil

Renã A. S. Robinson Department of Chemistry, Vanderbilt University, Nashville, TN, USA

Vanderbilt Memory and Alzheimer's Center, Vanderbilt University Medical Center, Nashville, TN, USA

Vanderbilt Institute of Chemical Biology, Vanderbilt University, Nashville, TN, USA

Davide Seripa Geriatric Unit, Fondazione IRCCS Casa Sollievo della Sofferenza, Foggia, Italy

Liming Shen College of Life Science and Oceanography, Shenzhen University, Shenzhen, People's Republic of China

Licia C. Silva-Costa Laboratory of Neuroproteomics, Department of Biochemistry and Tissue Biology, Institute of Biology, University of Campinas (UNICAMP), Campinas, SP, Brazil

Danijela Stanisic Laboratory of Chemical Biology, Department of Organic Chemistry, Institute of Chemistry, University of Campinas (UNICAMP), Campinas, São Paulo, Brazil

Kaitlyn E. Stepler Department of Chemistry, Vanderbilt University, Nashville, TN, USA

Nicholas Szoko Cleveland Clinic Lerner College of Medicine of Case Western Reserve University, Cleveland, OH, USA

Sho Takatori Laboratory of Neuropathology and Neuroscience, Graduate School of Pharmaceutical Sciences, The University of Tokyo, Tokyo, Japan

Ljubica Tasic Laboratory of Chemical Biology, Department of Organic Chemistry, Institute of Chemistry, University of Campinas (UNICAMP), Campinas, São Paulo, Brazil

Taisuke Tomita Laboratory of Neuropathology and Neuroscience, Graduate School of Pharmaceutical Sciences, The University of Tokyo, Tokyo, Japan

Christoph W. Turck Proteomics and Biomarkers, Department of Translational Research in Psychiatry, Max Planck Institute of Psychiatry, Munich, Germany

Marzena Ułamek-Kozioł Laboratory of Ischemic and Neurodegenerative Brain Research, Mossakowski Medical Research Centre, Polish Academy of Sciences, Warsaw, Poland

First Department of Neurology, Institute of Psychiatry and Neurology, Warsaw, Poland

Wenbo Wang Laboratory of Neuropathology and Neuroscience, Graduate School of Pharmaceutical Sciences, The University of Tokyo, Tokyo, Japan

Sijian Xia College of Life Science and Oceanography, Shenzhen University, Shenzhen, People's Republic of China

Fang Yao College of Life Science and Oceanography, Shenzhen University, Shenzhen, People's Republic of China

Huajie Zhang College of Life Science and Oceanography, Shenzhen University, Shenzhen, People's Republic of China

Wei Zhang Translational Psychiatry Program, Department of Psychiatry and Behavioral Sciences, McGovern Medical School, The University of Texas Health Science Center at Houston (UTHealth), Houston, TX, USA

Yuxi Zhao College of Life Science and Oceanography, Shenzhen University, Shenzhen, People's Republic of China

Danqing Zhao Department of Obstetrics and Gynecology, Affiliated Hospital of Guizhou Medical University, Guiyang, People's Republic of China

Chapter 1
The Potential of 'Omics to Link Lipid Metabolism and Genetic and Comorbidity Risk Factors of Alzheimer's Disease in African Americans

Kaitlyn E. Stepler and Renã A. S. Robinson

Abstract Alzheimer's disease (AD) disproportionately affects African Americans (AAs) and Hispanics, who are more likely to have AD than non-Hispanic Whites (NHWs) and Asian Americans. Racial disparities in AD are multifactorial, with potential contributing factors including genetics, comorbidities, diet and lifestyle, education, healthcare access, and socioeconomic status. Interestingly, comorbidities such as hypertension, type 2 diabetes mellitus, and cardiovascular disease also impact AAs. It is plausible that a common underlying molecular basis to these higher incidences of AD and comorbidities exists especially among AAs. A likely common molecular pathway that is centrally linked to AD and these noted comorbidities is alterations in lipid metabolism. Several genes associated with AD risk—most notably, the ε4 allele of the apolipoprotein E (APOE) gene and several mutations in the ATP-binding cassette transporter A7 (ABCA7) gene—are linked to altered lipid metabolism, especially in AAs. This review explores the role of lipid metabolism in AD broadly, as well as in other comorbidities that are prevalent in AAs. Because there are gaps in our understanding of the molecular basis of higher incidences of AD in AAs, 'omics approaches such as proteomics and lipidomics are presented as potential methods to improve our knowledge in these areas.

Keywords Lipid metabolism · Alzheimer's disease · Proteomics · African Americans · Comorbidities · Lipidomics

K. E. Stepler
Department of Chemistry, Vanderbilt University, Nashville, TN, USA

R. A. S. Robinson (✉)
Department of Chemistry, Vanderbilt University, Nashville, TN, USA

Vanderbilt Memory and Alzheimer's Center, Vanderbilt University Medical Center, Nashville, TN, USA

Vanderbilt Institute of Chemical Biology, Vanderbilt University, Nashville, TN, USA
e-mail: rena.as.robinson@vanderbilt.edu

P. C. Guest (ed.), *Reviews on Biomarker Studies in Psychiatric and Neurodegenerative Disorders*, Advances in Experimental Medicine and Biology 1118,
https://doi.org/10.1007/978-3-030-05542-4_1

1.1 Introduction

Alzheimer's disease (AD) is a neurodegenerative disorder marked by accumulation of extracellular amyloid beta (Aβ) plaques and intracellular hyperphosphorylated tau tangles [also referred to as senile plaques and neurofibrillary tangles, (NFTs)] in the brain, which lead to neuronal dysfunction and death. Other primary hallmarks of the disease include mitochondrial dysfunction, decreased synaptic plasticity, compromised blood-brain barrier, and oxidative stress. According to the Alzheimer's Association, approximately 5.7 million Americans have AD [1], although this number is not equally spread among different subgroups of the population. AD disproportionately affects certain racial subgroups, which is alarming considering that underrepresented minorities will comprise a larger proportion of both the entire older population and the population of AD sufferers by 2050 [2–4]. AD and related dementia prevalence are both higher in African Americans (AAs) and Hispanics than in non-Hispanic Whites (NHWs) and Asian Americans [3, 5–7]. In terms of incidence, AAs are 2–3 and Hispanics are about 1.5 times as likely to develop AD and related dementias compared to NHWs [8, 9], and AAs have a 65% higher risk than Asian Americans [10]. AAs and Hispanics also have a higher prevalence of cognitive impairment in adults aged 55 and older compared to NHWs [3].

Although it is well-established that racial disparities exist in AD, there are many contributing factors such as socioeconomic aspects, genetics, and comorbidities, and whether or not there are molecular underpinnings related to these remains unknown. Socioeconomic factors include education level, healthcare access, and willingness to seek care and treatment which are noted in AD as having differences between racial subgroups [7, 9]. AAs are less likely than NHWs to seek care for symptoms of mild cognitive impairment (MCI), a preliminary stage of AD [11], and are less likely than NHWs to receive AD pharmacotherapy treatment (e.g., cholinesterase inhibitors or memantine) upon disease diagnosis [2, 12]. Prevalence of genetic risk factors such as the ε4 allele of the apolipoprotein E (APOE) gene and various single-nucleotide polymorphisms (SNPs) in the ATP-binding cassette transporter A7 (ABCA7) gene and comorbidities, such as cardiovascular disease and type 2 diabetes mellitus (T2DM) that increase risk of AD, also contribute to racial disparities in AD. Prevalence of comorbidities is suggested to be a larger contributing factor than genetics [8].

One common molecular pathway that affects both genetic and comorbidity factors in AD is alterations in lipid metabolism. Significant evidence links dysregulation of lipid metabolism to AD [13–15]. Lipids play an integral role in AD pathogenesis through their interaction with Aβ, particularly in cell membranes and lipid rafts that can promote Aβ aggregation and disrupt membrane integrity (Fig. 1.1) [16]. Lipid rafts can affect amyloid precursor protein (APP) processing leading to an increase in Aβ production [16, 17]. Impaired cholesterol metabolism has been implicated in tau hyperphosphorylation processes and leads to increased oxidative stress, inflammation, phospholipase activation, and vascular dysfunction (Fig. 1.1) [16, 18]. Lipids contribute to neuronal dysfunction and dystrophy, disruption of the autophagy/lysosomal system, increased apoptosis, and compromised membrane function [19, 20].

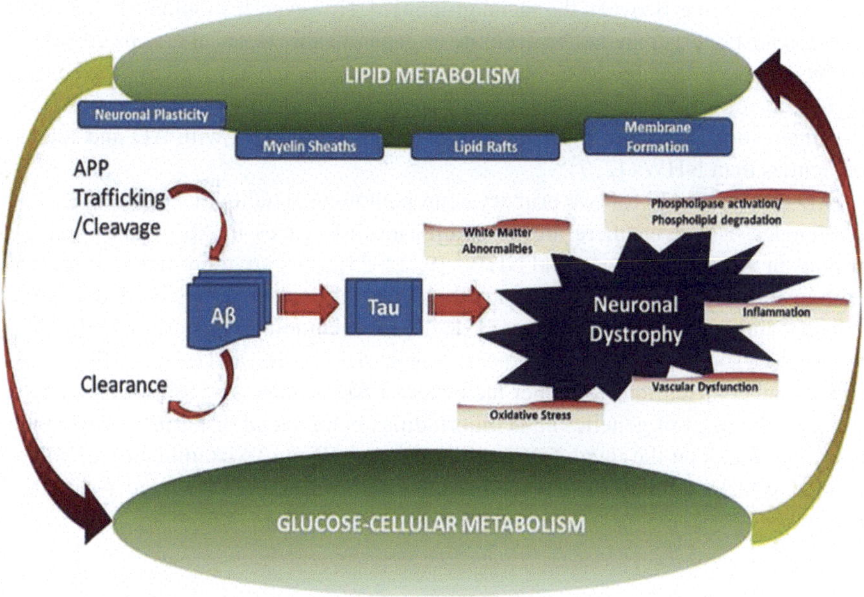

Fig. 1.1 Relationship of glucose-cellular metabolism and lipid metabolism in AD pathogenesis (source: see Ref. [20])

It is clear that lipid metabolism is important in AD pathogenesis and, as we will describe in this review, especially important in comorbidities affecting AAs. Mechanisms of lipid metabolism in AD are not fully understood, and their role in racial disparities of AD is unexplored. Fortunately, 'omics approaches are powerful methods with which to study molecules in lipid metabolism pathways in this context. 'Omics approaches, which include genomics, proteomics, metabolomics, and lipidomics, allow thousands of molecules to be investigated simultaneously and can give a systematic insight to changes in lipid metabolism in tissues. Such insight is important to help with tailored AD prevention, early diagnosis, and personalized treatment strategies for racial groups with high incidences of AD.

1.2 Potential Roles of Lipid Metabolism in AD Racial Disparities

The differences in AD prevalence and incidence among AAs and NHWs noted above are significant and can be partially evaluated by disease presentation and cognitive performance. AAs and Hispanics are more likely to present with more severe symptoms than NHWs [2, 9], and AAs are more likely to present at an earlier age of onset than NHWs [2]. Several studies demonstrate that AAs tend to score lower on cognitive tests than NHWs [2, 5, 21], although the rates of cognitive decline are similar between the racial groups [5]. However, it should be highlighted that diagnostic tests may not be accurate or generalizable to racial groups outside of NHWs. For example, the Mini-Mental State Examination was found to have a much higher rate of false-positive diagnosis of cognitive impairment in AAs than NHWs [9]. Interestingly, despite these differences, AAs live longer with AD and related dementias than NHWs [2, 7].

AD is well-known for two primary neuropathological hallmarks: Aβ plaques and tau tangles. However, there are no significant differences in Aβ plaques and tau tangles in the brains of AAs and NHWs [9, 22–24]. AAs are more likely to present with mixed AD and other dementia pathologies, particularly Lewy body dementia, infarcts, and cerebrovascular disease [22, 25]. These findings suggest that other factors such as socioeconomic differences, genetic risk factors, and comorbidities have substantial contributions to higher incidence of AD in AAs. The following sections examine the roles of genetics and comorbidities in the racial disparities of AD with a specific focus on the involvement of lipid metabolism. An examination of socioeconomic factors is beyond the scope of this review, and we refer readers to other reviews [3, 26, 27].

1.2.1 Comorbidities

Comorbidities describe health conditions that can increase an individual's risk for diseases, such as AD. Traumatic brain injury, stroke, dyslipidemia/hypercholesterolemia, cardiovascular disease, T2DM, obesity, and hypertension (HTN) all increase risk of AD (Fig. 1.2) [2, 6, 28–32]. These comorbidities also disproportionately affect AAs compared to NHWs. Findings from the Atherosclerosis Risk in Communities (ARIC) study suggest racial disparities in brain aging may be due to differences in risk factor presence, severity, and control [5]. Notably, alterations in lipid metabolism are common in AD and these comorbidities in AAs, which suggests that lipid metabolism may be an important underlying cause of racial disparities of AD.

1.2.1.1 Dyslipidemia

Dyslipidemia is a group of lipid disorders that present due to genetic predisposition or underlying events, such as insulin resistance, excess weight, and hypothyroidism. These events result in abnormal levels of high-density lipoprotein (HDL), low-density lipoprotein (LDL), very-low-density lipoprotein (VLDL), and triglycerides [32]. Clinically, dyslipidemia is noted by total cholesterol ≥240 mg/dL, LDL ≥160 mg/dL, HDL ≤40 mg/dL, or the use of lipid-lowering medications [33] and includes high cholesterol levels (hereafter referred to as hypercholesterolemia).

The connection between AD and cholesterol has been firmly established (Fig. 1.3) and discussed in the literature on multiple occasions [13–15, 34]. In the

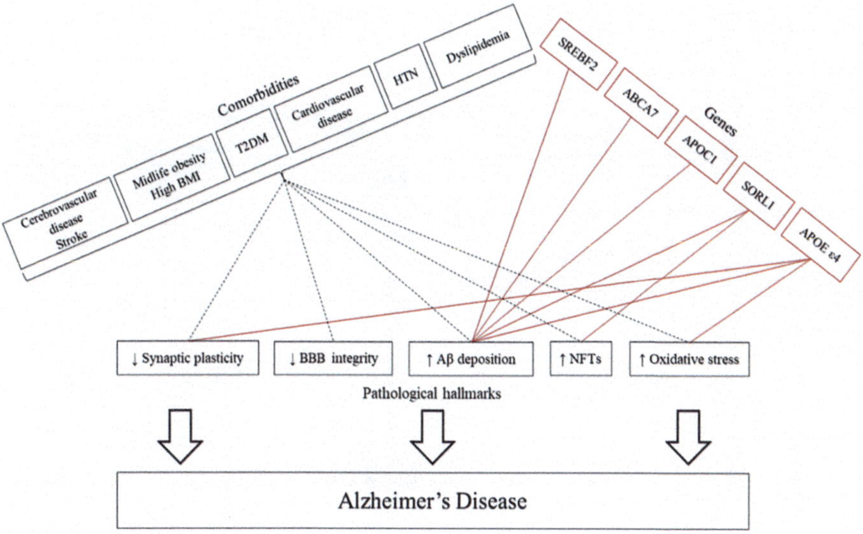

Fig. 1.2 The connections among AD risk factors (genes in red and comorbidities in black) with altered lipid metabolism and AD pathological hallmarks

brain, cholesterol is produced in astrocytes and is transported to neurons by apoE, a process necessary for neurons to form functional synapses [17]. Both increased and decreased levels of cholesterol in the brain have been suggested to contribute to Aβ production [16], and cholesterol accumulates in both Aβ plaques and tau tangles [17]. Brain cholesterol is synthesized de novo, is present in several forms including cholesterol esters, and greatly impacts Aβ monomer formation and aggregation. Peripheral cholesterol levels which come from new synthesis or dietary uptake are also implicated in AD. AD patients have peripheral lipoprotein profiles (i.e., high total plasma cholesterol and LDL and low HDL level) similar to dyslipidemia [32, 35]. This is likely due to a compromised blood-brain barrier (BBB) in AD which would allow HDL to leak from the periphery into the brain, whereas VLDL/LDL is not transported across this barrier (Fig. 1.3).

Dyslipidemia was associated with increased Aβ plaque burden in a Japanese population, even when adjusting for APOE ε4 genotype [31]. In a cohort of AAs without ε4 alleles, higher mean serum total cholesterol levels were observed in individuals with AD compared to those that were cognitively normal [36]. Consistent with these observations, hypercholesterolemia has been associated with AD [17, 30] and in midlife is associated with increased MCI risk later in life [11]. Transgenic mouse studies suggest that the mechanism by which hypercholesterolemia increases AD risk is acceleration of Aβ deposition in the brain [32]. On the other hand, in a meta-analysis of modifiable risk factors for AD, there was no association between high cholesterol levels and incidence of AD [37]. This could be due to the necessity for APOE ε4 dependence.

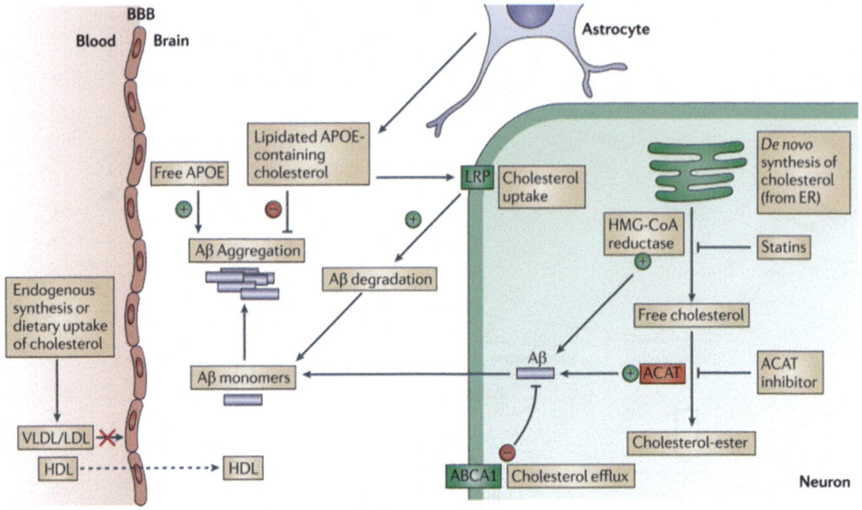

Fig. 1.3 Role of cholesterol in Aβ pathology in AD. ApoE is also included in this figure as it is a cholesterol transporter. Proteins highlighted in green decrease Aβ pathology; proteins highlighted in red increase Aβ pathology. Abbreviations: *ACAT* acyl-CoA:cholesterol acyltransferase 1 (also known as sterol *O*-acyltransferase), *LRP* LDL receptor-related protein, *HMG-CoA* 3-hydroxy-3-methylglutaryl-CoA. Source: See Ref. [34]

Contradictory findings regarding cholesterol levels in AAs and NHWs exist. In an ARIC study, AAs and NHWs did not have different cholesterol levels in midlife (50–60 years old) [38], consistent with a report that lipid profiles between AAs and NHWs are not significantly different on a national level [33]. However, over the age of 45, there is higher incidence of dyslipidemia in AAs compared to NHWs [39]. This incidence is more noticeable in older age groups (i.e., 65–74 years old). Others have found that AAs have higher levels of hypercholesterolemia than NHWs [11]. Together, the discrepancy in these reports requires more evaluation of cholesterol and dyslipidemia levels in AAs but, more importantly, highlights that the role of dyslipidemia in AAs with AD warrants further investigation.

1.2.1.2 HTN

Dyslipidemia is also a risk factor for HTN. Serum cholesterol levels correlate with both systolic and diastolic blood pressure in individuals with HTN [40]. HTN is independently associated with increased cognitive decline, MCI, and AD [5, 11, 17], and high systolic blood pressure is associated with an increased risk of AD [37]. Midlife HTN is associated with increased AD and dementia risk later in adult life [41]. Higher systolic blood pressure increases the odds of brain infarcts and is associated with an increased number of NFTs in postmortem brain tissue [42]. HTN among AAs is likely to increase risk for certain neurovascular pathologies, such as cerebral amyloid angiopathy, white matter lesions, and vascular endothelial damage [6]. The Systolic Blood Pressure Intervention Trial Memory and Cognition in Decreased Hypertension study recently discovered that reducing systolic blood pressure to less than 120 mmHg decreased MCI risk by 19% in a study cohort that was 30% AA and 10% Hispanic. Importantly, this study implicates treatment of HTN as an effective measure to prevent dementia and AD in multiple racial groups [43, 44].

It is well-known that HTN is more prevalent in AAs than in NHWs [5, 6, 9, 11, 45, 46]. The prevalence of HTN, including diagnosed and undiagnosed cases, in AA men is 42.4% and in AA women is 45% in the United States, and these are 10–12% higher than those for NHWs and Mexican Americans [33]. HTN occurs at an earlier age of onset in AAs [46, 47]. Interestingly, AAs are less likely to have their blood pressure under control when compared to NHWs and Hispanics, despite the fact that AAs are more aware of their HTN and take medications [33, 45, 48]. Increases in systolic blood pressure increase risk of stroke, congestive heart failure, and end-stage renal disease in AAs [46]. Furthermore, in individuals with AD, there is a higher prevalence of HTN in AAs compared to NHWs [6, 21, 25].

The American Heart Association states that HTN is the "most potent risk to cardiovascular health of African Americans" [33]. There is an apparent intersection of HTN and dyslipidemia as risk factors for cardiovascular disease and AD such that progress made with mechanistic understanding of both of these risk factors can have a positive impact on human health, especially among AAs.

1.2.1.3 Obesity

Obesity can lead to inflammation in the brain, compromised BBB integrity, and changes in neuronal structure, synaptic plasticity, and memory [32]. Both low and high body mass indices have been associated with cognitive impairment and dementia [2]. A higher body mass index (BMI, ≥30 kg/m^2) in midlife is associated with increased dementia and AD risk, presumably due to increased amyloid deposition [23]. On the other hand, a higher BMI in late life reduces risk of cognitive decline and dementia [41]. A high BMI and obesity are more prevalent in AAs [2], and AAs have higher obesity rates than NHWs at various ages [33, 38]. BMI has stronger effects on other conditions, especially diabetes, metabolic syndrome, and HTN in AAs [2]. Highlighting these aspects of obesity is necessary as high BMI and obesity, HTN, and T2DM are comorbidities of AD.

1.2.1.4 T2DM

Comprehensive discussions of the connections between T2DM, lipid metabolism, and AD can be found in several reviews [49–52]. Briefly, early in the pathogenesis of T2DM, insulin resistance causes lipid accumulation in skeletal muscle and the liver [51]. On the other hand, chronic elevation of free fatty acid levels in plasma can cause insulin resistance and development of T2DM [50]. Circulating triglycerides affect insulin transport across the BBB [49]. T2DM has been associated with

AD [30, 32, 52] and confers significantly increased AD risk [32, 41]. T2DM results in lower insulin and insulin receptor levels, which impairs synaptic function and decreases memory formation [32]. Prediabetes and T2DM are associated with increased cognitive decline [53], vascular dementia, and compromised BBB integrity [49]. A dysfunctional BBB could result in increased brain insulin levels and thus prevent Aβ clearance and degradation.

The associations between T2DM and AD are unaffected by the presence of the APOE ε4 allele [54]. However, the APOE ε4 allele has been linked to decreased expression of insulin-degrading enzyme, which could increase brain insulin levels and contribute to AD pathology as aforementioned. T2DM also increases risk of progression from MCI to dementia [18, 32] and is associated with cognitive impairment. Cognitive impairment is more severe with longer duration of diabetes, poorer control, presence of complications, and comorbid HTN or depression [49]. T2DM is associated with lower cognitive scores at baseline and at a 6-year follow-up in a multiethnic cohort [55], and in another cohort coexisting T2DM was found to accelerate AD pathogenesis [30].

There have been conflicting findings regarding the association of T2DM with AD pathology in the brain [18, 49]. Although systemic insulin resistance has been associated with brain Aβ via PET imaging [56], most studies found no relationships between AD neuropathology and T2DM [57, 58]. In a cross-sectional study of older Brazilian adults, although there was no overall association between AD neuropathology and T2DM, a higher NFT burden was detected when both T2DM and the APOE ε4 allele were present [57]. However, T2DM has been linked to cerebral infarcts, cerebrovascular pathology [59–61], and stroke [54].

T2DM is more prevalent in AAs [2, 5, 33, 62, 63] and Hispanics [6, 9] than NHWs. Prevalence estimates for T2DM in AAs range from 1.4 to 2.3 times higher than in NHWs [11, 63]. According to the National Health and Nutrition Examination Survey, the prevalence of combined diagnosed and undiagnosed T2DM is 21.8% in AAs and 11.3% in NHWs, and over one-third of the cases in AAs were undiagnosed [33]. T2DM was associated with greater cognitive decline in AAs [64], while a study of the Minority Aging Research Study and Memory and Aging Project cohorts found similar effects of diabetes on cognition in AAs and NHWs [65]. Glucose levels in AAs with T2DM were significantly higher than those in AAs who did not develop dementia. These levels then declined prior to dementia diagnosis, while glucose levels remained stable in NHWs with T2DM [66, 67]. Although most evidence supports the existence of racial disparities in T2DM, one study did not find an association between T2DM and race [57].

1.2.1.5 Vascular Diseases

Vascular diseases, which encompass cardiovascular disease, heart disease, atherosclerosis, vascular dementia, and cerebrovascular disease, also involve dysregulated lipid metabolism and thus should be briefly addressed. Vascular pathology increases dementia risk [68]. In a study of MCI and cognitively unimpaired individuals, increased vascular risk factors—measured by Framingham Stroke Risk Profile score taking into account age, systolic blood pressure, anti-HTN medications, T2DM, smoking, cardiovascular disease history, and atrial fibrillation—were associated with increased cognitive decline [69]. Vascular risk factors increased risk of conversion from MCI to AD, and treatment of these risk factors reduced risk of conversion [70]. History of coronary artery disease and myocardial infarction are also associated with higher dementia rates and more Aβ plaques in the brain [6]. Cerebrovascular disease is more commonly comorbid with AD than other neurodegenerative diseases and when combined can manifest in earlier clinical symptoms of dementia [71]. Surprisingly, in a meta-analysis, stroke was found to have no effect or a negative effect on AD risk [37], which is in contrast to a study reporting that stroke increases AD risk especially in the presence of HTN and T2DM [29].

AAs and Hispanics have a 2.4 and 2 times higher incidence of stroke compared to NHWs [6], and stroke mortality rates for AAs are also 4.5 times higher than in NHWs [33]. These racial differences are exacerbated in younger age groups such as 45–59-year-olds in AAs [72, 73]. There are conflicting reports on whether or not stroke increases the likelihood of developing AD and related dementias in AAs [3, 6, 74]. Atherosclerosis is associated with increased risk of AD [37] and is commonly detected in the brain [75]. A larger proportion of dementia cases are attributed to vascular dementia in AAs and Asian Americans/Pacific Islanders than NHWs and Hispanics [3, 6].

Overall, the prior sections grossly demonstrate that comorbidities greatly influence risk of AD, incidence of AD especially among AAs, and disease pathogenesis. Several of the comorbidities discussed clearly implicate lipid metabolism as a primary feature of the comorbid disease and of AD.

1.2.2 Genetics

Genetic factors are well-known to play a role in AD risk (Table 1.1) and most likely also contribute to racial disparities in AD. Although there is no evidence for a separate genomic region with a different contribution to age-related cognitive decline between NHWs and AAs [76], risk genes such as APOE, ABCA7, and others have been discovered. Frequency of risk alleles and single-nucleotide polymorphisms (SNPs) for these genes, as well as strength of their association with AD, varies among racial groups, and in some cases, there are SNPs and genes that are only associated with AD for a given racial group. Many of these genetic risk factors for AD are related to lipid metabolism (Fig. 1.4), which will be discussed in detail in the remainder of this section.

1.2.2.1 APOE ε4

The APOE gene codes for the protein apolipoprotein E (apoE), which is one of the most abundant lipoproteins in the central nervous system. The primary function of apoE is to maintain lipid and cholesterol homeostasis in the brain, which it

Table 1.1 Genes known to increase risk of AD

Gene	Disease-associated SNP/allele	References
ABCA1[a]	SNPs rs2230806, rs4149313, rs2230805, rs2230808	Wavrant-De Vrièze et al. [77]; Koldamova et al. [78]; Fehér et al. [79]
ABCA7[a]	SNPs rs11550680, rs142076058, rs3764647, rs3752239, rs3764650, rs3752246, rs78117248, rs4147929	Aikawa et al. [80]; Almeida et al. [81]; Hollingworth et al. [82]; Naj et al. [83]; Cuyvers et al. [84]; Lambert et al. [85]
APOC1[a]	Insertion/deletion polymorphism rs11568822, H2 allele	Zhou et al. [86]; Petit-Turcotte et al. [87]; Ki et al. [88]
APOD[a]	Intron 1 polymorphism	Desai et al. [89]
APOE[a]	ε4 allele	Barnes and Bennett [2]; El Gaamouch et al. [16]; Martins et al. [15]; Zhao et al. [90]
BIN1[a]	SNPs rs55636820, rs7561528, rs744373	Reitz et al. [91]; Reitz and Mayeux [92]; Hollingworth et al. [82]; Naj et al. [83]; Seshadri et al. [93]
CD2AP	SNP rs9349407	Naj et al. [83]
CD33	SNPs rs3826656, rs3865444	Bertram et al. [94]; Hollingworth et al. [82]; Naj et al. [83]
CLU[a]	SNPs rs11136000, rs1532278	Lambert et al. [95]; Naj et al. [83]; Harold et al. [96]; Seshadri et al. [93]
CR1	SNPs rs3818361, rs6656401, rs6701713	Hollingworth et al. [82]; Lambert et al. [95]; Naj et al. [83]
EPHA1	SNPs rs11771145, rs11767557	Hollingworth et al. [82]; Naj et al. [83]; Seshadri et al. [93]
MS4A gene cluster	SNP rs610932 in MS4A6A SNP rs670139 in MS4A4E SNP rs4938933 in MS4A4A	Hollingworth et al. [82]; Naj et al. [83]
PICALM[a]	SNPs rs561655, rs3851179	Reitz and Mayeux [92]; Harold et al. [96]; Seshadri et al. [93]
SORL1[a]	SNPs rs2298813, rs2070045, rs668387, rs689021, rs641120, rs1784933, rs3824966, rs12285364	Rogaeva et al. [97]; Lee et al. [98]; Chou et al. [99]
SIGMAR1[a]	Long runs of homozygosity in Chr4q313, 15q24.1, 3p21.31 regions	Ghani et al. [100]
SREBF2[a]	SNP rs2269657	Picard et al. [101]

[a]Related to lipid metabolism

accomplishes via phospholipid and cholesterol transport. ApoE is responsible for delivery of phospholipids and cholesterol to neurons and various receptors for utilization or clearance [15–17]. The transport of cholesterol to neurons is particularly important for synapse formation and neuronal functioning [17]. In addition to carrying lipids, apoE also has the ability to bind Aβ and aid in its clearance from the brain (Fig. 1.3) [15].

While the APOE ε2 allele has a protective effect against AD [15–17, 90, 102], the APOE ε4 allele is one of the strongest genetic risk factors for AD [2, 15–17, 90, 102]. The ε4 allele exerts a dose-dependent effect on AD risk. One ε4 allele confers

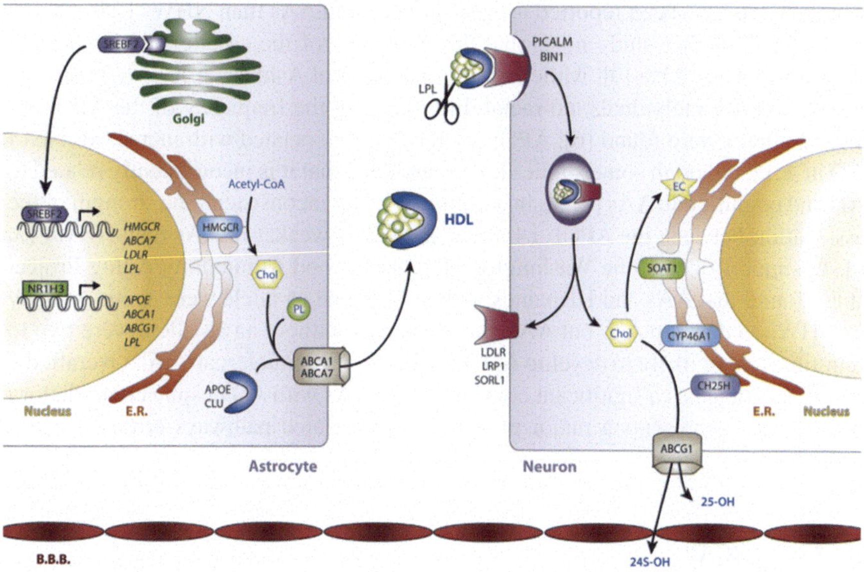

Fig. 1.4 Representation of cholesterol metabolism in the brain, including several AD risk genes and their gene products (source: see Ref. [101])

2 to 3 times greater risk of developing AD, while two ε4 alleles leads to a 12 times greater risk for AD [103]. The ε4 allele has been associated with increased Aβ accumulation and deposition in the brain and cerebral vessels [90, 102, 104], as well as increased tau tangles [2, 18]. Interestingly, individuals with T2DM and the APOE ε4 allele had more of both types of neuropathology than individuals with neither or only one of these factors [105]. APOE ε4 has also been associated with an earlier age of onset of AD [16, 102], more rapid rate of cognitive decline [16], decreased cognitive performance, and increased memory decline [106]. The ε2, ε3, and ε4 alleles lead to three corresponding protein isoforms of apoE, which have different abilities to carry lipids and bind Aβ. ApoE4 is less effective at binding and clearing Aβ from the brain than apoE2 and apoE3 [15, 16, 32, 107]. Additionally, the apoE4 isoform increases both Aβ production and fibril formation compared to the other two isoforms [90, 104]. The apoE4 isoform also suppresses synaptic protein expression, which impairs synapse transmission and plasticity and could contribute to synapse dysfunction and loss that occurs early in AD pathology [90, 102]. ApoE4 is less protective against oxidative stress than the other isoforms and leads to increased dysfunction of cholinergic neurons in AD [15, 16]. APOE ε4 has also recently been associated with vascular pathologies, such as cerebral microbleeds, white matter lesions, and coronary artery calcification, which provides a link between genetic and comorbid risk factors for AD [108].

Disparities in APOE allele frequency and association with AD have been noted among racial groups. Multiple studies have determined that there is an increased frequency of the ε4 allele in AA populations compared to NHWs [2, 25], and the ε4/ε4 genotype has also been reported as more common in AAs than NHWs [25]. On the other hand, another study notes that the frequency of the ε4 allele varies in AAs from about 17 to 21% [9], while in a Florida study of Ashkenazi Jewish, Hispanic, NHW, and AA individuals, no racial differences in the frequency of the APOE ε4 and ε2 alleles were found [6]. APOE ε4 has been associated with increased risk of AD in AAs, although some studies have concluded that it is inconsistently related to AD and cognition in AAs [2, 3]. In addition to being inconsistent, the strength of the association between the APOE ε4 allele and AD is weaker in AAs than NHWs [3, 6]. In support of this, the Washington Heights-Inwood Community Aging Project study found that AAs and Hispanics with at least one ε4 allele were about as likely as NHWs to develop AD, but AAs and Hispanics without an ε4 allele were two to four times more likely to develop AD than their NHW counterparts [6]. Overall, the APOE ε4 allele has a significant contribution to AAs with AD despite contradictory findings and represents a major protein in lipid-related pathways critical for AD pathogenesis.

1.2.2.2 ABCA7

The ABCA7 protein, coded for by the ABCA7 gene, is a transmembrane protein that is an important transporter of lipids through the cell membrane energetically driven by ATP binding [80]. ABCA7 is particularly important for cholesterol and phospholipid efflux mediated by apolipoproteins and through these interactions functions in the biosynthesis of HDL and is involved in the lipidation of apoE [16, 91]. Additionally, there are direct links between ABCA7 and AD, as ABCA7 can affect APP transport through the cell membrane and is involved in the Aβ clearance pathway [80, 91]. For example, overexpression of ABCA7 leads to decreased Aβ levels. On the other hand, deletion of ABCA7 accelerates Aβ production (likely by allowing APP endocytosis), impairs Aβ clearance, and increases Aβ plaque burden [16, 80]. The changes in ABCA7 function due to AD risk mutations are not entirely known. The ABCA7 deletion rs142076058 may result in a frameshift mutation that either causes synthesis of an aberrant protein or nonsense-mediated decay of the truncated RNA transcript [109]. Either of these consequences would likely result in reduced ABCA7 levels and have similar effects to those of the ABCA7 deletion described above. An additional study hypothesizes that the rs3764650 SNP decreases ABCA7 expression before AD onset which increases AD risk, and after AD onset ABCA7 expression increases [110].

Many SNPs within the ABCA7 gene have been associated with AD (Table 1.2), although premature termination codon mutations and loss-of-function mutations have also been noted [80]. AD risk conferred by ABCA7 has been confirmed specifically in AAs, and the association is stronger and more widespread in AAs than NHWs [2, 91, 111]. A genome-wide association study (GWAS) supports this in that SNP rs115550680 is significantly associated with AD in AAs with an effect size (70–80%) comparable to that of APOE ε4 [91, 92]. Notably, many of the ABCA7 SNPs are present and confer risk in one specific racial group. For example, a dele-

Table 1.2 ABCA7 SNPs associated with AD in AAs

SNP	Populations associated with AD	Sources
rs3752239	AAs	Aikawa et al. [80]
rs3752246	Multiple racial groups	Naj et al. [83]
rs3764647	AAs	Aikawa et al. [80]
rs3764650	NHWs, AAs	Almeida et al. [81]; Hohman et al. [111]
rs115550680	AAs	Reitz et al. [91]; Reitz and Mayeux [112]
rs142076058	AAs, rare in NHWs	Cukier et al. [109]; Aikawa et al. [80]

tion in ABCA7 (rs142076058) is significantly associated with AD, which was commonly identified in AA cases and controls (15.2 and 9.74%, respectively) but only in 0.12% of NHWs [109]. There are more SNPs (rs3764647, rs142076058, rs3752239) that have only been associated with AD in AAs, while rs3764650 is associated with AD in both NHWs [80, 81] and in individuals with more African ancestry [111]. These findings provide clear evidence that ABCA7 is likely a contributing factor to the racial disparities in AD, as ABCA7 is a stronger genetic risk factor for AD in AAs than NHWs.

1.2.2.3 Other Genes

In addition to APOE and ABCA7, other AD risk genes include bridging integrator 1 (BIN1), clusterin (CLU), phosphatidylinositol binding clathrin assembly protein (PICALM), sortilin-related receptor 1, ABCA1, L(DLR) class A-type repeats containing (SORL1), apolipoprotein D (APOD), apolipoprotein C1 (APOC1), sigma non-opioid intracellular receptor 1 (SIGMAR1), and sterol regulatory element binding transcription factor 2 (SREBF2) [79, 86, 89, 92, 98, 100, 101]. For the purpose of this review, only those genes related to lipid metabolism will be discussed. Lipid-related AD risk genes involve pathways such as regulation of Aβ formation (ABCA7, APOE, ABCA1, CLU, PICALM, SORL1), regulation of NFT assembly (APOE, CLU, SORL1), and protein-lipid complex assembly (ABCA7, ABCA1, APOC1, APOE, BIN1) [113]. BIN1, CLU, and PICALM are less common risk genes. Clusterin, also known as apolipoprotein J, has functions that parallel those of apoE, while both BIN1 and PICALM are involved in receptor-mediated endocytosis, which is important for lipid internalization and transport including transport mediated by apoE and clusterin [114]. In a GWAS study of AAs, AD-associated SNPs in BIN1, CLU, and PICALM had a 10–20% increase in AD risk and significantly lower effect sizes than in NHWs [92]. This result is confounded by another GWAS study that shows CLU and PICALM are not associated with AD in AAs but that BIN1 is associated with SNPs in AAs that differ from those in NHWs [91]. ABCA1 has functions similar to ABCA7 in apolipoprotein transport and Aβ clearance. In an AD mouse model, deficiency of ABCA1 increases Aβ deposition in the brain and decreases apoE levels, while ABCA1 overexpression decreases Aβ deposition [115, 116]. Interestingly, the effect of ABCA1 on Aβ was APOE genotype-dependent as

ABCA1 deficiency reduced Aβ clearance in APOE4 mice but not in APOE3 mice [115]. Several SNPs in ABCA1 have been associated with AD in various populations including NHWs, Swedish, and Chinese [77–79], although no data were available for AA groups. SORL1 is involved in the regulation of lipoprotein lipase trafficking, APP processing and trafficking, and tau cellular processes, and its expression is reduced in AD brains [97, 99, 117]. Various SNPs and haplotypes in SORL1 have been associated with AD in NHWs, Hispanics, AAs, and Asians, although these genetic variations are not the same across racial groups [97–99].

Other potential genetic risk factors for AD in AAs exist. ApoD, similar in function to apoE, has four polymorphisms in the APOD gene unique to individuals of African ancestry and is associated with increased risk of AD [89]. ApoC1 is known to interfere with apoE receptor interactions and thus decreases clearance of lipoproteins containing apoE [87]. ApoC1 also can inhibit the cholesteryl ester transfer protein and activate lecithin-cholesterol acyltransferase and thus cholesterol esterification. Although the APOC1 gene is located adjacent to the APOE gene, the H2 allele of APOC1 was found to be a risk factor for AD independent of APOE ε4. These two alleles occur frequently together in AD populations and combined produce higher AD risk than either allele individually [87, 88]. Decreased levels of apoC1 mRNA with the H2 allele and increased apoC1 protein levels have been reported in AD [87]. Additionally, an insertion/deletion polymorphism in the APOC1 gene was found to increase AD risk in NHWs, Asians, and Caribbean Hispanics, but not in AAs [86].

Long runs of homozygosity in genes in AAs are associated with AD. The most notable of these genes is SIGMAR1 which encodes a protein that functions in lipid transport from the ER and helps to regulate various cellular functions via regulation of biosynthesis of lipid microdomains in the membrane [100]. Lastly, SREBF2 is a protein involved in lipid homeostasis and cholesterol biosynthesis with activity affected by brain cholesterol levels. SNP rs2269657 of SREBF2 was associated with AD pathological biomarkers and gene expression levels in the Alzheimer's Disease Neuroimaging Initiative cohort, which includes individuals of multiple ethnic backgrounds [101]. In summary, there are several genes critical to lipid metabolism, which have reported risk for AD, especially in AAs, and these warrant further investigation.

1.3 'Omics Approaches to Study Lipid Metabolism in AD

Review of the current literature involving lipid metabolism and AD reveals gaps that remain in our molecular understanding of AD, particularly across different racial groups. Some studies have focused on one or a few genes, proteins, and lipids in lipid metabolism pathways in AD. For example, a study on plasma levels of total cholesterol, HDL, LDL, and triglycerides revealed that higher midlife total cholesterol and triglycerides levels were associated with increased 20-year cognitive decline in NHWs but not in AAs [118]. In another study, apolipoproteins J, E, A-I, and C-III and their subspecies were measured with ELISA in plasma samples from the Ginkgo Evaluation of Memory Study cohort [119]. Higher apoE and apoA-I levels were associated with lower Aβ deposition and lower hippocampal volume, respectively. Lower plasma apoE levels were associated with higher Aβ deposition [120]. Heart-type fatty acid-binding protein was increased in cerebrospinal fluid (CSF) of AD patients [121]. Western blots have been used to study phospholipase D1 in AD brain tissue as well, which found elevated levels [122]. Neurodegenerative markers in AD brain tissue were recently measured in AAs although these markers are not related to lipid metabolism [123].

The limited focus of one or a few targeted species prevents a comprehensive picture of the changes occurring in the lipid metabolism pathway in AD from being gained. 'Omics approaches including genomics, proteomics, metabolomics, lipidomics, and transcriptomics enable comprehensive analyses of their respective molecule classes and can fill these gaps in our understanding of lipid metabolism. It is especially important to note that a majority of 'omics studies in AD are focused on NHWs or other majority populations and grossly exclude AAs. This presents an opportunity for the field to ensure that 'omics studies are more racially inclusive.

Discovery-based 'omics approaches are used for broad studies that can help with disease understanding and biomarker discovery [124, 125]. Several studies have utilized discovery-based proteomics to study AD for various purposes, such as to examine changes in the overall proteome in aging and MCI or AD [126–140] and to study proteins associated with oxidative stress [141–145]. Discovery-based proteomics of AD and Parkinson's disease brain tissue samples identified a combined total of 11,840 proteins [146]. The power of such studies is deep proteome coverage, which leads to insights on many biological pathways in health and disease. We refer the reader to several recent reviews on proteomics studies in AD [147–149].

On the other hand, 'omics approaches can also be targeted, enabling the focused study of a few to a few hundred species. Targeted proteomics has been used to analyze CSF samples in AD for potential biomarkers in several instances [148, 150–152]. Targeted proteomics in the brain identified several proteins with expression levels that correlated with Aβ and tau pathology [149, 153–156]. Targeted proteomics has also been used to specifically study proteins involved in lipid metabolism, mainly apolipoproteins in blood and blood-based bio-specimens [151–159]. One targeted proteomics assay was developed to analyze 12 apolipoproteins in serum or plasma samples, which identified significant effects of gender and use of lipid-lowering medications on apolipoprotein levels [157]. Few examples of targeted proteomics applied to study lipid metabolism in AD exist. A two-dimensional gel electrophoresis analysis quantified three fatty acid-binding proteins in AD and observed decreased levels in AD brain [160]. Serum protein analysis showed that there was no significant difference in apoE levels in MCI patients and controls [161]. Targeted lipidomics has been used to study changes in ceramide levels in human neurons in response to a neurotoxic signaling glycerophospholipid [162]. Targeted metabolomics has been used to measure 188 lipid and metabolite species in plasma samples from a subset of mostly AA participants in the ARIC study [163]. The metabolomics results from this study are particularly interesting because ten of

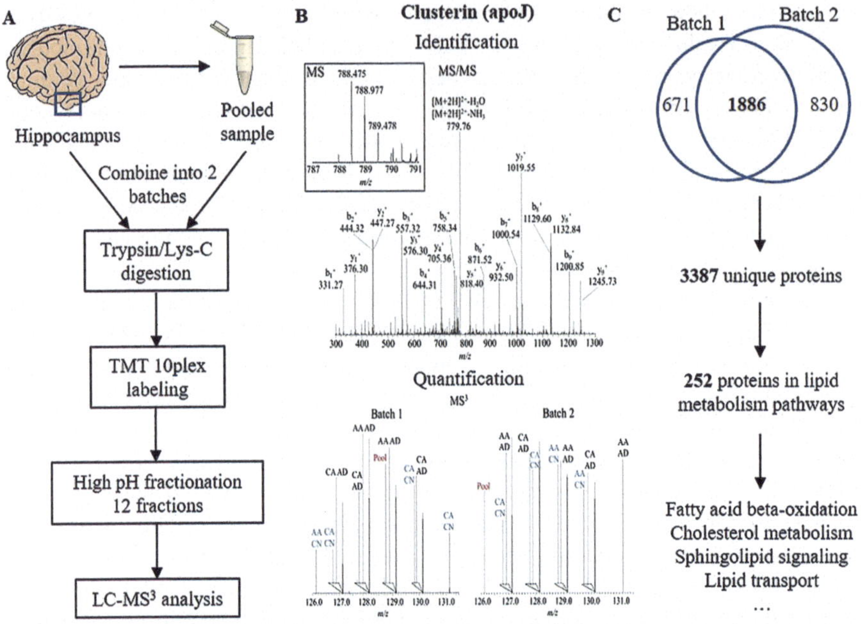

Fig. 1.5 Summary of our hippocampal proteomics experimental workflow (**a**), example MS, MS/MS, and MS^3 spectra from a lipid metabolism protein (clusterin) (**b**), and an overview of our protein identifications from these analyses (**c**) whereby two independent batches of TMT-tagged samples were analyzed. Abbreviations: *TMT* tandem mass tags, *LC* liquid chromatography, *MS* mass spectrometry. Unique proteins were identified in either or both batches of samples

the species that had been previously found to be predictive of MCI or dementia in NHWs were not predictive of either condition in this mostly AA cohort.

Multi-omics, or the integration of multiple types of 'omics data, can provide a more complete picture of the biological system being studied and allow determination of changes potentially associated with disease pathogenesis, biomarker discovery, and therapeutic targets [164–167]. Several studies have applied such methods in investigations of AD by using proteomics in combination with genome and transcriptome data to identify potential pathways or networks that may contribute to AD pathogenesis [168–170].

Our laboratory has demonstrated the use of multi-omics, specifically proteomics and lipidomics, to study lipid metabolism in AD across racial groups (manuscripts in preparation). In a pilot AD study, we have analyzed postmortem hippocampal tissues using a quantitative proteomics workflow (Fig. 1.5a). The workflow applies tandem mass tagging (TMT) of tryptic peptides to liquid chromatography-mass spectrometry (LC-MS), tandem MS, and MS^3 on an Orbitrap Fusion Lumos mass spectrometer. This comprehensive workflow enables peptide identification from the MS/MS spectra and accurate relative quantification from the MS^3 spectra (Fig. 1.5b). From this study we identified over 3300 proteins, which included 252 lipid-related proteins (Fig. 1.5c). These lipid-related proteins encompass many pathways including lipid transport, cholesterol metabolism, sphingolipid signaling, and others (Fig. 1.5c), and it should be noted that we detected many apolipoprotein species such as apoJ (Fig. 1.5b). We are currently evaluating protein expression differences between cognitively normal and AD groups and racial subgroups such as NHWs and AAs. Investigations of other brain regions using this proteomics workflow are ongoing. Plasma proteomics analysis of AA AD patients compared to NHWs will also be beneficial to identify peripheral changes in lipid metabolism in AD, and such studies are underway.

We are also developing robust platforms to study lipid pathways in AD plasma samples including both discovery and targeted approaches using LC-MS/MS on a QExactive HF mass spectrometer. In a pilot discovery-based lipidomics study of NHW cognitively normal individuals and AD patients, we tentatively identified over 2000 lipid compounds in positive ionization mode and over 1350 lipid compounds in negative ionization mode. The majority of these identified lipids are glycerosphingolipids, glycerolipids, and sphingolipids. Several lipid compounds identified in our study have been previously established as differentially expressed in AD, and our results change in a consistent manner (Fig. 1.6) [171–174].

These 'omics experiments have enabled many proteins and lipids to be measured simultaneously in AD and will provide a comprehensive study of various lipid pathways in AD. Additionally, these studies are among the first to apply 'omics approaches to study AD in AAs and may give insights into the molecular differences between NHWs and AAs in AD.

1.4 Conclusions

This review has demonstrated that many genetic and comorbidity risk factors for AD are interconnected and especially important in increasing our understanding of racial disparities in AD. As discussed, HTN, obesity, and other comorbidities such as T2DM and vascular diseases are especially prevalent in AAs. AD risk factors that involve lipid metabolism point to alterations in lipid transport, cholesterol synthesis, lipid homeostasis, and others as key pathways that could contribute to higher incidence of AD in AAs. However, it is also evident that the molecular mechanisms underlying these differences are still not fully known. Filling these gaps will advance understanding of the molecular basis for racial disparities in AD and potentially lead to improved AD prevention, diagnosis, and treatment strategies tailored to these high-risk populations. 'Omics techniques can help us gain a complete understanding of changes in lipid metabolism occurring in AD and should be explored in AAs, Hispanics, and other racial groups.

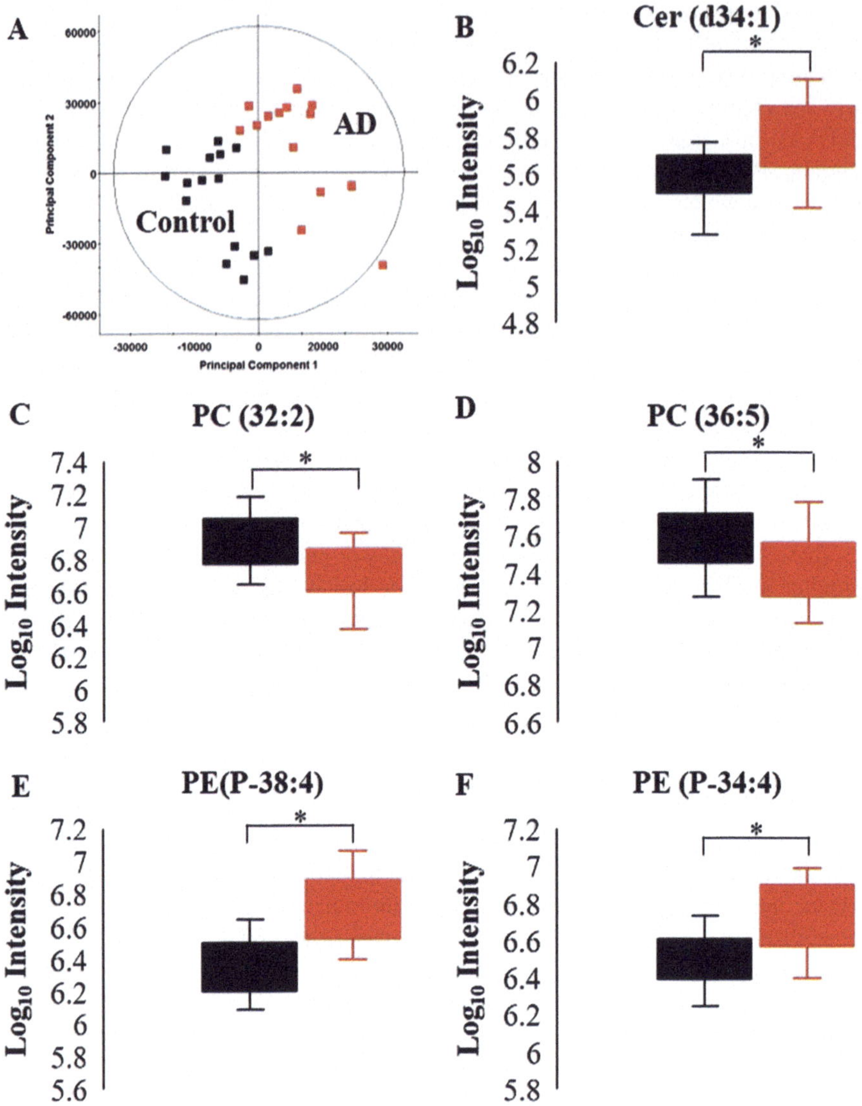

Fig. 1.6 Principal component analysis plot showing the clustering of cognitively normal and AD groups, including triplicate measurements (**a**) and box plots showing average intensities in control (cognitively normal) and AD groups for five lipids that have been previously found to be differentially expressed in AD (see Refs. [171–174]): ceramide d34:1 (**b**), phosphatidylcholine 32:2 (**c**), phosphatidylcholine 36:5 (**d**), phosphatidylethanolamine P-38:4 (**e**), and phosphatidylethanolamine P-34:4 (**f**). For all box plots, $N = 5$ per group; black is control group; red is AD. * indicates $p < 0.05$

Acknowledgments The authors acknowledge funding from the Alzheimer's Association (AARGD-17-533405), Vanderbilt University Start-Up Funds, the University of Pittsburgh Alzheimer Disease Research Center funded by the National Institutes of Health and National Institute on Aging (P50AG005133, RASR), and the Vanderbilt Institute of Chemical Biology (fellowship, KES). We would like to thank Mostafa J. Khan for sharing his lipidomics data.

References

1. Alzheimer's Association (2018) 2018 Alzheimer's disease facts and figures. Alzheimers Dement 14:367–429
2. Barnes LL, Bennett DA (2014) Alzheimer's disease in African Americans: risk factors and challenges for the future. Health Aff (Millwood) 33:580–586
3. Lines L, Sherif NA, Wiener J (2014) Racial and ethnic disparities among individuals with Alzheimer's disease in the United States: a literature review. RTI Press, Research Triangle Park, NC. https://pdfs.semanticscholar.org/7ed5/b9ed14cd3e002df546134e76766d01c-5c4aa.pdf
4. Matthews KA, Xu W, Gaglioti AH, Holt JB, Croft JB, Mack D et al (2018) Racial and ethnic estimates of Alzheimer's disease and related dementias in the United States (2015–2060) in adults aged ≥65 years. Alzheimers Dement. https://doi.org/10.1016/j.jalz.2018.06.3063 [Epub ahead of print]
5. Gottesman RF, Fornage M, Knopman DS, Mosley TH (2015) Brain aging in African-Americans: the atherosclerosis risk in communities (ARIC) experience. Curr Alzheimer Res 12:607–613
6. Manly JJ, Mayeux R (2004) Ethnic differences in dementia and Alzheimer's disease. In: Anderson NB, Bulatao RA, Cohen B (eds) Critical perspectives on racial and ethnic differences in health in late life. National Academies Press, Washington, DC. ASIN: B00FBZPHCU
7. Mehta KM, Yeo GW (2017) Systematic review of dementia prevalence and incidence in United States race/ethnic populations. Alzheimers Dement 13:72–83
8. Alzheimer's Association (2017) Alzheimer's disease facts and figures. Alzheimers Dement 13:325–373
9. Chin AL, Negash S, Hamilton R (2011) Diversity and disparity in dementia: the impact of ethnoracial differences in Alzheimer disease. Alzheimer Dis Assoc Disord 25:187–195
10. Mayeda ER, Glymour MM, Quesenberry CP, Whitmer RA (2015) Inequalities in dementia incidence between six racial and ethnic groups over 14 years. Alzheimers Dement 12:216–224
11. Burke SL, Cadet T, Maddux M (2017) Chronic health illnesses as predictors of mild cognitive impairment among African American older adults. J Natl Med Assoc 110(4):314–325
12. Gilligan AM, Malone DC, Warholak TL, Armstrong EP (2012) Racial and ethnic disparities in Alzheimer's disease pharmacotherapy exposure: an analysis across four state Medicaid populations. Am J Geriatr Pharmacother 10:303–312
13. Liu Q, Zhang J (2014) Lipid metabolism in Alzheimer's disease. Neurosci Bull 30:331–345
14. Gamba P, Testa G, Sottero B, Gargiulo S, Poli G, Leonarduzzi G (2012) The link between altered cholesterol metabolism and Alzheimer's disease. Ann N Y Acad Sci 1259:54–64
15. Martins IJ, Berger T, Sharman MJ, Verdile G, Fuller SJ, Martins RN (2009) Cholesterol metabolism and transport in the pathogenesis of Alzheimer's disease. J Neurochem 111:1275–1308
16. El Gaamouch F, Jing P, Xia J, Cai D (2016) Alzheimer's disease risk genes and lipid regulators. J Alzheimers Dis 53:15–29
17. Burns M, Duff K (2002) Cholesterol in Alzheimer's disease and tauopathy. Ann N Y Acad Sci 977:367–375

18. Sato N, Morishita R (2015) The roles of lipid and glucose metabolism in modulation of β-amyloid, tau, and neurodegeneration in the pathogenesis of Alzheimer disease. Front Aging Neurosci 7:199. https://doi.org/10.3389/fnagi.2015.00199
19. Torres M, Busquets X, Escribá PV (2016) Brain lipids in the pathophysiology and treatment of Alzheimer's disease. In: Moretti DV (ed) Update on dementia. InTech, Rijeka, pp 127–167. https://doi.org/10.5772/64757
20. Wong MW, Braidy N, Poljak A, Pickford R, Thambisetty M, Sachdev PS (2017) Dysregulation of lipids in Alzheimer's disease and their role as potential biomarkers. Alzheimers Dement 13:810–827
21. Zamrini E, Parrish JA, Parsons D, Harrell LE (2004) Medical comorbidity in black and white patients with Alzheimer's disease. South Med J 97:2–6
22. Barnes LL, Leurgans S, Aggarwal NT, Shah RC, Arvanitakis Z, James BD et al (2015) Mixed pathology is more likely in black than white decedents with Alzheimer dementia. Neurology 85:528–534
23. Gottesman RF, Schneider AC, Zhou Y, Coresh J, Green E, Gupta N et al (2017) Association between midlife vascular risk factors and estimated brain amyloid deposition. JAMA 317:1443–1450
24. Wilkins CH, Grant EA, Schmitt SE, McKeel DW, Morris JC (2006) The neuropathology of Alzheimer disease in African American and white individuals. Arch Neurol 63:87–90
25. Graff-Radford NR, Besser LM, Crook JE, Kukull WA, Dickson DW (2016) Neuropathological differences by race from the National Alzheimer's coordinating center. Alzheimers Dement 12:669–677
26. Mortimer JA, Graves AB (1993) Education and other socioeconomic determinants of dementia and Alzheimer's disease. Neurology 43:S39–S44
27. Borenstein AR, Copenhaver CI, Mortimer JA (2006) Early-life risk factors for Alzheimer's disease. Alzheimer Dis Assoc Disord 20:63–72
28. Ramos-Cejudo J, Wisniewski T, Marmar C, Zetterberg H, Blennow K, de Leon MJ et al (2018) Traumatic brain injury and Alzheimer's disease: the cerebrovascular link. EBioMedicine 28:21–30
29. Honig LS, Tang MX, Albert S, Costa R, Luchsinger J, Manly J et al (2003) Stroke and the risk of Alzheimer disease. Arch Neurol 60:1707–1712
30. Chakrabarti S, Khemka VK, Banerjee A, Chatterjee G, Ganguly A, Biswas A (2015) Metabolic risk factors of sporadic Alzheimer's disease: implications in the pathology, pathogenesis and treatment. Aging Dis 6:282–299
31. Matsuzaki T, Sasaki K, Hata J, Hirakawa Y, Fujimi K, Ninomiya T et al (2011) Association of Alzheimer disease pathology with abnormal lipid metabolism: the Hisayama study. Neurology 77:1068–1075
32. Nday CM, Eleftheriadou D, Jackson G (2017) Shared pathological pathways of Alzheimer's disease with specific comorbidities: current perspectives and interventions. J Neurochem 144(4):360–389
33. Carnethon MR, Pu J, Howard G, Albert MA, Anderson CAM, Bertoni AG et al (2017) Cardiovascular health in African Americans: a scientific statement from the American Heart Association. Circulation 136:e393–e423
34. Di Paolo G, Kim TW (2011) Linking lipids to Alzheimer's disease: cholesterol and beyond. Nat Rev Neurosci 12:284–296
35. Martins IJ (2015) Diabetes and cholesterol dyshomeostasis involve abnormal α-synuclein and amyloid beta transport in neurodegenerative diseases. Austin Alzheimer's and Parkinson's Disease. https://api.research-repository.uwa.edu.au/portalfiles/portal/17312757/Full_publication.pdf
36. Evans RM, Emsley CL, Gao S, Sahota A, Hall KS, Farlow MR et al (2000) Serum cholesterol, APOE genotype, and the risk of Alzheimer's disease: a population-based study of African Americans. Neurology 54:240–242
37. Xu W, Tan L, Wang H-F, Jiang T, Tan M-S, Tan L et al (2015) Meta-analysis of modifiable risk factors for Alzheimer's disease. J Neurol Neurosurg Psychiatry 86(12):1299–1306

38. Gonzalez HM, Tarraf W, Harrison K, Windham BG, Tingle J, Alonso A et al (2017) Midlife cardiovascular health and 20-year cognitive decline: atherosclerosis risk in communities study results. Alzheimers Dement 14(5):579–589
39. Howard G, Safford MM, Moy CS, Howard VJ, Kleindorfer DO, Unverzagt FW et al (2017) Racial differences in the incidence of cardiovascular risk factors in older black and white adults. J Am Geriatr Soc 65:83–90
40. Osuji CU, Omejua EG, Onwubuya EI, Ahaneku GI (2012) Serum lipid profile of newly diagnosed hypertensive patients in Nnewi, south-East Nigeria. Int J Hypertens 2012:710486. https://doi.org/10.1155/2012/710486
41. Barnes DE, Yaffe K (2011) The projected effect of risk factor reduction on Alzheimer's disease prevalence. Lancet Neurol 10:819–828
42. Arvanitakis Z, Capuano AW, Lamar M, Shah RC, Barnes LL, Bennett DA et al (2018) Late-life blood pressure association with cerebrovascular and Alzheimer disease pathology. Neurology 91(6):e517–e525
43. Williamson JW (2018) A randomized trial of intensive versus standard systolic blood pressure control and the risk of mild cognitive impairment and dementia: results from SPRINT MIND. Proceedings of Alzheimer's Association International Conference 2018, Chicago, IL, USA. ID 27525
44. Nasrallah IM (2018) A randomized trial of intensive versus standard systolic blood pressure control on brain structure: results from SPRINT MIND MRI. Proceedings of Alzheimer's Association International Conference 2018, Chicago, IL, USA. ID 27526
45. Fuchs FD (2011) Why do black Americans have higher prevalence of hypertension? An enigma still unsolved. Hypertension 57:379–380
46. Lackland DT (2014) Racial differences in hypertension: implications for high blood pressure management. Am J Med Sci 348:135–138
47. Muntner P, He J, Cutler JA, Wildman RP, Whelton PK (2004) Trends in blood pressure among children and adolescents. JAMA 291:2107–2113
48. Redmond N, Baer HJ, Hicks LS (2011) Health behaviors and racial disparity in blood pressure control in the National Health and nutrition examination survey. Hypertension 57:383–389
49. Arnold SE, Arvanitakis Z, Macauley-Rambach SL, Koenig AM, Wang HY, Ahima RS et al (2018) Brain insulin resistance in type 2 diabetes and Alzheimer disease: concepts and conundrums. Nat Rev Neurol 14(3):168–181
50. Boden G, Laakso M (2004) Lipids and glucose in type 2 diabetes: what is the cause and effect? Diabetes Care 27:2253–2259
51. Savage DB, Petersen KF, Shulman GI (2007) Disordered lipid metabolism and the pathogenesis of insulin resistance. Physiol Rev 87:507–520
52. Schilling MA (2016) Unraveling Alzheimer's: making sense of the relationship between diabetes and Alzheimer's disease. J Alzheimers Dis 51:961–977
53. Marseglia A, Fratiglioni L, Kalpouzos G, Wang R, Bäckman L, Xu W (2018) Prediabetes and diabetes accelerate cognitive decline and predict microvascular lesions: a population-based cohort study. Alzheimers Dement Aug 13. doi: 10.1016/j.jalz.2018.06.3060. [Epub ahead of print]
54. Cheng D, Noble J, Tang MX, Schupf N, Mayeux R, Luchsinger JA (2011) Type 2 diabetes and late-onset Alzheimer's disease. Dement Geriatr Cogn Disord 31:424–430
55. Bangen KJ, Gu Y, Gross AL, Schneider BC, Skinner JC, Benitez A et al (2015) Relationship between type 2 diabetes mellitus and cognitive change in a multiethnic elderly cohort. J Am Geriatr Soc 63:1075–1083
56. Willette AA, Johnson SC, Birdsill AC, Sager MA, Christian B, Baker LD et al (2015) Insulin resistance predicts brain amyloid deposition in late middle-aged adults. Alzheimers Dement 11:504–510.e1
57. Pimentel dos Santos Matioli MN, Suemoto CK, Rodriguez RD, Farias DS, da Silva MM, Paraizo Leite RE et al (2017) Diabetes is not associated with Alzheimer's disease neuropathology. J Alzheimers Dis 60:1035–1043

58. Heitner J, Dickson D (1997) Diabetics do not have increased Alzheimer-type pathology compared with age-matched control subjects: a retrospective postmortem immunocytochemical and histofluorescent study. Neurology 49:1306–1311
59. Arvanitakis Z, Schneider JA, Wilson RS, Li Y, Arnold SE, Wang Z et al (2006) Diabetes is related to cerebral infarction but not to AD pathology in older persons. Neurology 67:1960–1965
60. Abner EL, Nelson PT, Kryscio RJ, Schmitt FA, Fardo DW, Woltjer RL et al (2016) Diabetes is associated with cerebrovascular but not Alzheimer's disease neuropathology. Alzheimers Dement 12:882–889
61. Ahtiluoto S, Polvikoski T, Peltonen M, Solomon A, Tuomilehto J, Winblad B et al (2010) Diabetes, Alzheimer disease, and vascular dementia: a population-based neuropathologic study. Neurology 75:1195–1202
62. Brancati FL, Kao W, Folsom AR, Watson RL, Szklo M (2000) Incident type 2 diabetes mellitus in African American and white adults: the atherosclerosis risk in communities study. JAMA 283:2253–2259
63. Marshall M (2005) Diabetes in African Americans. Postgrad Med J 81:734–740
64. Mayeda ER, Haan MN, Neuhaus J, Yaffe K, Knopman DS, Sharrett AR et al (2014) Type 2 diabetes and cognitive decline over 14 years in middle-aged African Americans and whites: the ARIC brain MRI study. Neuroepidemiology 43:220–227
65. Arvanitakis Z, Bennett DA, Wilson RS, Barnes LL (2010) Diabetes and cognitive systems in older black and white persons. Alzheimer Dis Assoc Disord 24:37–42
66. Hendrie HC, Zheng M, Lane KA, Ambuehl R, Purnell C, Li S et al (2018) Changes of glucose levels precede dementia in African-Americans with diabetes but not in Caucasians. Alzheimers Dement 14(12):1572-1579
67. Hendrie HC, Zheng M, Li W, Lane K, Ambuehl R, Purnell C et al (2017) Glucose level decline precedes dementia in elderly African Americans with diabetes. Alzheimers Dement 13:111–118
68. Azarpazhooh MR, Avan A, Cipriano LE, Munoz DG, Sposato LA, Hachinski V (2017) Concomitant vascular and neurodegenerative pathologies double the risk of dementia. Alzheimers Dement 14:148–156
69. Jefferson AL, Hohman TJ, Liu D, Haj-Hassan S, Gifford KA, Benson EM et al (2015) Adverse vascular risk is related to cognitive decline in older adults. J Alzheimers Dis 44:1361–1373
70. Li J, Wang YJ, Zhang M, Xu ZQ, Gao CY, Fang CQ et al (2011) Vascular risk factors promote conversion from mild cognitive impairment to Alzheimer disease. Neurology 76:1485–1491
71. Toledo JB, Arnold SE, Raible K, Brettschneider J, Xie SX, Grossman M et al (2013) Contribution of cerebrovascular disease in autopsy confirmed neurodegenerative disease cases in the National Alzheimer's coordinating Centre. Brain 136:2697–2706
72. Gorelick PB (1998) Cerebrovascular disease in African Americans. Stroke 29:2656–2664
73. Morgenstern LB, Spears WD, Goff DC, Grotta JC, Nichaman MZ (1997) African Americans and women have the highest stroke mortality in Texas. Stroke 28:15–18
74. Sandberg G, Stewart W, Smialek J, Troncoso JC (2001) The prevalence of the neuropathological lesions of Alzheimer's disease is independent of race and gender. Neurobiol Aging 22:169–175
75. Arvanitakis Z, Leurgans SE, Fleischman DA, Schneider JA, Rajan KB, Pruzin JJ et al (2018) Memory complaints, dementia, and neuropathology in older blacks and whites. Ann Neurol 83:718–729
76. Raj T, Chibnik LB, McCabe C, Wong A, Replogle JM, Yu L et al (2017) Genetic architecture of age-related cognitive decline in African Americans. Neurol Genet 3(1):e125. https://doi.org/10.1212/NXG.0000000000000125
77. Vrièze FW-D, Compton D, Womick M, Arepalli S, Adighibe O, Li L et al (2007) ABCA1 polymorphisms and Alzheimer's disease. Neurosci Lett 416:180–183
78. Koldamova R, Fitz NF, Lefterov I (2010) The role of ATP-binding cassette transporter A1 in Alzheimer's disease and neurodegeneration. Biochim Biophys Acta 1801:824–830

79. Fehér Á, Giricz Z, Juhász A, Pákáski M, Janka Z, Kálmán J (2018) ABCA1 rs2230805 and rs2230806 common gene variants are associated with Alzheimer's disease. Neurosci Lett 664:79–83
80. Aikawa T, Holm ML, Kanekiyo T (2018) ABCA7 and pathogenic pathways of Alzheimer's disease. Brain Sci 8(2):E27. https://doi.org/10.3390/brainsci8020027
81. Almeida JFF, Dos Santos LR, Trancozo M, de Paula F (2018) Updated meta-analysis of BIN1, CR1, MS4A6A, CLU, and ABCA7 variants in Alzheimer's disease. J Mol Neurosci 64(3):471–477
82. Hollingworth P, Harold D, Sims R, Gerrish A, Lambert JC, Carrasquillo MM et al (2011) Common variants at ABCA7, MS4A6A/MS4A4E, EPHA1, CD33 and CD2AP are associated with Alzheimer's disease. Nat Genet 43:429–435
83. Naj AC, Jun G, Beecham GW, Wang LS, Vardarajan BN, Buros J et al (2011) Common variants at MS4A4/MS4A6E, CD2AP, CD33 and EPHA1 are associated with late-onset Alzheimer's disease. Nat Genet 43:436–441
84. Cuyvers E, De Roeck A, Van den Bossche T, Van Cauwenberghe C, Bettens K, Vermeulen S et al (2015) Mutations in ABCA7 in a Belgian cohort of Alzheimer's disease patients: a targeted resequencing study. Lancet Neurol 14:814–822
85. Lambert JC, Ibrahim-Verbaas CA, Harold D, Naj AC, Sims R, Bellenguez C et al (2013) Meta-analysis of 74,046 individuals identifies 11 new susceptibility loci for Alzheimer's disease. Nat Genet 45:1452–1458
86. Zhou Q, Zhao F, Lv Z-P, Zheng C-G, Zheng W-D, Sun L et al (2014) Association between APOC1 polymorphism and Alzheimer's disease: a case-control study and meta-analysis. PLoS One 9(1):e87017. https://doi.org/10.1371/journal.pone.0087017
87. Petit-Turcotte C, Stohl SM, Beffert U, Cohn JS, Aumont N, Tremblay M et al (2001) Apolipoprotein C-I expression in the brain in Alzheimer's disease. Neurobiol Dis 8:953–963
88. Ki C-S, Na DL, Kim DK, Kim HJ, Kim J-W (2002) Genetic association of an apolipoprotein C-I (APOC1) gene polymorphism with late-onset Alzheimer's disease. Neurosci Lett 319:75–78
89. Desai PP, Hendrie HC, Evans RM, Murrell JR, DeKosky ST, Kamboh MI (2003) Genetic variation in apolipoprotein D affects the risk of Alzheimer disease in African-Americans. Am J Med Genet B Neuropsychiatr Genet 116B:98–101
90. Zhao N, Liu C-C, Qiao W, Bu G (2017) Apolipoprotein E, receptors, and modulation of Alzheimer's disease. Biol Psychiatry 83(4):347–357
91. Reitz C, Jun G, Naj A, Rajbhandary R, Vardarajan BN, Wang L-S et al (2013) Variants in the ATP-binding cassette transporter (ABCA7), apolipoprotein E ε4, and the risk of late-onset Alzheimer disease in African Americans. JAMA 309:1483–1492
92. Reitz C, Mayeux R (2014) Genetics of Alzheimer's disease in Caribbean Hispanic and African American populations. Biol Psychiatry 75:534–541
93. Seshadri S, Fitzpatrick AL, Ikram MA, DeStefano AL, Gudnason V, Boada M et al (2010) Genome-wide analysis of genetic loci associated with Alzheimer disease. JAMA 303:1832–1840
94. Bertram L, Lange C, Mullin K, Parkinson M, Hsiao M, Hogan MF et al (2008) Genome-wide association analysis reveals putative Alzheimer's disease susceptibility loci in addition to APOE. Am J Hum Genet 83:623–632
95. Lambert J-C, Heath S, Even G, Campion D, Sleegers K, Hiltunen M et al (2009) Genome-wide association study identifies variants at CLU and CR1 associated with Alzheimer's disease. Nat Genet 41:1094–1099
96. Harold D, Abraham R, Hollingworth P, Sims R, Gerrish A, Hamshere ML et al (2009) Genome-wide association study identifies variants at CLU and PICALM associated with Alzheimer's disease. Nat Genet 41:1088–1093
97. Rogaeva E, Meng Y, Lee JH, Gu Y, Kawarai T, Zou F et al (2007) The neuronal sortilin-related receptor SORL1 is genetically associated with Alzheimer's disease. Nat Genet 39:168–177

98. Lee JH, Cheng R, Schupf N, Manly J, Lantigua R, Stern Y et al (2007) The association between genetic variants in SORL1 and Alzheimer's disease in an urban, multiethnic, community-based cohort. Arch Neurol 64:501–506
99. Chou C-T, Liao Y-C, Lee W-J, Wang S-J, Fuh J-L (2016) SORL1 gene, plasma biomarkers, and the risk of Alzheimer's disease for the Han Chinese population in Taiwan. Alzheimers Res Ther 8(1):53. https://doi.org/10.1186/s13195-016-0222-x
100. Ghani M, Reitz C, Cheng R, Vardarajan BN, Jun G, Sato C et al (2015) Association of long runs of homozygosity with Alzheimer disease among African American individuals. JAMA Neurol 72:1313–1323
101. Picard C, Julien C, Frappier J, Miron J, Théroux L, Dea D et al (2018) Alterations in cholesterol metabolism-related genes in sporadic Alzheimer's disease. Neurobiol Aging 66:180.e1–180.e9. https://doi.org/10.1016/j.neurobiolaging.2018.01.018
102. Bales KR (2010) Brain lipid metabolism, apolipoprotein E and the pathophysiology of Alzheimer's disease. Neuropharmacology 59:295–302
103. Girard H, Potvin O, Nugent S, Dallaire-Theroux C, Cunnane S, Duchesne S (2017) Faster progression from MCI to probable AD for carriers of a single-nucleotide polymorphism associated with type 2 diabetes. Neurobiol Aging 64:157.e11–157.e17. https://doi.org/10.1016/j.neurobiolaging.2017.11.013
104. Pirttila T, Soininen H, Heinonen O, Lehtimaki T, Bogdanovic N, Paljarvi L et al (1996) Apolipoprotein E (apoE) levels in brains from Alzheimer disease patients and controls. Brain Res 722:71–77
105. Peila R, Rodriguez BL, Launer LJ (2002) Type 2 diabetes, APOE gene, and the risk for dementia and related pathologies: the Honolulu-Asia aging study. Diabetes 51:1256–1262
106. Corlier F, Hafzalla G, Faskowitz J, Kuller LH, Becker JT, Lopez OL et al (2018) Systemic inflammation as a predictor of brain aging: contributions of physical activity, metabolic risk, and genetic risk. NeuroImage 172:118–129
107. Hall K, Murrell J, Ogunniyi A, Deeg M, Baiyewu O, Gao S et al (2006) Cholesterol, APOE genotype, and Alzheimer disease: an epidemiologic study of Nigerian Yoruba. Neurology 66:223–227
108. Lin Y-F, Smith AV, Aspelund T, Betensky RA, Smoller JW, Gudnason V et al (2018) Genetic overlap between vascular pathologies and Alzheimer's dementia and potential causal mechanisms. Alzheimers Dement Sep 19. doi: 10.1016/j.jalz.2018.08.002. [Epub ahead of print]
109. Cukier HN, Kunkle BW, Vardarajan BN, Rolati S, Hamilton-Nelson KL, Kohli MA et al (2016) ABCA7 frameshift deletion associated with Alzheimer disease in African Americans. Neurol Genet 2(3):e79. https://doi.org/10.1212/NXG.0000000000000079
110. Vasquez JB, Fardo DW, Estus S (2013) ABCA7 expression is associated with Alzheimer's disease polymorphism and disease status. Neurosci Lett 556:58–62
111. Hohman TJ, Cooke-Bailey JN, Reitz C, Jun G, Naj A, Beecham GW et al (2016) Global and local ancestry in African-Americans: implications for Alzheimer's disease risk. Alzheimers Dement 12:233–243
112. Reitz C, Mayeux R (2014) Alzheimer disease: epidemiology, diagnostic criteria, risk factors and biomarkers. Biochem Pharmacol 88:640–651
113. Han Z, Huang H, Gao Y, Huang Q (2017) Functional annotation of Alzheimer's disease associated loci revealed by GWASs. PLoS One 12(6):e0179677. https://doi.org/10.1371/journal.pone.0179677
114. Jones L, Harold D, Williams J (2010) Genetic evidence for the involvement of lipid metabolism in Alzheimer's disease. Biochim Biophys Acta 1801:754–761
115. Fitz NF, Cronican AA, Saleem M, Fauq AH, Chapman R, Lefterov I et al (2012) Abca1 deficiency affects Alzheimer's disease-like phenotype in human ApoE4 but not in ApoE3-targeted replacement mice. J Neurosci 32:13125–13136
116. Wahrle SE, Jiang H, Parsadanian M, Kim J, Li A, Knoten A et al (2008) Overexpression of ABCA1 reduces amyloid deposition in the PDAPP mouse model of Alzheimer disease. J Clin Invest 118:671–682

117. Felsky D, Szeszko P, Yu L, Honer WG, De Jager PL, Schneider JA et al (2013) The SORL1 gene and convergent neural risk for Alzheimer's disease across the human lifespan. Mol Psychiatry 19:1125–1132
118. Power MC, Rawlings A, Sharrett AR, Bandeen-Roche K, Coresh J, Ballantyne CM et al (2017) Association of midlife lipids with 20-year cognitive change: a cohort study. Alzheimers Dement 14(2):167–177
119. Koch M, DeKosky ST, Fitzpatrick AL, Furtado JD, Lopez OL, Kuller LH et al (2018) Apolipoproteins and Alzheimer's pathophysiology. Alzheimer's Dement (Amst) 10:545. https://doi.org/10.1016/j.dadm.2018.07.001
120. Hughes TM, Lopez OL, Evans RW, Kamboh MI, Williamson JD, Klunk WE et al (2014) Markers of cholesterol transport are associated with amyloid deposition in the brain. Neurobiol Aging 35:802–807
121. Chiasserini D, Biscetti L, Eusebi P, Salvadori N, Frattini G, Simoni S et al (2017) Differential role of CSF fatty acid binding protein 3, α-synuclein, and Alzheimer's disease core biomarkers in Lewy body disorders and Alzheimer's dementia. Alzheimers Res Ther 9(1):52. https://doi.org/10.1186/s13195-017-0276-4
122. Krishnan B, Kayed R, Taglialatela G (2018) Elevated phospholipase D isoform 1 in Alzheimer's disease patients' hippocampus: relevance to synaptic dysfunction and memory deficits. Alzheimers Dement (N.Y.) 4:89–102
123. Ferguson SA, Panos JJ, Sloper D, Varma V (2017) Neurodegenerative markers are increased in postmortem BA21 tissue from African Americans with Alzheimer's disease. J Alzheimers Dis 59:57–66
124. Moya-Alvarado G, Gershoni-Emek N, Perlson E, Bronfman FC (2015) Neurodegeneration and Alzheimer's disease. What can proteomics tell us about the Alzheimer's brain? Mol Cell Proteomics 15(2):409–425
125. Robinson RAS, Amin B, Guest PC (2017) Multiplexing biomarker methods, proteomics and considerations for Alzheimer's disease. Adv Exp Med Biol 974:21–48
126. Andreev VP, Petyuk VA, Brewer HM, Karpievitch YV, Xie F, Clarke J et al (2012) Label-free quantitative LC-MS proteomics of Alzheimer's disease and normally aged human brains. J Proteome Res 11:3053–3067
127. Begcevic I, Kosanam H, Martinez-Morillo E, Dimitromanolakis A, Diamandis P, Kuzmanov U et al (2013) Semiquantitative proteomic analysis of human hippocampal tissues from Alzheimer's disease and age-matched control brains. Clin Proteomics 10(1):5. https://doi.org/10.1186/1559-0275-10-5
128. Evans AR, Gu L, Guerrero R, Robinson RAS (2015) Global cPILOT analysis of the APP/PS-1 mouse liver proteome. Proteomics Clin Appl 9:872–884
129. Fania C, Arosio B, Capitanio D, Torretta E, Gussago C, Ferri E et al (2017) Protein signature in cerebrospinal fluid and serum of Alzheimer's disease patients: the case of apolipoprotein A-1 proteoforms. PLoS One 12(6):e0179280. https://doi.org/10.1371/journal.pone.0179280
130. Hondius DC, van Nierop P, Li KW, Hoozemans JJM, van der Schors RC, van Haastert ES et al (2016) Profiling the human hippocampal proteome at all pathologic stages of Alzheimer's disease. Alzheimers Dement 12:654–668
131. Lopez MF, Mikulskis A, Kuzdzal S, Bennett DA, Kelly J, Golenko E et al (2005) High-resolution serum proteomic profiling of Alzheimer disease samples reveals disease-specific, carrier-protein–bound mass signatures. Clin Chem 51:1946–1954
132. Manavalan A, Mishra M, Feng L, Sze SK, Akatsu H, Heese K (2013) Brain site-specific proteome changes in aging-related dementia. Exp Mol Med 45:e39. https://doi.org/10.1038/emm.2013.76
133. Minjarez B, Calderon-Gonzalez KG, Rustarazo ML, Herrera-Aguirre ME, Labra-Barrios ML, Rincon-Limas DE et al (2016) Identification of proteins that are differentially expressed in brains with Alzheimer's disease using iTRAQ labeling and tandem mass spectrometry. J Proteome 139:103–121
134. Muenchhoff J, Poljak A, Song F, Raftery M, Brodaty H, Duncan M et al (2015) Plasma protein profiling of mild cognitive impairment and Alzheimer's disease across two independent cohorts. J Alzheimers Dis 43:1355–1373

135. Musunuri S, Wetterhall M, Ingelsson M, Lannfelt L, Artemenko K, Bergquist J et al (2014) Quantification of the brain proteome in Alzheimer's disease using multiplexed mass spectrometry. J Proteome Res 13:2056–2068
136. Neuner SM, Wilmott LA, Hoffmann BR, Mozhui K, Kaczorowski CC (2017) Hippocampal proteomics defines pathways associated with memory decline and resilience in normal aging and Alzheimer's disease mouse models. Behav Brain Res 322:288–298
137. Sultana R, Boyd-Kimball D, Cai J, Pierce WM, Klein JB, Merchant M et al (2007) Proteomics analysis of the Alzheimer's disease hippocampal proteome. J Alzheimers Dis 11:153–164
138. Tsuji T, Shiozaki A, Kohno R, Yoshizato K, Shimohama S (2002) Proteomic profiling and neurodegeneration in Alzheimer's disease. Neurochem Res 27:1245–1253
139. Zahid S, Oellerich M, Asif AR, Ahmed N (2014) Differential expression of proteins in brain regions of Alzheimer's disease patients. Neurochem Res 39:208–215
140. Gu L, Evans AR, Robinson RAS (2015) Sample multiplexing with cysteine-selective approaches: cysDML and cPILOT. J Am Soc Mass Spectrom 26:615–630
141. Aluise CD, Robinson RAS, Beckett TL, Murphy MP, Cai J, Pierce WM et al (2010) Preclinical Alzheimer disease: brain oxidative stress, Aβ peptide & proteomics. Neurobiol Dis 39:221–228
142. Aluise CD, Robinson RAS, Cai J, Pierce WM, Markesbery WR, Butterfield DA (2011) Redox proteomics analyses of brains from subjects with amnestic mild cognitive impairment compared to brains from subjects with preclinical Alzheimer's disease: insights into memory loss in MCI. J Alzheimers Dis 23:257–269
143. Castegna A, Aksenov M, Thongboonkerd V, Klein JB, Pierce WM, Booze R et al (2002) Proteomic identification of oxidatively modified proteins in Alzheimer's disease brain. Part II: Dihydropyrimidinase-related protein 2, alpha-enolase and heat shock cognate 71. J Neurochem 82:1524–1532
144. Reed TT, Pierce WM Jr, Turner DM, Markesbery WR, Butterfield DA (2009) Proteomic identification of nitrated brain proteins in early Alzheimer's disease inferior parietal lobule. J Cell Mol Med 13:2019–2029
145. Robinson RAS, Joshi G, Huang Q, Sultana R, Baker AS, Cai J et al (2011) Proteomics analysis of brain proteins in APP/PS-1 human double mutant knock-in mice with increasing amyloid β-peptide deposition: insights into the effects of in vivo treatment with N-acetylcysteine as a potential therapeutic intervention in mild cognitive impairment and Alzheimer disease. Proteomics 11:4243–4256
146. Ping L, Duong DM, Yin L, Gearing M, Lah JJ, Levey AI et al (2018) Global quantitative analysis of the human brain proteome in Alzheimer's and Parkinson's disease. Sci Data 5:180036. https://doi.org/10.1038/sdata.2018.36
147. Brinkmalm A, Portelius E, Ohrfelt A, Brinkmalm G, Andreasson U, Gobom J et al (2015) Explorative and targeted neuroproteomics in Alzheimer's disease. Biochim Biophys Acta 1854:769–778
148. Korolainen MA, Nyman TA, Aittokallio T, Pirttila T (2010) An update on clinical proteomics in Alzheimer's research. J Neurochem 112:1386–1414
149. Papassotiropoulos A, Fountoulakis M, Dunckley T, Stephan DA, Reiman EM (2006) Genetics, transcriptomics, and proteomics of Alzheimer's disease. J Clin Psychiatry 67:652–670
150. Paterson RW, Heywood WE, Heslegrave AJ, Magdalinou NK, Andreasson U, Sirka E et al (2016) A targeted proteomic multiplex CSF assay identifies increased malate dehydrogenase and other neurodegenerative biomarkers in individuals with Alzheimer's disease pathology. Transl Psychiatry 6(11):e952. https://doi.org/10.1038/tp.2016.194
151. Heywood WE, Galimberti D, Bliss E, Sirka E, Paterson RW, Magdalinou NK et al (2015) Identification of novel CSF biomarkers for neurodegeneration and their validation by a high-throughput multiplexed targeted proteomic assay. Mol Neurodegener 10:64. https://doi.org/10.1186/s13024-015-0059-y

152. Begcevic I, Brinc D, Brown M, Martinez-Morillo E, Goldhardt O, Grimmer T et al (2018) Brain-related proteins as potential CSF biomarkers of Alzheimer's disease: a targeted mass spectrometry approach. J Proteome 182:12–20
153. Yu L, Petyuk VA, Gaiteri C, Mostafavi S, Young-Pearse T, Shah RC et al (2018) Targeted brain proteomics uncover multiple pathways to Alzheimer's dementia. Ann Neurol 84(1):78–88
154. Brinkmalm G, Sjödin S, Simonsen AH, Hasselbalch SG, Zetterberg H, Brinkmalm A et al (2018) A parallel reaction monitoring mass spectrometric method for analysis of potential CSF biomarkers for Alzheimer's disease. Proteomics Clin Appl 12(1). https://doi.org/10.1002/prca.201700131
155. Chang RY, Etheridge N, Dodd PR, Nouwens AS (2014) Targeted quantitative analysis of synaptic proteins in Alzheimer's disease brain. Neurochem Int 75:66–75
156. Oeckl P, Metzger F, Nagl M, von Arnim CA, Halbgebauer S, Steinacker P et al (2016) Alpha-, beta-, and gamma-synuclein quantification in cerebrospinal fluid by multiple reaction monitoring reveals increased concentrations in Alzheimer's and Creutzfeldt-Jakob disease but no alteration in synucleinopathies. Mol Cell Proteomics 15:3126–3138
157. Dittrich J, Adam M, Maas H, Hecht M, Reinicke M, Ruhaak LR et al (2018) Targeted online SPE-LC-MS/MS assay for the quantitation of 12 apolipoproteins from human blood. Proteomics 18(3–4). https://doi.org/10.1002/pmic.201700279
158. Chen J, Wang M, Turko IV (2012) Mass spectrometry quantification of clusterin in the human brain. Mol Neurodegener 7:41. https://doi.org/10.1186/1750-1326-7-41
159. Henderson CM, Bollinger JG, Becker JO, Wallace JM, Laha TJ, MacCoss MJ et al (2017) Quantification by nano liquid chromatography parallel reaction monitoring mass spectrometry of human apolipoprotein A-I, apolipoprotein B, and hemoglobin A1c in dried blood spots. Proteomics Clin Appl 11(7–8). https://doi.org/10.1002/prca.201600103
160. Cheon MS, Kim SH, Fountoulakis M, Lubec G (2003) Heart type fatty acid binding protein (H-FABP) is decreased in brains of patients with down syndrome and Alzheimer's disease. J Neural Transm Suppl 67:225–234
161. Ijsselstijn L, Papma JM, Dekker LJ, Calame W, Stingl C, Koudstaal PJ et al (2013) Serum proteomics in amnestic mild cognitive impairment. Proteomics 13:2526–2533
162. Kennedy MA, Moffat TC, Gable K, Ganesan S, Niewola-Staszkowska K, Johnston A et al (2016) A signaling lipid associated with Alzheimer's disease promotes mitochondrial dysfunction. Sci Rep 6:19332. https://doi.org/10.1038/srep19332
163. Li D, Misialek JR, Boerwinkle E, Gottesman RF, Sharrett AR, Mosley TH et al (2017) Prospective associations of plasma phospholipids and mild cognitive impairment/dementia among African Americans in the ARIC neurocognitive study. Alzheimers Dement (Amst) 6:1–10
164. Hasin Y, Seldin M, Lusis A (2017) Multi-omics approaches to disease. Genome Biol 18(1):83. https://doi.org/10.1186/s13059-017-1215-1
165. Karahalil B (2016) Overview of systems biology and omics technologies. Curr Med Chem 23:4221–4230
166. Pimplikar SW (2017) Multi-omics and Alzheimer's disease: a slower but surer path to an efficacious therapy? Am J Physiol Cell Physiol 313:C1–C2
167. Tosto G, Reitz C (2016) Use of "omics" technologies to dissect neurologic disease. Handb Clin Neurol 138:91–106
168. Jaeger PA, Lucin KM, Britschgi M, Vardarajan B, Huang R-P, Kirby ED et al (2016) Network-driven plasma proteomics expose molecular changes in the Alzheimer's brain. Mol Neurodegener 11:31. https://doi.org/10.1186/s13024-016-0095-2
169. Seyfried NT, Dammer EB, Swarup V, Nandakumar D, Duong DM, Yin L et al (2017) A multi-network approach identifies protein-specific co-expression in asymptomatic and symptomatic Alzheimer's disease. Cell Syst 4:60–72.e4
170. Zhang Q, Ma C, Gearing M, Wang PG, Chin LS, Li L (2018) Integrated proteomics and network analysis identifies protein hubs and network alterations in Alzheimer's disease. Acta Neuropathol Commun 6(1):19. https://doi.org/10.1186/s40478-018-0524-2

171. Whiley L, Sen A, Heaton J, Proitsi P, García-Gómez D, Leung R et al (2014) Evidence of altered phosphatidylcholine metabolism in Alzheimer's disease. Neurobiol Aging 35:271–278
172. Klavins K, Koal T, Dallmann G, Marksteiner J, Kemmler G, Humpel C (2015) The ratio of phosphatidylcholines to lysophosphatidylcholines in plasma differentiates healthy controls from patients with Alzheimer's disease and mild cognitive impairment. Alzheimers Dement (Amst) 1:295–302
173. Li D, Misialek JR, Boerwinkle E, Gottesman RF, Sharrett AR, Mosley TH et al (2016) Plasma phospholipids and prevalence of mild cognitive impairment and/or dementia in the ARIC neurocognitive study (ARIC-NCS). Alzheimers Dement (Amst) 3:73–82
174. Mielke MM, Bandaru VVR, Haughey NJ, Rabins PV, Lyketsos CG, Carlson MC (2010) Serum sphingomyelins and ceramides are early predictors of memory impairment. Neurobiol Aging 31:17–24

Chapter 2
The Role of Biomarkers in Alzheimer's Disease Drug Development

Jeffrey Cummings

Abstract Biomarkers have a key role in Alzheimer's disease (AD) drug development. Biomarkers can assist in diagnosis, demonstrate target engagement, support disease modification, and monitor for safety. The amyloid (A), tau (T), neurodegeneration (N) Research Framework emphasizes brain imaging and CSF measures relevant to disease diagnosis and staging and can be applied to drug development and clinical trials. Demonstration of target engagement in Phase 2 is critical before advancing a treatment candidate to Phase 3. Trials with biomarker outcomes are shorter and smaller than those required to show clinical benefit and are important to understanding the biological impact of an agent and inform go/no-go decisions. Companion diagnostics are required for safe and effective use of treatments and may emerge in AD drug development programs. Complementary biomarkers inform the use of therapies but are not mandatory for use. Biomarkers promise to de-risk AD drug development, attract sponsors to AD research, and accelerate getting new drugs to those with or at risk for AD.

Keywords Alzheimer's disease · Drug development · Clinical trials · Biomarker · Positron-emission tomography · Amyloid · Tau · Neurodegeneration

2.1 The Role of Biomarkers in Alzheimer's Disease Drug Development

Alzheimer's disease (AD) is a neurodegenerative disorder that progressively compromises cognition, function, and behavior [1, 2]. AD becomes more common in the elderly and is reaching epidemic proportions with the graying of the global population. The frequency of AD doubles in frequency every 5 years after the age of 60 [3]. An estimated current 35 million victims worldwide will grow to over 130 million by

J. Cummings (✉)
Cleveland Clinic Lou Ruvo Center for Brain Health, Las Vegas, NV, USA
e-mail: cumminj@ccf.org

P. C. Guest (ed.), *Reviews on Biomarker Studies in Psychiatric and Neurodegenerative Disorders*, Advances in Experimental Medicine and Biology 1118,
https://doi.org/10.1007/978-3-030-05542-4_2

2050 [4]. The cost of AD to the global economy will increase from its estimated 818 billion US dollars (USD) in 2015 to 2 trillion USD by 2030 [5]. To address this growing public health crisis, it is critical that treatments that defer the onset, slow the progression, or improve the symptoms of AD be identified.

There is high failure rate of AD drug development; there have been no new drug approvals for AD since 2003, and the failure rate in development programs exceeds 99% [6]. To advance new therapies for AD, it is imperative that the vulnerabilities of the drug development process be identified and addressed. The improvement must span target identification, drug screening and optimization, use and interpretation of animal models of AD, and the clinical trial process [7]. The risk of AD drug development is high, and biomarkers represent a promising means of reducing the risk and increasing the likelihood of technical success. Understanding of AD is improving rapidly, and key biological events are being identified. In some cases these events are accompanied by biomarkers measurable by brain imaging or in the cerebrospinal fluid (CSF) or blood. The use of these biomarkers to improve the drug development process can de-risk AD drug development. This contribution describes the increasing role of biomarkers in AD drug development.

Several new advances relevant to biomarkers are included in this review. The amyloid (A), tau (T), neurodegeneration (N) Research Framework uses biomarkers to diagnose AD [8]. These same biomarkers can also serve important roles in drug development including demonstrating target engagement or providing support for disease modification [9]. The US Food and Drug Administration (FDA) developed an AD staging system that facilitates trials in patients with preclinical and prodromal AD and emphasizes the potential role for biomarkers in drug development in early AD [10]. This staging system and the use of biomarkers is described and accelerated approval of new treatments are discussed. The use of biomarkers in both disease-modifying and symptomatic drug development is presented.

2.2 Overview of Biomarkers in AD Drug Development

A biomarker is a characteristic that is objectively measured and evaluated as an indicator of normal biological processes, pathologic processes, or biological responses to a therapeutic intervention [11]. Biomarkers help characterize the baseline state, a disease process, or a response to treatment. Biomarkers include measures of genes, "omics" technologies (genomics, transcriptomics, proteomics, metabolomics, lipidomics), imaging, blood, electrocardiograms, or evaluations of organ function (e.g., liver functions, etc.) [12, 13]. The National Institutes of Health (NIH) developed an extensive glossary of biomarker-related terms—Biomarkers, EndpointS, and other Tools (BEST) resource—to provide a shared vocabulary for biomarker discussions [14].

Table 2.1 presents an overview of the roles played by biomarkers in AD drug development. The principal uses of biomarkers include demonstrating the presence of AD-type pathological changes with CSF measures or amyloid positron-emission

Table 2.1 Roles of biomarkers in AD drug development with examples

Role in trial	Biology identified	Fluid biomarker	Brain imaging
Diagnosis and participant identification	Presence of AD-type pathological changes	Low CSF Aβ42 or CSF Aβ42/t-tau ratio or Aβ42/p-tau ratio	Positive amyloid imaging
Target engagement	Reduction of amyloid production	Reduced Aβ42 production as shown by SILK	
	Removal of aggregated Aβ		Reduced Aβ aggregation as shown by amyloid imaging
Support for disease modification	Reduction of measures of neurodegeneration compared to placebo	Reduced CSF t-tau	Drug-placebo difference in favor of active treatment for FDG PET hypometabolism or MRI atrophy
Analytic stratification	Identification of ApoE-4 carrier status	ApoE genotype	
Adverse effect monitoring	Effects on the liver or blood	Blood liver function tests, complete blood count	
	Production of ARIA by monoclonal antibodies		MRI monitoring for ARIA

Aβ42 amyloid beta protein 42 amino acid length fragment, *AD* Alzheimer's disease, *ApoE* apolipoprotein E, *ARIA* amyloid-related imaging abnormalities, *CSF* cerebrospinal fluid, *FDG* fluorodeoxyglucose, *MRI* magnetic resonance imaging, *PET* positron-emission tomography, *SILK* stable isotope labeling kinetics

tomography (PET) for inclusion in AD trials, demonstrating target engagement by the candidate therapy, generating supportive evidence of disease modification, informing analytic stratification, and monitoring of adverse effects of treatment. Means of scoring biomarkers to increase confidence in their role in drug development have been proposed but not yet widely adopted [15].

2.3 A,T,N Framework for Alzheimer's Disease Diagnosis and Characterization

The A,T,N Research Framework uses biomarkers to diagnose and characterize AD [8]. Amyloid measures include amyloid PET (Fig. 2.1) and CSF amyloid beta (Aβ) protein; tau measures include tau PET (Fig. 2.2) and CSF phosphorylated tau (p-tau); neurodegeneration is reflected in atrophy on magnetic resonance imaging (MRI) (Fig. 2.3), CSF levels of total tau (t-tau), or fluorodeoxyglucose (FDG) PET (Fig. 2.4). In this approach, reduced N in the treatment groups compared to the placebo group is the object of disease-modifying therapy (DMT) [16, 17].

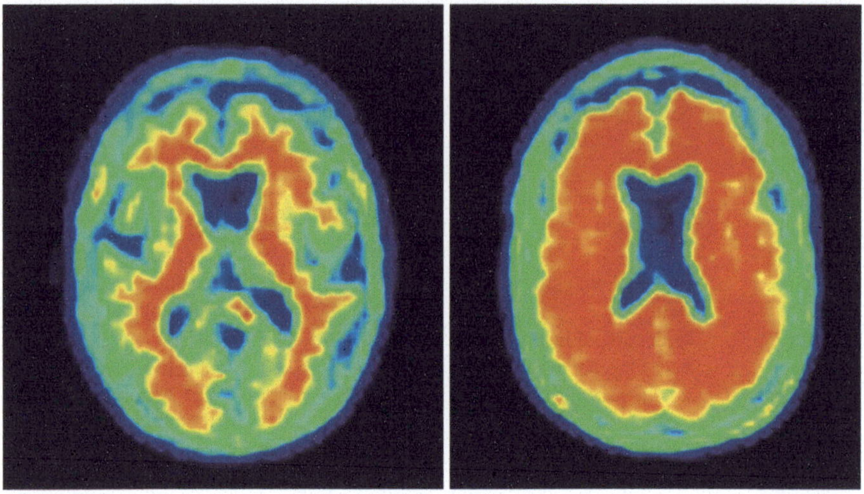

Fig. 2.1 Normal (left) and abnormal (right) amyloid PET

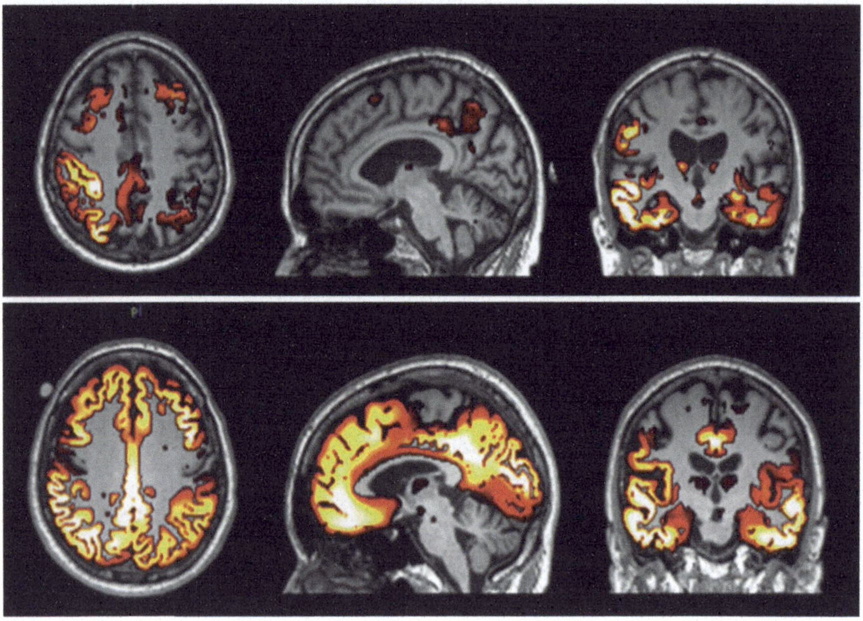

Fig. 2.2 Low tau (above) and high tau (below) PET aligned with MRI (images courtesy of Dawn Matthews)

Reductions in aggregated Aβ on amyloid PET or changes in CSF Aβ42 demonstrate impact on A, and drug-placebo differences in aggregated tau on tau PET or CSF p-tau establish effects on T. Amyloid PET measures the aggregated, deposited fibrillar, insoluble form of Aβ, and CSF amyloid is a measure of the soluble

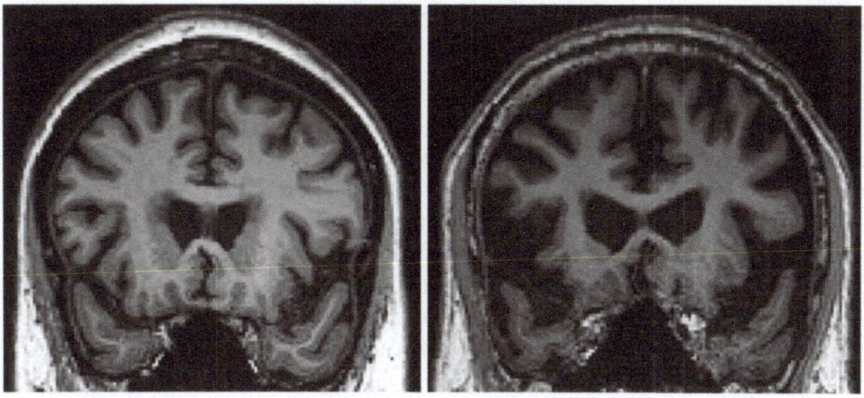

Fig. 2.3 Early AD (left) and late AD (right) MRI. The scan on the right shows whole brain atrophy and ventricular enlargement (images courtesy of Karthik Sreenivasan)

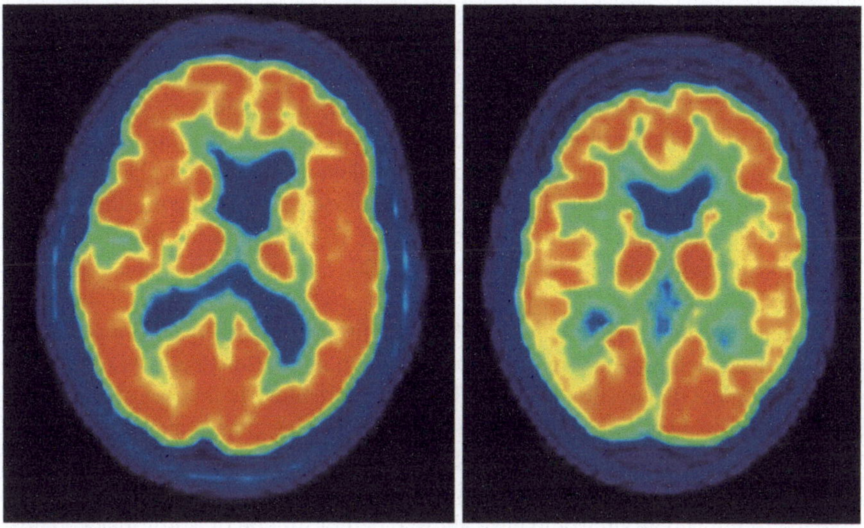

Fig. 2.4 Normal (left) and abnormal (right) fluorodeoxyglucose PET scans

monomeric form of the peptide. Similarly, tau PET measures the fibrillar deposited form of the tau protein, and CSF p-tau is the soluble form of the tau protein. Oligomeric Aβ and oligomeric tau may represent the neurotoxic form of these peptides and do not have currently accepted measures that have been shown to be useful in trials. Drug-placebo differences in A and T would represent important effects on AD biology. They are markers of intermediate steps of the biological changes leading to cell death and do not themselves represent evidence of disease modification. Evidence linking these biomarkers to neuronal loss might allow them to function as surrogate markers of N; this evidence is lacking. A and T are currently best regarded as target engagement biomarkers.

2.4 Biomarkers for Participant Selections

Participation in AD treatment trials requires that the patient have AD. The clinical diagnosis of AD dementia is approached using the 1984 criteria of the National Institute of Neurological and Communicative Disorders and Stroke and the Alzheimer's Disease and Related Disorders Association (NINCDS-ADRDA) [18] or the 2011 criteria of the National Institute on Aging-Alzheimer's Association (NIA-AA) [19]. Recent studies with amyloid imaging show that a substantial portion of individuals diagnosed with these criteria lack biomarker evidence of AD. Using the cohort of the Alzheimer's Disease Neuroimaging Initiative (ADNI), Landau and colleagues found that 15% of patients diagnosed clinically with AD dementia had amyloid PET and CSF findings incompatible with the diagnosis [20]. Similarly, among patients diagnosed clinically with mild AD dementia, Sevigny and coworkers found that 25% failed to show abnormal amyloid levels on amyloid PET [21]. These findings demonstrate that the clinical diagnosis of AD is insufficient to establish a secure diagnosis or be certain of the associated pathology. Measures of A are critical to supporting the diagnosis of AD and providing the rationale for anti-AD therapy. Patients that are amyloid negative have slower progression than those with AD dementia even if they have evidence of neurodegeneration on MRI; these individuals have been labeled as suspected non-Alzheimer pathology (SNAP) [22]. SNAP patients, if included in trial populations, will decrease the rate of change in the placebo group and compromise the ability to demonstrate a drug-placebo difference in the trial. Thus, amyloid biomarkers are needed to show the presence of the AD pathology substrate and to optimize the rate of decline in the placebo group. These considerations apply to trials of both DMTs and symptomatic cognition enhancers.

A recent drug development program for idalopirdine—a 5-HT6 antagonist targeting cognitive enhancement—recruited patients with mild-to-moderate AD dementia based on clinical diagnostic criteria without a confirmatory biomarker. A subgroup of the patients had known amyloid status, and this group declined significantly faster on the Alzheimer's Disease Assessment Scale-Cognitive Subscale (ADAS-Cog) than patients with negative or unknown amyloid status [23]. The greater decline in the biomarker-enriched group would have allowed demonstration of a drug-placebo difference with smaller sample sizes if the agent had been efficacious. This is an example of how diagnostic confirmation of an AD diagnosis can be important in development programs for symptomatic as well as disease-modifying agents.

Mild cognitive impairment (MCI) is an etiologically nonspecific syndrome comprised of several entities associated with cognitive impairment including AD, precursor phases of other dementias such as dementia with Lewy bodies (DLB) and frontotemporal dementia (FTD), depression, and other unrecognized states. This heterogeneity is manifest in the longitudinal outcomes of MCI that include progression to AD, progression to other types of dementias, recovery to normal cognition, and continuation in the MCI state [24]. Bangen found that 37% of amnestic MCI patients did not have brain amyloid by PET assessment indicating that they did not have AD as the key associated pathology [25]. Similarly, Wisse and colleagues [22] reported that 36% of the MCI population they assessed lacked positive findings on amyloid imaging. Of MCI patients who progressed to dementia, 29% were found to have non-AD diagnoses as the primary cause of dementia at autopsy [26]. These studies suggest that at least one-third of patients with MCI do not meet biomarker criteria for prodromal AD using the criteria of the International Work Group [27]. As discussed above, the absence of the pathological changes of AD indicates that the substrate of many AD therapies is absent, and the decline in the placebo group on which power calculations and sample sizes are based becomes less predictable. Biomarker confirmation of the presence of AD pathological changes should be pursued in both DMT and symptomatic therapy drug development programs for prodromal AD or MCI due to AD [27, 28].

Preclinical AD participants do not evidence cognitive abnormalities (although they may have decline from past cognitive performance levels) and can be identified only through the use of biomarkers. Primary and secondary prevention trials can target this population. Cognitively normal individuals with normal amyloid PET or normal CSF levels of Aβ are subjects for primary prevention trials; those with normal cognition and evidence of abnormal brain amyloid can be participants in secondary prevention trials. Of 353 ADNI subjects over age 65 with normal cognition, 45% (160) had normal CSF Aβ42 levels and negative amyloid PET, 47% had abnormal CSF Aβ42 and abnormal amyloid PET, and 7% (26) had abnormal CSF Aβ42 and normal amyloid PET [29].

A meta-analysis of studies of amyloid PET in those with normal cognition shows that the rate of amyloid positivity increases from 10.4% in those 50–55 years old to 43.8% among those 90+ years old [30]. Figure 2.5 shows the prevalence of positive amyloid imaging by age and establishes the expected screen fail and screen positive rate for cognitively normal individuals if no other screening criteria are employed. The rate of amyloid positivity is increased twofold in apolipoprotein E (ApoE) 4 gene carriers [31].

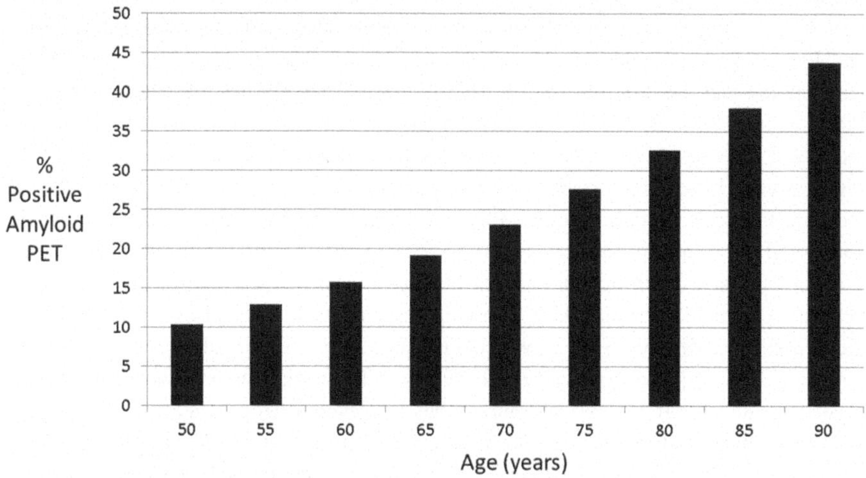

Fig. 2.5 Prevalence of amyloid PET positivity by age (data from Jansen et al. [30])

Together these observations indicate that biomarkers are required for diagnostic confidence in preclinical, prodromal, and dementia trials. Amyloid biomarkers confirm the diagnosis and provide confidence in the predicted decline of the placebo group and the ability to predict effect sizes and sample size requirements. Amyloid biomarkers establish the presence of the target pathology for anti-amyloid trials. Amyloid biomarkers show that the patient has AD and not some other unknown pathology that could create a neuronal environment with an unknown therapeutic response to anti-amyloid treatments. Establishing a firm diagnosis of AD by demonstrating abnormal brain amyloid metabolism or aggregation is important for non-amyloid therapies directed at tau, neuroprotection, inflammation, or other AD-related pathologies. In these trials, amyloid PET or CSF Aβ function as inclusion criteria and not as outcomes. Despite the advantages of confirming the diagnosis of AD with biomarkers, a recent review of current AD trials and the AD pipeline showed that 11 of 25 DMT trials in Phase 3 and 21 of 38 DMT trials in Phase 2 did not require evidence of amyloidosis at baseline [32]. Failure to show a drug-placebo difference in these trials will leave unresolved the question of whether there was lack of efficacy of the test agent or an insufficient number of trial participants with AD.

2.5 Biomarkers of Target Engagement

In drug development programs, Phase 2a is typically used to establish proof of concept (POC) and Phase 2b to determine the doses to be advanced to Phase 3. POC can be based on a clinical response or on a biomarker response or some combination of clinical and biomarker outcomes. Symptomatic agents with detectable clinical responses such as the cholinesterase inhibitors can achieve POC in Phase 2 with clinical measures [33–35]. Symptomatic agents improve cognition above baseline and can show cognitive and global benefit without substantial decline in the placebo group. DMTs target slowing of progression in comparison to the decline observed in the participants in the placebo arm of the study. This typically takes a large trial and long observation period (12 months to 5 years depending on the stage of the participants and the outcomes chosen). Such large long studies are characteristic of Phase 3, not Phase 2. This created the "Phase 2 conundrum" in AD drug development and was resolved by some sponsors by advancing agents from Phase 1 to Phase 3 with little effort to show a drug-placebo efficacy difference or understand the safety issues of the agent in Phase 2 [36]. The Phase 2 conundrum can be addressed by using biomarkers as the principal readout with attention to the directional responses of clinical measures but not requiring demonstration of clinical benefit. Agents with biomarkers showing responses in early phase drug development are more likely to be advanced to later phase of development and to be approved [37, 38]. Drug-placebo differences in biomarker measures can be shown in smaller shorter trials. No biomarker has achieved surrogate status in AD drug development with definite evidence that a change in the biomarker predicts a clinical benefit. Nevertheless, a Phase 3 program can be de-risked by acquiring a repertoire of biomarkers and clinical measures that provide a "weight of evidence" argument supporting drug efficacy and inform the go-no, go decision process at end of Phase 2.

Phase 2 clinical trials should show target engagement and establish the dose(s) to be advanced to Phase 3. Unless target engagement is established, it is impossible to distinguish between a drug that failed to engage the target and a failed trial (usually poorly conducted) as the interpretation of a trial showing no drug-placebo difference [39]. Target engagement can be shown by receptor occupancy or proof of pharmacology. Receptor occupancy is more often used in development programs for symptomatic agents; proof of pharmacology is applicable to both symptomatic and disease-modifying drug development. Symptomatic agents can be advanced on the basis of clinical response, but demonstrating target engagement provides information relevant to brain penetration, dose optimization, and efficacy. DMTs should not be advanced to Phase 3 without evidence of target engagement in Phase 2. There are relatively few well-developed target engagement biomarkers. Sponsors should require development of biomarkers to show target engagement (fluid or imaging) to advance a drug development program.

2.6 Fluid Biomarkers of Target Engagement

Proof of pharmacology for enzyme inhibitors can be shown by stable isotope labeling kinetic (SILK) studies [40]. When used to measure amyloid protein synthesis, this technique involves labeling an amino acid in peripheral blood and then using mass spectrometry to determine when it appears (synthesis rate) in the amyloid protein and disappears (amyloid clearance rate) from the CSF. SILK has been used to show that amyloid clearance is decreased in late-onset AD and that this may be an important contributor to amyloid aggregation in this form of the disease [41]. In the autosomal dominant form of AD (ADAD), production of Aβ40 and Aβ42 was 24% higher in mutation carriers than noncarriers, and the fractional turnover rate of Aβ42 was 65% faster in mutation carriers [42]. These observations show that late-onset AD (LOAD) reflects reduced amyloid clearance; ADAD reflects amyloid over-production.

In drug development programs for inhibitors of enzymes involved in Aβ synthesis, SILK can be used to assess the short-term impact of inhibition of amyloid production. When testing a gamma-secretase inhibitor (semagacestat; LY 450139), patients received an oral dose of the inhibitor at the start of a 9-h infusion of labeled leucine; CSF sampling occurred continuously over the next 36 h. Hours 0–12 were used to calculate Aβ synthesis and hours 24–36 to calculate Aβ clearance. There was a dose-dependent decrease in Aβ production ranging from 47% to 84% with no effect on Aβ clearance [43]. SILK showed proof of pharmacology for inhibition of Aβ production over 36 h. A 16-week study of the agent showed no decrease in CSF Aβ at the end of the exposure period suggesting that short-term inhibition of gamma-secretase may not predict long-term inhibition [44].

Verubecestat, a β-site amyloid precursor protein cleaving enzyme 1 (BACE 1) inhibitor, decreased CSF levels of Aβ over a 7-day continuous sampling period [45], and the levels remained 70–80% lower than placebo-treated patients after an 18-month treatment period [46]. In this case, short-term observations were consistent with long-term data. There was no associated cognitive or functional benefit from prolonged Aβ reductions in these patients with mild-to-moderate AD.

Gamma-secretase inhibitors decrease the production of Aβ42 and increase the production of shorter amyloid fragments. A rise in Aβ15–16 is a target engagement biomarker for gamma-secretase inhibitors showing that the enzyme activity has been decreased and pharmacologic consequences are measurable. This can be shown acutely using short-term measures [47].

Labeled leucine has also been used to interrogate tau protein kinetics [48]. No treatment-related effects of tau agents have been reported using this approach. This technique will provide insight into tau therapeutics and impact of drugs on tau dynamics.

CSF p-tau and Aβ1–42 can serve as pharmacodynamic measures of target engagement. P-tau is increased in AD and is hypothesized to reflect either cell death with release of the neurofibrillary tangle-associated p-tau protein into the CSF or leakage of p-tau from the extracellular space into the CSF during the prion-like transfer of p-tau from cell-to-cell [49]. T-tau measures are considered measures of neurodegeneration. Table 2.2 shows the results of trials where p-tau was measured. Patients on active treatment observed with bapineuzumab, gantenerumab, and semagacestat had modest reductions in p-tau in some studies. Most trials showed no effect on p-tau levels.

Table 2.2 Effects on p-tau and t-tau of recent major clinical trials

Agent	N in the CSF portion of the study	Percent change p-tau compared to placebo	Percent change T-tau compared to placebo	MOA
AN1792 [107]	21	Not reported	Reduced in the active treatment group compared to the pbo group ($p < 0.001$)	Anti-amyloid vaccine
Antioxidant* trial [108]	66	No drug-placebo difference	No drug-placebo difference	Antioxidant
Semagacestat [97]	47	P-tau increased 16% in the pbo group and declined by 8% and 4%, respectively, in the low- and high-dose groups ($p = <0.001$)	No drug-pbo difference	GSI
Bapineuzumab; ApoE carrier study [109]	212	5.8 pg/ml decrease in treatment group; 0.95 pg/ml increase in the pbo group ($p = 0.0.005$)	Not reported	mAb

(continued)

Table 2.2 (continued)

Agent	N in the CSF portion of the study	Percent change p-tau compared to placebo	Percent change T-tau compared to placebo	MOA
Bapineuzumab; ApoE noncarrier study [109]	188	No drug-pbo difference in the 5 mg/kg treatment group; for the 10 mg/kg group, there was a 8.18 pg/ml decrease in the treatment group and a 1.98 pg/ml in the placebo group ($p = 0.009$)	Not reported	
Solanezumab; Expedition [51]	45	No drug-pbo difference	No drug-pbo difference	mAb
Solanezumab; Expedition II [51]	76	No drug-pbo difference	No drug-pbo difference	mAb
Gantenerumab [69]	209	−5.61% change from baseline for 105 mg dose ($p = <0.001$ compared to pbo); −7.15% change from baseline for 225 mg dose ($p = <0.001$ compared to pbo)	−1.08% change from baseline for 105 mg dose ($p = 0.05$ compared to pbo); −2.91% change from baseline for 225 mg dose ($p = 0.02$ compared to pbo)	mAb
IVIG [110]		No drug-placebo difference	No drug-placebo difference	
Solanezumab; Expedition III [52]	258	No drug-pbo difference	No drug-pbo difference	mAb
Crenezumab [53]		No drug-pbo difference	No drug-pbo difference	mAb
Verubecestat [46]	111	Decrease of 0.42% in the pbo group and 5.86% in the 40 mg group (not significant)	Increase of 7.52% in the pbo group and 3.35% in the 40 mg group (not significant)	BACE 1 inhibitor
Azeliragon (TTP488)		No drug-placebo difference	No drug-placebo difference	RAGE inhibitor

BACE β-site amyloid precursor protein cleaving enzyme 1, *GSI* gamma-secretase inhibitor, *IVIG* intravenous immunoglobulin, *mAb* monoclonal antibody, *MOA* mechanism of action, *pbo* placebo, *RAGE* receptor for advance glycation end products

*Antioxidant, 3 arm trial comparing placebo to 800 IU/day of vitamin E (α-tocopherol) plus 500 mg/day of vitamin C plus 900 mg/day of α-lipoic acid (E/C/ALA) or 400 mg of coenzyme Q 3 times/day

CSF Aβ1–42 is monitored as a pharmacodynamic outcome in trials of anti-amyloid agents (Table 2.3). Reduction in Aβ1–42 or a decrease in the Aβ42/40 ratio is a diagnostic hallmark of AD and correlates with increased amyloid plaque burden on amyloid PET [50]. Further reduction in CSF levels of Aβ1–42 has been achieved with BACE inhibitors. Verubecestat, for example, produced 80% reduction in Aβ1–42 levels after 18 months of treatment [46]. Solanezumab produced increased levels of CSF Aβ1–42 in Expedition/Expedition II [51] and in Expedition III [52]. Crenezumab was associated with decreased CSF Aβ compared to placebo [53]. The effect of other amyloid-related treatment mechanisms on CSF Aβ1–42 levels is less predictable. For example, it might be anticipated that agents that reduce amyloid aggregation would increase the monomeric form of Aβ1–42 measured in the CSF and decrease the oligomeric form [54]. Resolution of these issues awaits further empirical evidence.

Inflammation is increasingly recognized as a critical process of AD neurobiology [55]. Several biomarkers for inflammation have been proposed and may eventually be included as biomarkers in AD clinical trials, especially those where the test agent targets or affects inflammation. In CSF, C-reactive protein (CRP) (decreased) and TREM-2 (increased) differ between normal elderly controls and those with AD dementia [56]. CRP is also reduced in plasma of those with AD dementia compared to those with MCI or normal cognition [57]. Correlations between levels of CRP and cognition or disease progression are weak. Elevated CSF levels of the pro-inflammatory cytokine TNF-alpha and decreased levels of the anti-inflammatory cytokine TGF-β are associated with an increased risk of progression from MCI to AD dementia suggesting that inflammation is playing a role in this early phase of symptomatic disease progression [55]. Chemokines are also altered in AD including elevated levels of monocyte chemoattractant protein-1 [58]. Microglial activation is part of the inflammatory process and stimulates astrocytic expression of YKL-40. Increased levels of YKL-40 are evident in CSF and blood in AD dementia [59]. Microglial activation can be assessed with PET imaging using ligands binding to microglial proteins; elevated microglial activity has been shown in medial temporal, occipital, and parietal lobes in those with AD dementia [60]. Reduction of inflammatory markers may be a means of tracking anti-inflammatory effects of AD therapies.

Isoprostanes are prostaglandin isomers produced from polyunsaturated fatty acids in lipid membranes by free radicals and comprise an index of oxidative injury measurable in the CSF and plasma [61, 62]. Isoprostanes increase in the course of normal aging as well as in AD. F2-isoprostanes have the best measurement performance characteristics of assays of oxidative damage [61] and may be used to reflect decreased membrane injury associated with AD-related treatments that provide neuroprotection [63].

Cyclic guanosine monophosphate (cGMP)-specific phosphodiesterase (PDE) inhibitors are candidates for cognitive enhancement in AD by increasing brain cGMP levels. PDE inhibitors have shown proof of pharmacology by raising CSF levels of cGMP after oral dosing. With BI 409306, a PDE9 inhibitor, maximum CSF

Table 2.3 CSF amyloid (Aβ42) reduction in clinical trials of anti-amyloid agents

Agent	Phase	Population	Change compared to placebo	N in the CSF study	Duration of treatment	MOA
AN1792	2	Mild-to-moderate	No drug-placebo difference	21	12 months[a]	Anti-amyloid vaccine
Semagacestat [111]	2	Mild-to-moderate	No drug-placebo difference	43	14 weeks	Gamma-secretase inhibitor
Azeliragon (TTP488) [112]	2	Mild-to-moderate	No drug-placebo difference	65	10 weeks	RAGE inhibitor
Antioxidant[b] trial [108]	2	Mild-to-moderate	No drug-placebo difference	66	16 weeks	Antioxidant
Avagacestat [113]	2	Mild-to-moderate	Decrease to 66% of baseline in the high-dose group ($p = 0.03$ compared to pbo)	45	24 weeks	Gamma-secretase inhibitor
Semagacestat [97]	3	Mild-to-moderate	No drug-placebo difference	47	76 weeks[a]	Gamma-secretase inhibitor
Solanezumab; Expedition [51]	3	Mild-to-moderate	Total Aβ42 increased with no appreciable change in the placebo group (between-group difference, $p < 0.001$); levels of free Aβ42 did not change significantly	45	80 weeks	mAb
Avagacestat [114]	2	MCI	No drug-pbo difference	132	104 weeks	Gamma-secretase inhibitor
Gantenerumab [69]	3	Mild-to-moderate	No drug-pbo difference at either dose group	209	104 weeks	mAb
IVIG [110]	3	Mild-to-moderate	No drug-placebo difference	35	18 months	Polyclonal antibody
Solanezumab; Expedition III [52]	3	Mild	Free Aβ42 decreased more in the treatment group compared to the placebo group (−37.3 vs. −9.3; $p = 0.005$)	258	76 weeks	mAb
Crenezumab [53]	2	Mild-to-moderate	Aβ1–42 levels declined more in the crenezumab than pbo groups of −120.16 pg/ml (unadjusted $p = 0.017$) (300 mg SC cohort) and −170.50 pg/ml (unadjusted $p = 0.022$) (15 mg/kg IV cohort)	26	76 weeks	mAb
Verubecestat	3	Mild-to-moderate	<10% reductions in pbo group; 62.7–76.4% reductions in active treatment group	111	78 weeks	BACE 1 inhibitor

BACE β-site amyloid precursor protein cleaving enzyme 1, *IVIG* intravenous immunoglobulin, *mAb* monoclonal antibody, *MOA* mechanism of action, *RXR* retinoid X receptor

[a]AN1792 and semagacestat trials were interrupted before completion for safety reasons

[b]Antioxidant, 3 arm trial comparing placebo to 800 IU/day of vitamin E (α-tocopherol) plus 500 mg/day of vitamin C plus 900 mg/day of α-lipoic acid (E/C/ALA) or 400 mg of coenzyme Q 3 times/day

cGMP concentrations were achieved within 2–5 h, declining to baseline 10–14 h after dosing in a Phase 1 study with healthy volunteers. This is an example of using a proof-of-pharmacology biomarker in a development program for a symptomatic cognitive enhancing agent. The agent did not produce clinical benefit; other PDE inhibitors are in clinical development [32].

2.7 Imaging Biomarkers of Target Engagement

Receptor occupancy studies are valuable when there is a defined receptor for the test agent; this is most likely in development programs for symptomatic agents—cognitive enhancers or drugs to treat neuropsychiatric syndromes in neurodegenerative disorders. An example of investigation of receptor occupancy in a drug development program for neurodegenerative disease is the PET study of pimavanserin, a 5-HT2A inverse agonist. Ascending doses of pimavanserin were given to healthy volunteers, followed by radio-labeled [11C]*N*-methylspiperone ([11C]NMSP), a 5-HT2A ligand. Reduced NMSP binding was evident following 1 mg of pimavanserin and reached near maximal displacement with the 10–20 mg doses [64]. The radioligand PET study showed blood-brain barrier penetration, dose-related receptor occupancy, dose of maximal occupancy, and safety. Receptor occupancy studies are a means of showing target engagement in a drug development program. Pimavanserin is approved for hallucinations and delusions in Parkinson's disease psychosis [65], reduced symptoms of AD-related psychosis [66], and is being studied for dementia-related psychosis [67].

Amyloid and tau PET are biomarkers of target engagement but are not direct measures of cell loss and neurodegeneration. An example of reduction of fibrillar amyloid in a clinical trial emerged from the trial of the monoclonal antibody aducanumab [68]. There was a dose-dependent decrease in Aβ plaque evident at 6 and 12 months of exposure. There was a corresponding amelioration in progression at the highest doses as measured by the Clinical Dementia Rating Sum of Boxes (CDR-sb) and the Mini-Mental State Examination (MMSE) but not on the Neuropsychological Test Battery (NTB) or the Free and Cued Selective Reminding Test (FCSRT). Dose- and genotype-dependent amyloid-related imaging abnormalities (ARIA) were evident. This agent is now in Phase 3 clinical trials. Reduced plaque amyloid has been observed with the monoclonal antibodies gantenerumab and BAN-2401 [69] and with the small molecules verubecestat [46] and bexarotene [70].

Trials of a few other agents have shown reductions in brain plaque burden on amyloid PET but with no corresponding benefit for cognition or function (Table 2.4). These studies show that demonstration of target engagement in Phase 2 may not predict cognitive benefit in Phase 3. Figure 2.6 shows the relationship of A and T imaging to impact on N; ATN biomarkers are pharmacodynamic measures.

An underutilized opportunity to show target engagement and garner information supportive of proof of pharmacology is the use of functional MRI (fMRI) to explore circuit-level effects of Phase 2 interventions [71]. A,T,N are cellular- and tissue-

Table 2.4 Amyloid reduction in clinical trials of anti-amyloid agents as measured by amyloid PET standardized uptake ratio values (SUVR)

Agent	Phase	Population	Percent Reduction compared to placebo	N in the study	Duration of treatment	MOA
Bapineuzumab; ApoE carrier study [115]	3	Mild-to-moderate	Increase in pbo; no change in treatment group ($p = 0.004$)	75	71 weeks	mAb
Bapineuzumab; ApoE noncarrier study [115]	3	Mild-to-moderate	No drug-pbo difference	39	71 weeks	mAb
Gantenerumab [69]	3	Mild-to-moderate	4.8% reduction from baseline in the 225 mg dose group ($p = 0.01$ compared to pbo); no drug-pbo difference in the 105 mg dose group	55	100 weeks	mAb
Solanezumab; Expedition [51]	3	Mild-to-moderate	No drug-pbo difference	169	80 weeks	mAb
Solanezumab; Expedition II [51]	3	Mild-to-moderate	No drug-pbo difference	97	80 weeks	mAb
Solanezumab; Expedition III [52]	3	Mild	No drug-pbo difference	Not reported	76 weeks	mAb
Aducanumab	1	Mild	SUVR decreased significantly in the 3, 6, 10 mg/kg doses ($p = <0.001$); 20% reduction (mean score 1.44 at baseline to 1.16 at week 54) in the 10 mg/kg group	125	54 weeks	mAb
Semagacestat [97]	3	Mild-to-moderate	No drug-placebo difference	59	76 weeks[a]	Gamma-secretase inhibitor
Verubecestat	3	Mild-to-moderate				BACE 1 inhibitor
Bexarotene	2	Mild-to-moderate	7% reduction in ApoE-4 noncarriers ($p = 0.012$); no significant reduction in ApoE-4 carriers	20	4 weeks	RXR antagonist
IVIG [110]	3	Mild-to-moderate	20 drug-placebo difference	61	18 months	Polyclonal antibody

BACE β-site amyloid precursor protein cleaving enzyme 1, *IVIG* intravenous immunoglobulin, *mAb* monoclonal antibody, *MOA* mechanism of action, *MCI* mild cognitive impairment, *RAGE* receptor for advanced glycation end products, *RXR* retinoid X receptor

[a]The semagacestat study was terminated by the Data Safety Monitoring Board, and completion rates were 23–38%

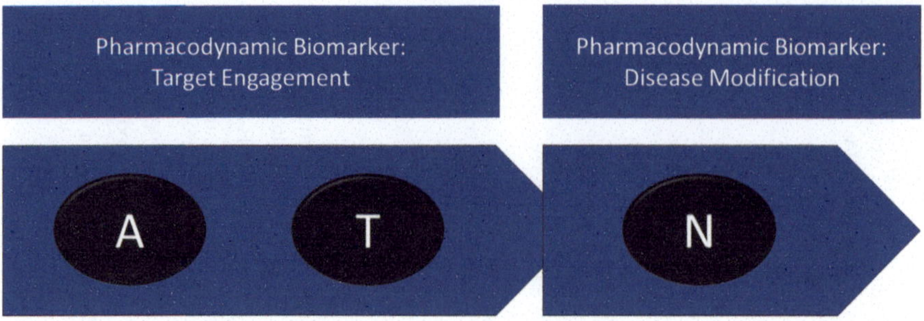

Fig. 2.6 Pharmacodynamic biomarkers. Tau (T) and amyloid (A) biomarkers function as target engagement biomarkers showing that an agent affects the brain protein; N biomarkers support disease modification if a drug-placebo difference is demonstrated

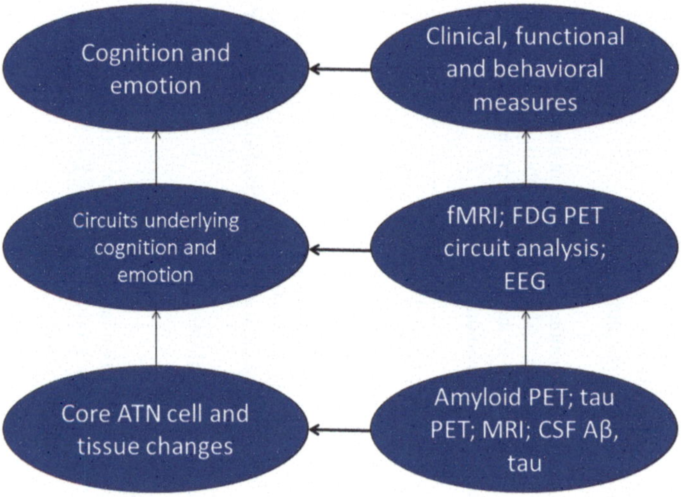

Fig. 2.7 Relationship of ATN (amyloid, tau, neurodegeneration) pathology to circuits that underlie human cognition and emotion is comprised in AD

level measures that reflect the core pathologic involvement of the brain in AD. Cognition and behavior are supported by complex brain circuits that are comprised of integrated nodes and connections [72] (Fig. 2.7). Aspects of these circuits are active at rest and differ from circuits that are dynamically engaged with specific cognitive activities. The activity pattern of resting state fMRI includes a posterior default mode network (DMN) and an anterior salience network (SN). The posterior DMN is comprised of a temporal-parietal network associated with memory and visuospatial function; the anterior SN includes frontal structures relevant to social-emotional function. In AD, the posterior DMN connectivity to posterior hippocampus and medial cingulo-parieto-occipital regions is diminished in contrast to intensified activity in the anterior SN [73, 74] (Fig. 2.8). Circuit function deterio-

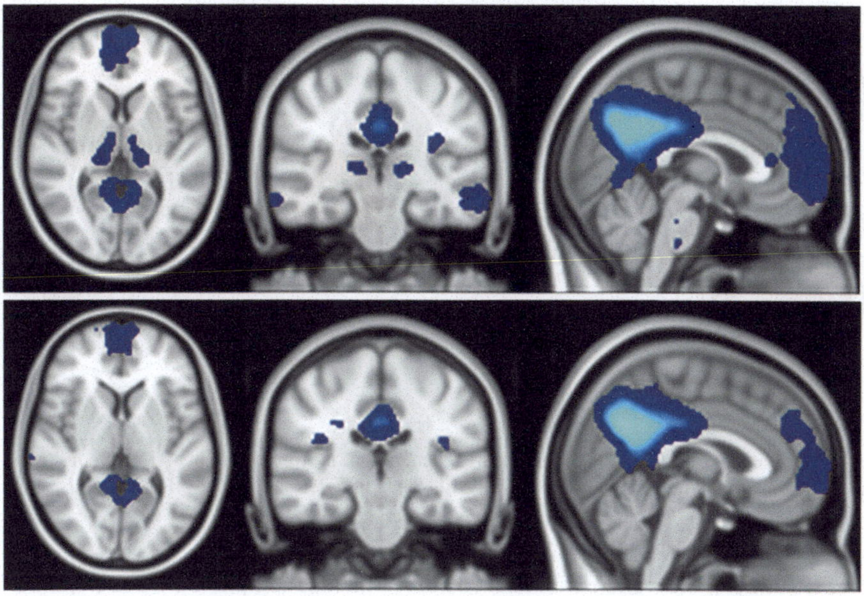

Fig. 2.8 Area of difference in default mode network (DMN) activation on functional MRI (fMRI) between cognitively normal amyloid-negative older adults and amyloid-positive individuals with mild cognitive impairment (MCI) from the AD Neuroimaging Initiative (ADNI) (figure courtesy of Zhengshi Yang)

rates in the preclinical phases of AD and continues to decline in prodromal and dementia phases [75]. Activated fMRI wherein patients perform cognitive tasks while in the MRI scanner show task-related regional activation and can also be used to investigate characteristic changes in AD [76]. Treatments that affect core A,T,N biology but do not impact circuit function are unlikely to produce a cognitive benefit compared to placebo. fMRI has had only limited application in multisite trials, but preliminary reliability studies support its implementation in Phase 2 trials [77, 78]. Treatment benefits demonstrated with fMRI would support circuit-level effects and increase confidence that the treatment will have cognitive benefit [71]. FDG PET circuit analyses and EEG may provide similar insights into circuit-level function but have not been explored in this context in multicenter studies [79].

2.8 Biomarkers Evidence of Disease Modification

A major role for biomarkers in current drug development programs is to provide evidence in support of disease modification for clinical trials of DMTs. Disease modification is defined as ameliorating the basic processes leading to cell death with a corresponding clinical benefit [16, 17]. To show that disease modification has occurred requires an impact on N (neurodegeneration) of the A,T,N Research

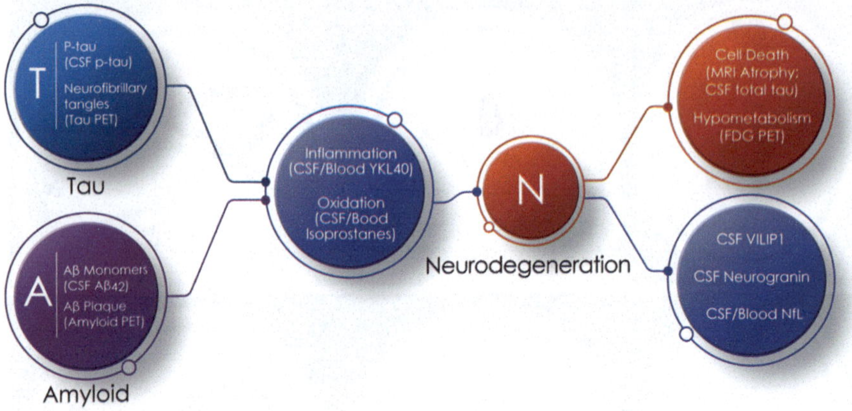

Fig. 2.9 The amyloid (A), tau (T), and neurodegeneration (N) framework of AD with consensus and emerging biomarkers (figure courtesy of Mike de la Flor)

Framework [8]. Recognized assessments of N include atrophy on MRI, hypometabolism on FDG PET, or increases in total tau in the CSF. Measures that are promising candidate biomarkers of N include neurofilament light (NfL), neurogranin, and visinin-like protein-1 (VILIP-1) (Fig. 2.9). NfL is an axonal protein appearing in CSF and plasma with neurodegeneration [80]. VILIP-1 is a neuronal calcium sensor protein previously used as a marker of acute ischemic stroke and found to be elevated in CSF of AD patients [81]. Neurogranin is a synaptic protein that is shed into CSF under circumstances of synaptic degeneration in AD [82]. NfL, total tau, and neurogranin have all been shown to correlate with regional cerebral atrophy, although the strength of correlations may vary [83].

MRI shows progressive atrophy in the course of AD with decline of whole brain and hippocampal volume and increasing ventricular size [84] (Fig. 2.3). Atrophy is correlated with cell loss [85] suggesting that interfering with neurodegeneration and cell loss should slow the rate of atrophy and create a drug-placebo difference at end of trials in favor of active treatment. Review of studies reporting MRI results and listed in Table 2.5 shows that this anticipated result has rarely been achieved. In most trials reporting MRI findings, there has been no drug-placebo difference at trial termination; in a few, the active treatment group has shown greater atrophy. It is uncertain whether the apparently greater shrinkage or "pseudo-atrophy" in the treatment group reflects amyloid removal or reduction of inflammation or if, at least with some treatment mechanisms, neurotoxicity and true increase in atrophy has occurred.

CSF total tau is included in the A,T,N Framework as a marker of N [8]. Tau protein is a microtubule-associated protein that has a critical role in intracellular transportation. It may become hyper-phosphorylated to p-tau in the process of forming neurofibrillary tangles where it becomes a marker of T, or it may appear directly in the CSF presumably as a product of cell death and reflecting N. CSF

Table 2.5 Volumetric MRI results in major recent clinical trials

Agent	Drug-placebo difference at the end of study	Duration of study
AN1792 [116]	Greater atrophy in the active treatment group (significant for whole brain and ventricular volume; not the hippocampal volume)	12 months[a]
Vitamin E or donepezil [117]	No drug-placebo difference in whole brain, ventricular, entorhinal cortex, or hippocampal volume with either treatment	36 months
Tramiprosate [118]	Dose-dependent preservation of hippocampal volume (post hoc analysis)	70 weeks
DHA [119]	No drug-placebo difference in whole brain volume, ventricular volume, or hippocampal volume	18 months
Valproate [111]	Greater atrophy in the treatment group compared to the placebo group (whole brain, ventricular, hippocampal)	24 months
Avagacestat [113]	No drug-pbo difference in whole brain, ventricular, or hippocampal volume	24 weeks
Semagacestat [97]	No drug-pbo difference in entorhinal cortex or hippocampal volume	76 weeks[a]
Bapineuzumab; ApoE carrier study [109]	No drug-pbo difference in whole brain volume	71 weeks
Bapineuzumab; ApoE noncarrier study [109]	No drug-pbo difference in whole brain volume	71 weeks
Solanezumab; Expedition [51]	No drug-pbo difference in whole brain or hippocampal volume	80 weeks
Solanezumab Expedition II [51]	No drug-pbo difference in whole brain or hippocampal volume	80 weeks
Azeliragon (TTP488) [120]	No drug-placebo difference in whole brain volume or hippocampal volume	18 months
Resveratrol [121]	Whole brain volume decreased and ventricular volume increased significantly in the treatment group compared to the placebo group	52 weeks
Avagacestat [114]	Greater atrophy rates were observed in the active treatment group for ventricular and whole brain volumes; differences were significant at weeks 24 and 56 but not at 104 (possibly due to small the number of patients remaining the study)	104 weeks
Avagacestat [114]	No drug-placebo difference in whole brain, ventricular, or hippocampal volume at study end; greater atrophy I the treatment group at weeks 24 and 56	104 weeks
Gantenerumab [69]	No drug-pbo difference in either dose group for whole brain, ventricular, or left-right hippocampal volume	100 weeks
IVIG [110]	No drug-placebo difference in whole brain volume, ventricular volume, or hippocampal volume	18 months
Solanezumab; Expedition III [52]	No drug-pbo difference in whole brain or ventricular volume	76 weeks
Crenezumab [52]	No drug-pbo difference in ventricular volume or whole brain volume	73 weeks
Verubecestat [46]	No significant drug-placebo difference in hippocampal volumes; numerically greater in the active treatment groups	78 weeks

BACE β-site amyloid precursor protein cleaving enzyme 1, *DHA* docosahexaenoic acid, *GSI* gamma-secretase inhibitor, *GSK* glycogen synthase kinase
IVIG intravenous immunoglobulin, *mAb* monoclonal antibody, *MOA* mechanism of action, *pbo* placebo
[a]AN1792 and semagacestat trials were stopped before planned completion

measures of p-tau and total tau have been collected in several major trials (Table 2.2). In a few trials, relatively small magnitude changes in either tau species have been observed. The interpretation of these changes is uncertain. Until recently, measurement of tau across or within laboratories produced inconsistent results, and the evolution of new techniques in tau measures will assist in using tau as an outcome in drug development [86].

Fluorodeoxyglucose PET is among the N measures of the A,T,N Framework (Figs. 2.4 and 2.9). The metabolic activity measured with FDG PET is largely reflective of synaptic activity and neuronal activating [8, 87], and hypometabolism is regarded as a reflection of synaptic compromise in the course of cell death. Relatively few AD clinical trials have included FDG PET as an outcome. Methodologies have evolved to suggest that it can be performed reliably as part of a multisite trial and have the ability to detect treatment effects with relatively small sample sizes [88, 89]. Brain metabolism can be increased with symptomatic treatments, and this potential confound must be considered in FDG studies [90]. Novel imaging biomarkers such as the synaptic vesicle glycoprotein 2A (SV2A) PET ligand, indicative of synaptic density, may present novel opportunities to document neurodegeneration [91, 92].

2.9 Biomarkers for Safety in AD Drug Development

Liver functions, blood counts, muscle enzymes, and electrocardiograms are key biomarkers for drug toxicity [93], and toxicity accounts for the termination of approximately 30% of drug development programs [94]. These measures are included in AD drug development programs. Liver toxicity has been observed with some BACE inhibitors and some 5-HT6 antagonists such as the Phase 2 studies of idalopirdine [95], and QTc prolongation was observed in a trial of citalopram for agitation in AD [96].

Off-target adverse events occurred in the course of gamma-secretase drug development with hypopigmentation, skin cancers, and cognitive and functional impairment relative to placebo [97]. The dermatologic changes are attributed to inhibition of NOTCH proteases; the cognitive and functional toxicity is of uncertain origin. Amyloid-related imaging abnormalities (ARIA) occur with some monoclonal antibodies, and monitoring these with MRI in the course of trials is critical to insuring the safety of these treatments [98].

2.10 FDA Classification of Biomarkers and Integration into Stages of Alzheimer's Disease

The FDA has proposed a staging system for AD beginning with biomarker-positive asymptomatic individuals (Stage 1); those with cognitive impairment measurable only with sensitive neuropsychological instruments and no functional impairment

(Stage 2); those with mild cognitive impairment and functional compromise measurable with sensitive instruments but not sufficient to meet criteria for dementia (Stage 3); and those with mild, moderate, and severe AD dementia (Stages 4, 5, 6) (Table 2.6) [10]. Biomarkers play an important role in this staging system particularly in the Stage 1 where there are no cognitive or functional changes, but biomarker changes indicative of AD pathological changes are present. Biomarkers that are abnormal in this early preclinical phase of the disease include positive amyloid imaging and low CSF Aβ42. The FDA Guidance provides for accelerated approval for a treatment at this stage based on a biomarker thought to be reasonably likely to predict clinical benefit and coupled with a post-approval plan to gather evidence on clinical outcomes. The FDA Guidance noted that no current AD biomarker can be regarded as a surrogate that reliably predicts clinical measure. Full approval would require demonstrating a drug-placebo difference on a clinical outcome. The FDA Guidance [10] indicates that in Stage 2, approval could be based on persuasive effects on neuropsychological measures supported by effects on the characteristic pathophysiological (biomarker) changes of AD. These approval discussions are independent of discussions of acceptable labeling of a new treatment where the designation of being a DMT will likely require robust effects on N-type biomarkers or effects on biomarkers known to predict N or to predict sustained cognitive and functional benefit.

The FDA divides biomarkers into categories of diagnostic, prognostic, predictive, response, and safety [99] (Table 2.7). Diagnostic biomarkers insure accurate diagnosis and allow categorization of a condition by the presence or absence of a specific pathophysiological state. Prognostic biomarkers indicate disease course and can be used to enrich populations to optimize establishing a drug-placebo difference. Predictive biomarkers assist in forecasting the response to treatment. Pharmacodynamic or activity biomarkers show that a biological response has occurred in an individual who received the therapeutic intervention. Pharmacodynamic biomarkers are used in Phase 2 studies to improve understanding of how to use a drug and to guide dose or regimen decisions for Phase 3. Evidence of disease modification also depends on pharmacodynamic biomarkers. Safety biomarkers are used to capture adverse events.

In AD drug development, evidence of amyloidosis is considered diagnostic of the AD pathological process; tau PET or MRI atrophy might serve a prognostic biomarkers of participants who will decline more rapidly; inflammatory markers might serve as a predictive biomarker for patients most likely to respond to anti-inflammatory therapies; both target engagement biomarkers and biomarkers of disease modification are pharmacodynamic biomarkers (Fig. 2.6); and use of MRI to monitor of ARIA is an example of a safety biomarker (Table 2.7).

2.11 Biomarker Qualification and Context of Use

Qualification refers to the FDA-defined process of reviewing drug development tools (DDTs) intended for use in multiple development programs [99]. DDTs include biomarkers, clinical outcome assessments, and animal models of drug

Table 2.6 Staging system for AD proposed by the FDA with means of achieving full approval or accelerated approval of drugs developed for early AD (FDA Guidance [10])

	Stage 1	Stage 2	Stage 3	Stage 4, 5, and 6
Biomarkers reflecting underlying AD pathophysiological changes	Positive	Positive	Positive	Positive
Cognition	Truly asymptomatic with no subjective complaint or detectable abnormalities on sensitive neuropsychological measures	Subtle detectable abnormalities on sensitive neuropsychological measures	Subtle or more apparent detectable abnormalities on sensitive neuropsychological measures	Mild, moderate, and severe AD dementia with worsening cognitive impairment
Function	No functional impairment	No functional impairment. The emergence of subtle functional impairment signals a transition to Stage 3	Mild but detectable functional impairment. The functional impairment in this stage is not severe enough to warrant a diagnosis of overt dementia	Mild, moderate, and severe AD dementia with worsening functional impairment
Clinical endpoints	A clinically meaningful benefit cannot be measured in these patients because there is no clinical impairment to assess (assuming that the duration of a trial is not sufficient to observe and assess the development of clinical impairment during the conduct of the trial)	Persuasive effect on sensitive measures of neuropsychological performance may provide adequate support for a marketing approval. A pattern of putatively beneficial effects demonstrated across multiple individual tests would increase the persuasiveness of the finding. A large magnitude of effect on sensitive measures of neuropsychological performance may also increase their persuasiveness	Favorable effect on cognitive and functional deficits. An integrated scale that adequately and meaningfully assesses both daily function and cognitive effects in early AD patients is acceptable as a single primary efficacy outcome measure	Co-primary approach to assessment of cognitive and functional (or global) measures

Alternate approach to clinical endpoints	Conduct a study of sufficient duration to allow the evaluation of the measures indicated for Stage 2 patients	A possible approach is to conduct a study of sufficient duration to allow the evaluation of the measures discussed for Stage 3 patients	Not specified in the guidance	Not specified in the guidance
Biomarker endpoints	An effect on the characteristic pathophysiologic changes of AD. A pattern of treatment effects seen across multiple individual biomarker measures would increase the persuasiveness of the putative effect	Supported by similarly persuasive effects on the characteristic pathophysiologic changes of AD	Not specified in the guidance	Not specified in the guidance
Full approval	An effect on the characteristic pathophysiologic changes of AD when the fundamental understanding of AD evolves sufficiently to establish surrogacy (no AD biomarkers currently meet the criteria for surrogacy)	The cognitive effects were found to be inherently clinically meaningful, either on face or because they reliably and inevitably are associated with functional benefit later in the course of the disease	Not specified in the guidance	Not specified in the guidance
Accelerated approval	An effect on the characteristic pathophysiologic changes of AD analyzed as a primary efficacy measure, may, serve as the basis for an accelerated approval (i.e., the biomarker effects would be found to be reasonably likely to predict clinical benefit, with a post-approval requirement for a study to confirm the predicted clinical benefit)	The cognitive effects were found to be reasonably likely to predict clinical benefit, with a post-approval requirement for a study to confirm the predicted clinical benefit required	Not specified in the guidance	Not specified in the guidance

Table 2.7 FDA terminology for biomarkers and identification of biomarkers in each category for AD drug development (adapted from Amur et al. [103])

FDA biomarker type	Examples for drug development	Examples from AD drug development
Diagnostic biomarkers	Patient selection	Positive amyloid imaging Low CSF Aβ42 or change in Aβ/tau or Aβ/p-tau ratio
Prognostic biomarkers	Stratify patients or enrich trials with patients likely to have disease	Tau PET to identify AD patients likely to have more rapid cognitive progression ApoE-4 carriers as a prognostic marker for ARIA in immunotherapy programs
Predictive biomarkers	Stratification Enrichment/inclusion criteria Enrichment/companion diagnostic	Use of tau PET to identify AD patients more likely to respond to anti-tau therapies
Response biomarkers	Pharmacodynamic biomarker as an indicator of intended drug activity Efficacy response biomarker as a surrogate for a clinical endpoint	Target engagement biomarkers (e.g., reduction in amyloid plaque in anti-amyloid programs) Markers of disease modification (e.g., drug-placebo differences in CSF total tau, FDG PET hypometabolism, or MRI atrophy)
Safety biomarkers	Biomarkers to detect adverse and off-target drug responses	MRI to monitor for ARIA in immunotherapy programs

ApoE apolipoprotein E, *ARIA* amyloid-related imaging abnormalities, *CSF* cerebrospinal fluid, *FDG* fluorodeoxyglucose, *MRI* magnetic resonance imaging, *PET* positron-emission tomography

development. Once qualified, drug developers can use the biomarker within the specific context of use (COU) for the qualified purpose as long as no new information that conflicts with the original basis for qualification has evolved. The qualification process is intended to expedite drug development by making publically available DDTs that can be widely employed. A qualified DDT can be relied on to have a specific interpretation in drug development and regulatory review. Qualification is a complex process that begins with an initiation request and DDT letter of intent. This stage (Stage 1) is followed by consultation and advice with an FDA Qualification Review Team (QRT) regarding the submitter's goals and the COU of the DDT, current understanding of the available data, identification of information gaps, discussion of additional information that may be needed, and construction of a plan for the qualification process (Stage 2). Stage 3 is comprised of review of the full qualification package [99]. The DDT COU proposal is reviewed by individual disciplines, and a final combined executive summary and recommendation are issued by the QRT. The qualification process is typically pursued by organizations such as the Coalition Against Major Disease (CAMD) of the Critical Path Institute [100]. No AD-related biomarkers have been advanced to qualification by the FDA. The European Medicines Agency (EMA) approved low CSF Aβ1–42 and high t-tau as qualified for identification of prodromal AD; CSF Aβ1–42, t-tau, and amyloid PET to enrich for subjects in trials of mild-to-moderate AD; and hippocampal volume for enrichment of trials in predementia stages of AD [101, 102].

DDT qualification is not necessary for use of a DDT within an individual drug development program, and use of a DDT in a program does not automatically qualify the DDT for the general COU. When qualified biomarkers are not available, the pharmaceutical developer engages with FDA to reach agreement on the use of a particular biomarker in the drug development program [103]. In the case of fluid biomarkers, the sponsor must present information on specified reagents, analytical validation, rigorous process standardization, procedures for sample collection and handling, measurement stability, environmental change tolerance, lot-to-lot variability, computational procedures, and validated, reliable, and accurate interpretation [13, 104]. The specific use of the biomarker in the development program must be specified with a program-specific COU. These discussions are confidential and do not become available to other sponsors.

2.12 Companion and Complementary Diagnostics

The reliance on biomarkers in the drug development process lends itself to the development of companion or complementary diagnostics. A companion diagnostic is an in vitro diagnostic device (IVD) required for the safe and effective use of a corresponding therapeutic. The companion test may identify those most likely to respond or to have side effects, monitor therapy, or identify those in whom the therapeutic product has been adequately studied and found to be safe and effective [99, 105]. When approved as a companion diagnostic, use of the IVD is mandatory before prescribing the drug or biologic. Extensive analytical and clinical validation is required for IVDs used as companion diagnostics [13]. A critical part of the companion diagnostic development is determining the cutoff level which determines that the test is abnormal and dichotomizes the population into those with normal or abnormal status [105]. If an anti-amyloid therapy is approved based on a clinical trial in which participants were defined by abnormal amyloid imaging, then amyloid PET may be identified as a companion diagnostic required for the safe and effective use of the therapeutic. Determining the standard uptake value ratio (SUVR) defining the participant as having brain amyloidosis will be required if the SUVR was used in the pivotal trial. Visual reads of amyloid PET would be required if that is the approach used in the trial.

Complementary diagnostic tests are not required for prescribing a therapeutic agent but can identify a biomarker-defined subset of patients that responds particularly well or aids in the risk/benefit assessments for individual patients [105, 106]. For example, an anti-amyloid monoclonal antibody might benefit both ApoE-4 carriers and noncarriers, but a higher rate of ARIA in carriers would play a role in the risk/benefit discussion with the potential treatment recipient. In this case, the ApoE genotype test would function as a complementary biomarker.

2.13 Summary

Biomarkers play a central role in AD drug development and are likely to become increasingly important as the biology of AD is more understood and the repertoire of biomarkers is expanded. Biomarkers can be used to assist in diagnosis, demonstrate target engagement, provide support for disease modification, and monitor for safety (Table 2.8). Diagnostic, predictive, prognostic, pharmacodynamic, and safety biomarkers have been identified. The A,T,N Research Framework integrates biomarkers into the process of AD diagnosis and can be applied to drug development and clinical trials. The FDA staging of AD facilitates drug development for predementia stages of AD and integrates biomarkers into the staging system. Companion and complementary biomarkers may be developed in concert with new therapeutics in drug development programs. Informed use of biomarkers promises to accelerate AD drug development and assist in bringing new therapies to those with or at risk for AD.

Table 2.8 Examples of biomarkers for each phase of AD drug development

Type of biomarker	Phase 1	Phase 2	Phase 3
Diagnostic biomarker (e.g., amyloid PET or CSF Aβ or Aβ/tau ratio)	If AD-spectrum patients are included in the Phase 1 program	All patients	All patients
Target engagement biomarker		Critical outcome of Phase 2 to allow progression to Phase 3	
Marker of disease modification			Critical outcome to allow labeling of the intervention as a DMT
Prognostic biomarker (e.g., ApoE-4 carrier status)		Important for analysis of outcomes and prediction of ARIA in immunotherapy programs	Important for analysis of outcomes and prediction of ARIA in immunotherapy programs
Safety biomarkers	Liver function and other laboratory tests, ECG, MRI to monitor for ARIA in immunotherapy programs	Liver function and other laboratory tests, ECG, MRI to monitor for ARIA in immunotherapy programs	Liver function and other laboratory tests, ECG, MRI to monitor for ARIA in immunotherapy programs

ApoE apolipoprotein, *ARIA* amyloid-related imaging abnormalities, *ECG* electrocardiography, *CSF* cerebrospinal fluid, *MRI* magnetic resonance imaging, *PET* positron-emission tomography

Acknowledgment J.C. acknowledges funding from the National Institute of General Medical Sciences (Grant: P20GM109025) and support from Keep Memory Alive.

Declaration of Interest: J.C. has provided consultation to Acadia, Avanir, BiOasis Technologies, Biogen, Boehringer-Ingelheim, Bracket, Eisai, Genentech, Grifols, Intracellular Therapies, Kyowa, Lilly, Lundbeck, Medavante, Merck, Nutricia, Otsuka, Pfizer, QR, Resverlogix, Samus, Servier, Suven, Takeda, Toyama, and United Neuroscience companies. J.C. acknowledges the funding from the National Institute of General Medical Sciences (Grant: P20GM109025) and support from Keep Memory Alive.

References

1. Masters CL, Bateman R, Blennow K, Rowe CC, Sperling RA, Cummings JL (2015) Alzheimer's disease. Nat Rev Dis Primers 1:15056. https://doi.org/10.1038/nrdp.2015.56
2. Scheltens P, Blennow K, Breteler MM, de Strooper B, Frisoni GB, Salloway S et al (2016) Alzheimer's disease. Lancet 388:505–517
3. Alzheimer's Association (2016) 2016 Alzheimer's disease facts and figures. Alzheimers Dement 12:459–509
4. Prince M, Wimo A, Guerchet M, Ali GC, Wu YT, Prina M et al (2015) Alzheimer's Disease International World Alzheimer Report 2015: the global impact of dementia, an analysis of prevalence, incidence, cost and trends. London, 2015. https://www.alz.co.uk/research/world-report-2015
5. Wimo A, Guerchet M, Ali GC, Wu YT, Prina AM, Winblad B et al (2017) The worldwide costs of dementia 2015 and comparisons with 2010. Alzheimers Dement 13:1–7
6. Cummings JL, Morstorf T, Zhong K (2014) Alzheimer's disease drug-development pipeline: few candidates, frequent failures. Alzheimers Res Ther 6:37–43
7. Cummings J, Ritter A, Zhong K (2018) Clinical trials for disease-modifying therapies in Alzheimer's disease: a primer, lessons learned, and a blueprint for the future. J Alzheimers Dis 64:S3–S22
8. Jack CR Jr, Bennett DA, Blennow K, Carrillo MC, Dunn B, Haeberlein SB et al (2018) NIA-AA Research framework: toward a biological definition of Alzheimer's disease. Alzheimers Dement 14:535–562
9. Cummings J (2018) The National Institute on Aging-Alzheimer's Association framework on Alzheimer's disease: application to clinical trials. Alzheimers Dement Jun 21. https://doi.org/10.1016/j.jalz.2018.05.006. [Epub ahead of print]
10. U.S. Food and Drug Administration (2018) Early Alzheimer's disease: developing drugs for treatment, guidance for industry. Office of Communications, Division of Drug Information, Center for Drug Evaluation and Research, Food and Drug Administration, Silver Spring, MD. https://www.fda.gov/downloads/Drugs/GuidanceComplianceRegulatoryInformation/Guidances/UCM596728.pdf
11. Biomarkers Definitions Working Group (2001) Biomarkers and surrogate endpoints: preferred definitions and conceptual framework. Clin Pharmacol Ther 69:89–95
12. U.S. Food and Drug Administration (2014) In vitro companion diagnostic devices: guidance for industry and Food and Drug Administration staff. Center for Drug Evaluation and Research, Division of Drug Information, Silver Spring, MD. https://www.fda.gov/downloads/MedicalDevices/DeviceRegulationandGuidance/GuidanceDocuments/UCM262327.pdf
13. Kraus VB (2018) Biomarkers as drug development tools: discovery, validation, qualification and use. Nat Rev Rheumatol 14:354–362
14. FDA-NIH Biomarker Working Group (2016) BEST (Biomarkers, EndpointS, and other Tools) resource. Food and Drug Administration and National Institutes of Health, Silver Spring, MD

15. Day M, Rutkowski JL, Feuerstein GZ (2009) Translational medicine--a paradigm shift in modern drug discovery and development: the role of biomarkers. Adv Exp Med Biol 655:1–12
16. Cummings J (2017) Disease modification and neuroprotection in neurodegenerative disorders. Transl Neurodegener 6:25. https://doi.org/10.1186/s40035-017-0096-2
17. Cummings JL, Fox N (2017) Defining disease modification for Alzheimer's disease clinical trials. J Prev Alzheimers Dis 4:109–115
18. McKhann G, Drachman D, Folstein M, Katzman R, Price D, Stadlan EM (1984) Clinical diagnosis of Alzheimer's disease: report of the NINCDS-ADRDA Work Group under the auspices of Department of Health and Human Services Task Force on Alzheimer's Disease. Neurology 34:939–944
19. McKhann GM, Knopman DS, Chertkow H, Hyman BT, Jack CR Jr, Kawas CH et al (2011) The diagnosis of dementia due to Alzheimer's disease: recommendations from the National Institute on Aging-Alzheimer's Association workgroups on diagnostic guidelines for Alzheimer's disease. Alzheimers Dement 7:263–269
20. Landau SM, Horng A, Fero A, Jagust WJ, Alzheimer's Disease Neuroimaging Initiative (2016) Amyloid negativity in patients with clinically diagnosed Alzheimer disease and MCI. Neurology 86:1377–1385
21. Sevigny J, Suhy J, Chiao P, Chen T, Klein G, Purcell D et al (2016) Amyloid PET screening for enrichment of early-stage Alzheimer disease clinical trials: experience in a phase 1b clinical trial. Alzheimer Dis Assoc Disord 30:1–7
22. Wisse LEM, Butala N, Das SR, Davatzikos C, Dickerson BC, Vaishnavi SN et al (2015) Suspected non-AD pathology in mild cognitive impairment. Neurobiol Aging 36:3152–3162
23. Ballard C, Atri A, Boneva N, Cummings JL, Frolich L, Molinuevo JL et al (2018) Predictors of Alzheimer disease progression for enrichment of clinical trials: analysis of placebo data from a phase 3 program (submitted)
24. Ellendt S, Vobeta B, Kohn N, Wagels L, Goerlich KS, Drexler E et al (2017) Predicting stability of mild cognitive impairment (MCI): findings of a community based sample. Curr Alzheimer Res 14:608–619
25. Bangen KJ, Clark AL, Werhane M, Edmonds EC, Nation DA, Evangelista N et al (2016) Cortical amyloid burden differences across empirically-derived mild cognitive impairment subtypes and interaction with APOE varepsilon4 genotype. J Alzheimers Dis 52:849–861
26. Jicha GA, Parisi JE, Dickson DW, Johnson K, Cha R, Ivnik RJ et al (2006) Neuropathologic outcome of mild cognitive impairment following progression to clinical dementia. Arch Neurol 63:674–681
27. Dubois B, Feldman HH, Jacova C, Hampel H, Molinuevo JL, Blennow K et al (2014) Advancing research diagnostic criteria for Alzheimer's disease: the IWG-2 criteria. Lancet Neurol 13:614–629
28. Albert MS, DeKosky ST, Dickson D, Dubois B, Feldman HH, Fox NC et al (2011) The diagnosis of mild cognitive impairment due to Alzheimer's disease: recommendations from the National Institute on Aging-Alzheimer's Association workgroups on diagnostic guidelines for Alzheimer's disease. Alzheimers Dement 7:270–279
29. Palmqvist S, Mattsson N, Hansson O, Alzheimer's Disease Neuroimaging Initiative (2016) Cerebrospinal fluid analysis detects cerebral amyloid-beta accumulation earlier than positron emission tomography. Brain 139:1226–1236
30. Jansen WJ, Ossenkoppele R, Knol DL, Tijms BM, Scheltens P, Verhey FR et al (2015) Prevalence of cerebral amyloid pathology in persons without dementia: a meta-analysis. JAMA 313:1924–1938
31. Insel PS, Palmqvist S, Mackin RS, Nosheny RL, Hansson O, Weiner MW et al (2016) Assessing risk for preclinical beta-amyloid pathology with APOE, cognitive, and demographic information. Alzheimers Dement (Amst) 4:76–84
32. Cummings J, Lee G, Ritter A, Zhong K (2018) Alzheimer's disease drug development pipeline: 2018. Alzheimers Dement (N Y) 4:195–214

33. Farlow M, Gracon SI, Hershey LA, Lewis KW, Sadowsky CH, Dolan-Ureno J (1992) A controlled trial of tacrine in Alzheimer's disease. The Tacrine Study Group. JAMA 268:2523–2529
34. Rogers SL, Farlow MR, Doody RS, Mohs R, Friedhoff LT (1998) A 24-week, double-blind, placebo-controlled trial of donepezil in patients with Alzheimer's disease. Donepezil Study Group. Neurology 50:136–145
35. Tan CC, Yu JT, Wang HF, Tan MS, Meng XF, Wang C et al (2014) Efficacy and safety of donepezil, galantamine, rivastigmine, and memantine for the treatment of Alzheimer's disease: a systematic review and meta-analysis. J Alzheimers Dis 41:615–631
36. Cummings JL (2008) Optimizing phase II of drug development for disease-modifying compounds. Alzheimers Dement 4:S15–S20
37. Cook D, Brown D, Alexander R, March R, Morgan P, Satterthwaite G et al (2014) Lessons learned from the fate of AstraZeneca's drug pipeline: a five-dimensional framework. Nat Rev Drug Discov 13:419–431
38. Morgan P, Brown DG, Lennard S, Anderton MJ, Barrett JC, Eriksson U et al (2018) Impact of a five-dimensional framework on R&D productivity at AstraZeneca. Nat Rev Drug Discov 17:167–181
39. Simon GM, Niphakis MJ, Cravatt BF (2013) Determining target engagement in living systems. Nat Chem Biol 9:200–205
40. Bateman RJ, Munsell LY, Morris JC, Swarm R, Yarasheski KE, Holtzman DM (2006) Human amyloid-beta synthesis and clearance rates as measured in cerebrospinal fluid in vivo. Nat Med 12:856–861
41. Mawuenyega KG, Sigurdson W, Ovod V, Munsell L, Kasten T, Morris JC et al (2010) Decreased clearance of CNS beta-amyloid in Alzheimer's disease. Science 330:1774. https://doi.org/ 10.1126/science.1197623
42. Potter R, Patterson BW, Elbert DL, Ovod V, Kasten T, Sigurdson W et al (2013) Increased in vivo amyloid-beta42 production, exchange, and loss in presenilin mutation carriers. Sci Transl Med 5:189ra77. https://doi.org/10.1126/scitranslmed.3005615
43. Bateman RJ, Siemers ER, Mawuenyega KG, Wen G, Browning KR, Sigurdson WC et al (2009) A gamma-secretase inhibitor decreases amyloid-beta production in the central nervous system. Ann Neurol 66:48–54
44. Fleisher AS, Raman R, Siemers ER, Becerra L, Clark CM, Dean RA et al (2008) Phase 2 safety trial targeting amyloid beta production with a gamma-secretase inhibitor in Alzheimer disease. Arch Neurol 65:1031–1038
45. Kennedy ME, Stamford AW, Chen X, Cox K, Cumming JN, Dockendorf MF et al (2016) The BACE1 inhibitor verubecestat (MK-8931) reduces CNS beta-amyloid in animal models and in Alzheimer's disease patients. Sci Transl Med 8:363ra150. https://doi.org/10.1126/scitranslmed.aad9704
46. Egan MF, Kost J, Tariot PN, Aisen PS, Cummings JL, Vellas B et al (2018) Randomized trial of verubecestat for mild-to-moderate Alzheimer's disease. N Engl J Med 378:1691–1703
47. Portelius E, Zetterberg H, Dean RA, Marcil A, Bourgeois P, Nutu M et al (2012) Amyloid-beta(1-15/16) as a marker for gamma-secretase inhibition in Alzheimer's disease. J Alzheimers Dis 31:335–341
48. Sato C, Barthelemy NR, Mawuenyega KG, Patterson BW, Gordon BA, Jockel-Balsarotti J et al (2018) Tau kinetics in neurons and the human central nervous system. Neuron 97:1284–1298.e7. https://doi.org/10.1016/j.neuron.2018.04.035
49. Coughlin D, Irwin DJ (2017) Emerging diagnostic and therapeutic strategies for tauopathies. Curr Neurol Neurosci Rep 17:72. https://doi.org/10.1007/s11910-017-0779-1
50. Lewczuk P, Matzen A, Blennow K, Parnetti L, Molinuevo JL, Eusebi P et al (2017) Cerebrospinal fluid Abeta42/40 corresponds better than Abeta42 to amyloid PET in Alzheimer's disease. J Alzheimers Dis 55:813–822
51. Doody RS, Thomas RG, Farlow M, Iwatsubo T, Vellas B, Joffe S et al (2014) Phase 3 trials of solanezumab for mild-to-moderate Alzheimer's disease. N Engl J Med 370:311–321

52. Honig LS, Vellas B, Woodward M, Boada M, Bullock R, Borrie M et al (2018) Trial of solanezumab for mild dementia due to Alzheimer's disease. N Engl J Med 378:321–330
53. Cummings JL, Cohen S, van Dyck CH, Brody M, Curtis C, Cho W et al (2018) ABBY: a phase 2 randomized trial of crenezumab in mild to moderate Alzheimer disease. Neurology 90:e1889–e1e97. https://doi.org/10.1212/WNL.0000000000005550
54. Lee SJ, Nam E, Lee HJ, Savelieff MG, Lim MH (2017) Towards an understanding of amyloid-beta oligomers: characterization, toxicity mechanisms, and inhibitors. Chem Soc Rev 46:310–323
55. Heneka MT, Carson MJ, El Khoury J, Landreth GE, Brosseron F, Feinstein DL et al (2015) Neuroinflammation in Alzheimer's disease. Lancet Neurol 14:388–405
56. Brosseron F, Traschutz A, Widmann CN, Kummer MP, Tacik P, Santarelli F et al (2018) Characterization and clinical use of inflammatory cerebrospinal fluid protein markers in Alzheimer's disease. Alzheimers Res Ther 10:25. https://doi.org/10.1186/s13195-018-0353-3
57. Yarchoan M, Louneva N, Xie SX, Swenson FJ, Hu W, Soares H et al (2013) Association of plasma C-reactive protein levels with the diagnosis of Alzheimer's disease. J Neurol Sci 333:9–12
58. Serpente M, Bonsi R, Scarpini E, Galimberti D (2014) Innate immune system and inflammation in Alzheimer's disease: from pathogenesis to treatment. Neuroimmunomodulation 21:79–87
59. Muszynski P, Groblewska M, Kulczynska-Przybik A, Kulakowska A, Mroczko B (2017) YKL-40 as a potential biomarker and a possible target in therapeutic strategies of Alzheimer's disease. Curr Neuropharmacol 15:906–917
60. Passamonti L, Rodriguez PV, Hong YT, Allinson KSJ, Bevan-Jones WR, Williamson D et al (2018) [(11)C]PK11195 binding in Alzheimer disease and progressive supranuclear palsy. Neurology 90:e1989–e1996
61. Galasko D, Montine TJ (2010) Biomarkers of oxidative damage and inflammation in Alzheimer's disease. Biomark Med 4:27–36
62. Montine TJ, Peskind ER, Quinn JF, Wilson AM, Montine KS, Galasko D (2011) Increased cerebrospinal fluid F2-isoprostanes are associated with aging and latent Alzheimer's disease as identified by biomarkers. NeuroMolecular Med 13:37–43
63. Kester MI, Scheffer PG, Koel-Simmelink MJ, Twaalfhoven H, Verwey NA, Veerhuis R et al (2012) Serial CSF sampling in Alzheimer's disease: specific versus non-specific markers. Neurobiol Aging 33:1591–1598
64. Nordstrom AL, Mansson M, Jovanovic H, Karlsson P, Halldin C, Farde L et al (2008) PET analysis of the 5-HT2A receptor inverse agonist ACP-103 in human brain. Int J Neuropsychopharmacol 11:163–171
65. Cummings J, Isaacson S, Mills R, Williams H, Chi-Burris K, Corbett A et al (2014) Pimavanserin for patients with Parkinson's disease psychosis: a randomised, placebo-controlled phase 3 trial. Lancet 383:533–540
66. Ballard C, Banister C, Khan Z, Cummings J, Demos G, Coate B et al (2018) Evaluation of the safety, tolerability, and efficacy of pimavanserin versus placebo in patients with Alzheimer's disease psychosis: a phase 2, randomised, placebo-controlled, double-blind study. Lancet Neurol 17:213–222
67. Cummings J, Ballard C, Tariot P, Owen R, Foff E, Youakim J et al (2018) Pimavanserin: potential treatment for dementia-related psychosis. J Prev Alzheimer Dis 5(4):253-258
68. Sevigny J, Chiao P, Bussiere T, Weinreb PH, Williams L, Maier M et al (2016) The antibody aducanumab reduces Abeta plaques in Alzheimer's disease. Nature 537:50–56
69. Ostrowitzki S, Lasser RA, Dorflinger E, Scheltens P, Barkhof F, Nikolcheva T et al (2017) A phase III randomized trial of gantenerumab in prodromal Alzheimer's disease. Alzheimers Res Ther 9:95. https://doi.org/10.1186/s13195-017-0318-y
70. Cummings JL, Zhong K, Kinney JW, Heaney C, Moll-Tudla J, Joshi A et al (2016) Double-blind, placebo-controlled, proof-of-concept trial of bexarotene in moderate Alzheimer's disease. Alzheimers Res Ther 8:4. https://doi.org/10.1186/s13195-016-0173-2
71. Cummings J, Zhong K, Cordes D (2017) Drug development in Alzheimer's disease: the role of default mode network assessment in phase II. US Neurol 13:67–69

72. Canter RG, Penney J, Tsai LH (2016) The road to restoring neural circuits for the treatment of Alzheimer's disease. Nature 539:187–196
73. Sorg C, Riedl V, Muhlau M, Calhoun VD, Eichele T, Laer L et al (2007) Selective changes of resting-state networks in individuals at risk for Alzheimer's disease. Proc Natl Acad Sci USA 104:18760–18765
74. Zhou J, Greicius MD, Gennatas ED, Growdon ME, Jang JY, Rabinovici GD et al (2010) Divergent network connectivity changes in behavioural variant frontotemporal dementia and Alzheimer's disease. Brain 133:1352–1367
75. McLaren DG, Sreenivasan A, Diamond EL, Mitchell MB, Van Dijk KR, Deluca AN et al (2012) Tracking cognitive change over 24 weeks with longitudinal functional magnetic resonance imaging in Alzheimer's disease. Neurodegener Dis 9:176–186
76. Dickerson BC, Sperling RA (2009) Large-scale functional brain network abnormalities in Alzheimer's disease: insights from functional neuroimaging. Behav Neurol 21:63–75
77. Atri A, O'Brien JL, Sreenivasan A, Rastegar S, Salisbury S, DeLuca AN et al (2011) Test-retest reliability of memory task functional magnetic resonance imaging in Alzheimer disease clinical trials. Arch Neurol 68:599–606
78. Sutton BP, Goh J, Hebrank A, Welsh RC, Chee MW, Park DC (2008) Investigation and validation of intersite fMRI studies using the same imaging hardware. J Magn Reson Imaging 28:21–28
79. Soares HD (2010) The use of mechanistic biomarkers for evaluating investigational CNS compounds in early drug development. Curr Opin Investig Drugs 11:795–801
80. Mattsson N, Andreasson U, Zetterberg H, Blennow K, Alzheimer's Disease Neuroimaging Initiative (2017) Association of plasma neurofilament light with neurodegeneration in patients with Alzheimer disease. JAMA Neurol 74:557–566
81. Babic Leko M, Borovecki F, Dejanovic N, Hof PR, Simic G (2016) Predictive value of cerebrospinal fluid visinin-like protein-1 levels for Alzheimer's disease early detection and differential diagnosis in patients with mild cognitive impairment. J Alzheimers Dis 50:765–778
82. Tarawneh R, D'Angelo G, Crimmins D, Herries E, Griest T, Fagan AM et al (2016) Diagnostic and prognostic utility of the synaptic marker neurogranin in Alzheimer disease. JAMA Neurol 73:561–571
83. Pereira JB, Westman E, Hansson O, Alzheimer's Disease Neuroimaging Initiative (2017) Association between cerebrospinal fluid and plasma neurodegeneration biomarkers with brain atrophy in Alzheimer's disease. Neurobiol Aging 58:14–29
84. Jack CR Jr, Shiung MM, Weigand SD, O'Brien PC, Gunter JL, Boeve BF et al (2005) Brain atrophy rates predict subsequent clinical conversion in normal elderly and amnestic MCI. Neurology 65:1227–1231
85. Whitwell JL, Josephs KA, Murray ME, Kantarci K, Przybelski SA, Weigand SD et al (2008) MRI correlates of neurofibrillary tangle pathology at autopsy: a voxel-based morphometry study. Neurology 71:743–749
86. Molinuevo JL, Blennow K, Dubois B, Engelborghs S, Lewczuk P, Perret-Liaudet A et al (2014) The clinical use of cerebrospinal fluid biomarker testing for Alzheimer's disease diagnosis: a consensus paper from the Alzheimer's Biomarkers Standardization Initiative. Alzheimers Dement 10:808–817
87. Dupont AC, Largeau B, Guilloteau D, Santiago Ribeiro MJ, Arlicot N (2018) The place of PET to assess new therapeutic effectiveness in neurodegenerative diseases. Contrast Media Mol Imaging 2018:7043578. https://doi.org/10.1155/2018/7043578
88. Alexander GE, Chen K, Pietrini P, Rapoport SI, Reiman EM (2002) Longitudinal PET evaluation of cerebral metabolic decline in dementia: a potential outcome measure in Alzheimer's disease treatment studies. Am J Psychiatry 159:738–745
89. Chen K, Langbaum JB, Fleisher AS, Ayutyanont N, Reschke C, Lee W et al (2010) Twelve-month metabolic declines in probable Alzheimer's disease and amnestic mild cognitive impairment assessed using an empirically pre-defined statistical region-of-interest: findings from the Alzheimer's Disease Neuroimaging Initiative. NeuroImage 51:654–664

90. Mega MS, Dinov ID, Porter V, Chow G, Reback E, Davoodi P et al (2005) Metabolic patterns associated with the clinical response to galantamine therapy: a fludeoxyglucose f 18 positron emission tomographic study. Arch Neurol 62:721–728
91. Mercier J, Provins L, Valade A (2017) Discovery and development of SV2A PET tracers: potential for imaging synaptic density and clinical applications. Drug Discov Today Technol 25:45–52
92. Chen MK, Mecca AP, Naganawa M, Finnema SJ, Toyonaga T, Lin SF et al (2018) Assessing synaptic density in Alzheimer disease with synaptic vesicle glycoprotein 2A positron emission tomographic imaging. JAMA Neurol 75(10):1215-1224
93. Kleinstreuer NC, Sullivan K, Allen D, Edwards S, Mendrick DL, Embry M et al (2016) Adverse outcome pathways: from research to regulation scientific workshop report. Regul Toxicol Pharmacol 76:39–50
94. Arrowsmith J, Miller P (2013) Trial watch: phase II and phase III attrition rates 2011–2012. Nat Rev Drug Discov 12:569. https://doi.org/10.1038/nrd4090
95. Wilkinson D, Windfeld K, Colding-Jorgensen E (2014) Safety and efficacy of idalopirdine, a 5-HT6 receptor antagonist, in patients with moderate Alzheimer's disease (LADDER): a randomised, double-blind, placebo-controlled phase 2 trial. Lancet Neurol 13:1092–1099
96. Porsteinsson AP, Drye LT, Pollock BG, Devanand DP, Frangakis C, Ismail Z et al (2014) Effect of citalopram on agitation in Alzheimer disease: the CitAD randomized clinical trial. JAMA 311:682–691
97. Doody RS, Raman R, Farlow M, Iwatsubo T, Vellas B, Joffe S et al (2013) A phase 3 trial of semagacestat for treatment of Alzheimer's disease. N Engl J Med 369:341–350
98. Sperling RA, Jack CR Jr, Black SE, Frosch MP, Greenberg SM, Hyman BT et al (2011) Amyloid-related imaging abnormalities in amyloid-modifying therapeutic trials: recommendations from the Alzheimer's Association Research Roundtable Workgroup. Alzheimers Dement 7:367–385
99. U.S. Food and Drug Administration (2014) Guidance for industry and FDA staff: qualification process for drug development tools. Division of Drug Information, Center for Drug Evaluation and Research, Food and Drug Administration, Silver Spring, MD. https://www.fda.gov/downloads/drugs/guidances/ucm230597.pdf
100. Romero K, de MM, Frank D, Anthony M, Neville J, Kirby L et al (2009) The Coalition Against Major Diseases: developing tools for an integrated drug development process for Alzheimer's and Parkinson's diseases. Clin Pharmacol Ther 86:365–367
101. Hill DL, Schwarz AJ, Isaac M, Pani L, Vamvakas S, Hemmings R et al (2014) Coalition Against Major Diseases/European Medicines Agency biomarker qualification of hippocampal volume for enrichment of clinical trials in predementia stages of Alzheimer's disease. Alzheimers Dement 10:421–429.e3. https://doi.org/10.1016/j.jalz.2013.07.003
102. Arneric SP, Batrla-Utermann R, Beckett L, Bittner T, Blennow K, Carter L et al (2017) Cerebrospinal fluid biomarkers for Alzheimer's disease: a view of the regulatory science qualification landscape from the Coalition Against Major Diseases CSF Biomarker Team. J Alzheimers Dis 55:19–35
103. Amur S, LaVange L, Zineh I, Buckman-Garner S, Woodcock J (2015) Biomarker qualification: toward a multiple stakeholder framework for biomarker development, regulatory acceptance, and utilization. Clin Pharmacol Ther 98:34–46
104. Gerlach CV, Derzi M, Ramaiah SK, Vaidya VS (2018) Industry perspective on biomarker development and qualification. Clin Pharmacol Ther 103:27–31
105. Jorgensen JT, Hersom M (2018) Clinical and regulatory aspects of companion diagnostic development in oncology. Clin Pharmacol Ther 103:999–1008
106. Scheerens H, Malong A, Bassett K, Boyd Z, Gupta V, Harris J et al (2017) Current status of companion and complementary diagnostics: strategic considerations for development and launch. Clin Transl Sci 10:84–92

107. Gilman S, Koller M, Black RS, Jenkins L, Griffith SG, Fox NC et al (2005) Clinical effects of Abeta immunization (AN1792) in patients with AD in an interrupted trial. Neurology 64:1553–1562
108. Galasko DR, Peskind E, Clark CM, Quinn JF, Ringman JM, Jicha GA et al (2012) Antioxidants for Alzheimer disease: a randomized clinical trial with cerebrospinal fluid biomarker measures. Arch Neurol 69:836–841
109. Salloway S, Sperling R, Fox NC, Blennow K, Klunk W, Raskind M et al (2014) Two phase 3 trials of bapineuzumab in mild-to-moderate Alzheimer's disease. N Engl J Med 370:322–333
110. Relkin NR, Thomas RG, Rissman RA, Brewer JB, Rafii MS, van Dyck CH et al (2017) A phase 3 trial of IV immunoglobulin for Alzheimer disease. Neurology 88:1768–1775
111. Fleisher AS, Truran D, Mai JT, Langbaum JB, Aisen PS, Cummings JL et al (2011) Chronic divalproex sodium use and brain atrophy in Alzheimer disease. Neurology 77:1263–1271
112. Sabbagh MN, Agro A, Bell J, Aisen PS, Schweizer E, Galasko D (2011) PF-04494700, an oral inhibitor of receptor for advanced glycation end products (RAGE), in Alzheimer disease. Alzheimer Dis Assoc Disord 25:206–212
113. Coric V, van Dyck CH, Salloway S, Andreasen N, Brody M, Richter RW et al (2012) Safety and tolerability of the gamma-secretase inhibitor avagacestat in a phase 2 study of mild to moderate Alzheimer disease. Arch Neurol 69:1430–1440
114. Coric V, Salloway S, van Dyck CH, Dubois B, Andreasen N, Brody M et al (2015) Targeting prodromal Alzheimer disease with avagacestat: a randomized clinical trial. JAMA Neurol 72:1324–1333
115. Liu E, Schmidt ME, Margolin R, Sperling R, Koeppe R, Mason NS et al (2015) Amyloid-beta 11C-PiB-PET imaging results from 2 randomized bapineuzumab phase 3 AD trials. Neurology 85:692–700
116. Fox NC, Black RS, Gilman S, Rossor MN, Griffith SG, Jenkins L et al (2005) Effects of Abeta immunization (AN1792) on MRI measures of cerebral volume in Alzheimer disease. Neurology 64:1563–1572
117. Petersen RC, Thomas RG, Grundman M, Bennett D, Doody R, Ferris S et al (2005) Vitamin E and donepezil for the treatment of mild cognitive impairment. N Engl J Med 352:2379–2388
118. Gauthier S, Aisen PS, Ferris SH, Saumier D, Duong A, Haine D et al (2009) Effect of tramiprosate in patients with mild-to-moderate Alzheimer's disease: exploratory analyses of the MRI sub-group of the Alphase study. J Nutr Health Aging 13:550–557
119. Quinn JF, Raman R, Thomas RG, Yurko-Mauro K, Nelson EB, Van Dyck C et al (2010) Docosahexaenoic acid supplementation and cognitive decline in Alzheimer disease: a randomized trial. JAMA 304:1903–1911
120. Galasko D, Bell J, Mancuso JY, Kupiec JW, Sabbagh MN, van DC et al (2014) Clinical trial of an inhibitor of RAGE-Abeta interactions in Alzheimer disease. Neurology 82:1536–1542
121. Turner RS, Thomas RG, Craft S, van Dyck CH, Mintzer J, Reynolds BA et al (2015) A randomized, double-blind, placebo-controlled trial of resveratrol for Alzheimer disease. Neurology 85:1383–1391

Chapter 3
Mitochondrial Involvement in Mental Disorders: Energy Metabolism and Genetic and Environmental Factors

Keiko Iwata

Abstract Mental disorders, such as major depressive disorder (MDD), bipolar disorder (BD), and schizophrenia (SZ), are generally characterized by a combination of abnormal thoughts, perceptions, emotions, behavior, and relationships with others. Multiple risk factors incorporating genetic and environmental susceptibility are associated with development of these disorders. Mitochondria have a central role in the energy metabolism, and the literature suggests energy metabolism abnormalities are widespread in the brains of subjects with MDD, BPD, and SZ. Numerous studies have shown altered expressions of mitochondria-related genes in these mental disorders. In addition, environmental factors for these disorders, such as stresses, have been suggested to induce mitochondrial abnormalities. Moreover, animal studies have suggested that interactions of altered expression of mitochondria-related genes and environmental factors might be involved in mental disorders. Further investigations into interactions of mitochondrial abnormalities with environmental factors are required to elucidate of the pathogenesis of these mental disorders.

Keywords Mental disorders · Mitochondria · Energy metabolism · Genetic factors · Environmental factors

3.1 Introduction

Mental disorders such as major depressive disorder (MDD), bipolar disorder (BD), and schizophrenia (SZ) are generally characterized by a combination of abnormal thoughts, perceptions, emotions, behavior, and relationships with others. MDD is characterized by sadness, loss of interest or pleasure, feelings of guilt or low self-worth, disturbed sleep or appetite, tiredness, and poor concentration. BD typically

K. Iwata (✉)
Venetian Institute of Molecular Medicine, Padova, Italy

Research Center for Child Mental Development, University of Fukui, Fukui, Japan
e-mail: kiwata@u-fukui.ac.jp

P. C. Guest (ed.), *Reviews on Biomarker Studies in Psychiatric and Neurodegenerative Disorders*, Advances in Experimental Medicine and Biology 1118,
https://doi.org/10.1007/978-3-030-05542-4_3

consists of both manic and depressive episodes separated by periods of normal mood. Manic episodes involve elevated or irritable moods, overactivity, pressure of speech, inflated self-esteem, and a decreased need for sleep. Psychoses, including SZ, are characterized by distortions in thinking, perception, emotions, language, sense of self, and behavior. Common psychotic experiences include hallucinations (hearing, seeing, or feeling things that are not there) and delusions (fixed false beliefs or suspicions that are firmly held even when there is evidence to the contrary). Multiple risk factors incorporating genetic and environmental susceptibility are associated with the development of these disorders. Based on twin studies, the heritability of MDD, BD, and SZ has been estimated about 40%, 75%, and 80%, respectively [1]. The way heritability has been estimated in twin studies means that gene-environment interactions involving environmental factors that are shared within a family are attributed to the genetic component and contribute to heritability estimates. Based on genetic data analyses, it becomes apparent that genetic variants account for a much smaller proportion of variance than the twin-based heritability estimates and genetic involvement is suggested to be 20–25% in these disorders [1].

3.2 Mitochondria

Mitochondria have a central role in the energy metabolism. This organelle plays a crucial role in adenosine 5′-triphosphate triphosphate (ATP) production through oxidative phosphorylation, a process that relies on the electron transport chain (ETC), composed of respiratory chain complexes I–IV and two intermediary substrates (coenzyme Q and cytochrome c) [2] (Fig. 3.1). These complexes are composed of numerous subunits encoded by both nuclear genes and mitochondrial DNA with the exception of complex II (encoded by nuclear genes only) (Fig. 3.1). The NADH and $FADH_2$ formed through the processes of glycolysis, fatty acid oxidation, and the citric acid cycle are energy-rich molecules that donate electrons to the ETC. Complexes I, III, and IV function as proton pumps that are driven by the free energy of coupled oxidation reactions. During the electron transfer, protons are pumped from the mitochondrial matrix into the intermembrane space. The proton gradient drives the ADP phosphorylation via the ATP synthase (complex V). In addition to energy metabolism, mitochondria are also involved in amino acid, lipid, and steroid metabolism and serve as Ca^{2+} buffers, sources of free radicals, and regulators of apoptosis.

3.3 Mitochondrial Abnormalities in Mental Disorders

Postmortem brain studies have shown mitochondrial abnormalities in SZ [3, 4]. A significant decrease in the volume density and count of mitochondria in oligodendroglial cells in the caudate nucleus and prefrontal area has been observed in the postmortem

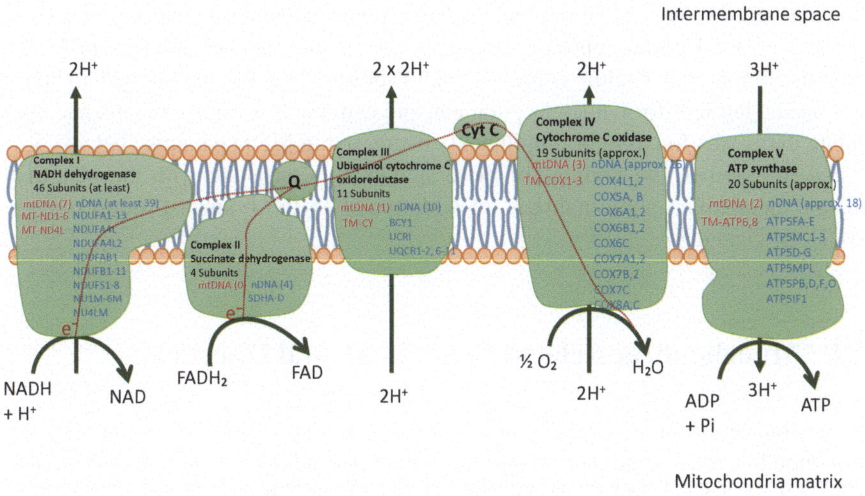

Fig. 3.1 Mitochondrial respiratory chain. For mammals, the respiratory chain consists of four enzyme complexes (complexes I–IV) and two intermediary substrates (coenzyme Q and cytochrome c). A proton gradient drives the ADP phosphorylation via ATP synthase (complex V)

brains of subjects with SZ compared to those of controls [3]. It has been also demonstrated that the caudate and the putamen of subjects with SZ contain significantly fewer mitochondrial profiles than those of controls [4]. Brain imaging studies have identified brain mitochondrial abnormalities in psychiatric disorders. Regional high-energy phosphorus compounds can be measured by phosphorus-31 magnetic resonance spectroscopy (^{31}P-MRS) neuroimaging. A decreased ratio of phosphomonoesters to phosphodiesters in subjects with BD was consistent with differences in membrane turnover, suggesting alterations in phospholipid metabolism and mitochondrial function in the brains of subjects with BD [5]. In SZ, many studies have shown alterations in membrane phospholipids, including decreased levels of free phosphomonoesters [6]. In addition, alterations in high-energy phosphate metabolism and regulation of oxidative phosphorylation have been found in the brains of subjects with MDD [7–12]. *N*-Acetylaspartate (NAA) synthesis correlates with mitochondrial metabolism in neurons, and the NAA concentration is used as an indicator of neuronal density and mitochondrial function [13]. In addition, studies using proton (^{1}H)-MRS have identified NAA deficits in the brains of subjects with MDD, BD, and SZ [14–16].

Glucose metabolism in the brain of mental disorders has been measured by positron emission tomography (PET) using ^{18}F-fluoro-deoxy-glucose (FDG). Using this approach, it has been shown that negative schizophrenia symptoms are associated with frontal, prefrontal, and anterior cingulate cortical hypoactivity and hypometabolism [17]. Similarly, positive schizophrenia symptoms have been linked to hypermetabolism in the temporolimbic system, including the amygdala, basal ganglia, and temporal cortical regions [17]. In BD, studies found a pattern of corticolimbic metabolism that was dysregulated accompanied by hypometabolism in frontal cortical regions [17]. In other studies, a reduced metabolic rate was observed in the prefrontal cortex, anterior cingulate cortex, and caudate nucleus in MDD patients compared to healthy controls [18]. In addition, glucose metabolism in bilateral insula, left lentiform nucleus putamen, and extranuclear, right caudate, and cingulate gyrus was found to be significantly decreased in MDD patients [19].

Taken together, the findings of these studies are consistent with the idea that energy metabolism abnormalities are widespread in the brains of subjects with MDD, BPD, and SZ.

3.4 Mitochondria-Related Genes in Mental Disorders

In postmortem brains of SZ subjects, reduced expression in a majority of mitochondria-related genes encoded by mitochondrial DNA, such as MT-ATP6, MT-ATP8, MT-CO3, MT-CYB, MT-ND1, MT-ND2, MT-ND3, MT-ND4, MT-ND5, and MT-ND6, has been observed [20]. In addition, the expression levels of nuclear genes encoding subunits of complex I (11 genes), complex III (4 genes), and complex IV (11 genes) have been found to be reduced in postmortem prefrontal or frontal cortical areas of subjects with SZ [21, 22]. Likewise, significantly lower mRNA and corresponding protein levels of complex I subunits have been demonstrated in the striatum in SZ patients [23]. In addition, studies using microarray analysis have revealed that NDUFS7 (complex I), NDUFS8 (complex I), UQCRC2 (complex III), COX5A (complex IV), COX6C (complex IV), ATP5C1 (complex V), ATP5J (complex V), and ATP5G3 (complex V) are present at decreased levels in the postmortem prefrontal cortex of subjects with BD [24]. Similarly, the mRNA of two isoforms of creatine kinase, which are involved in synthesis and metabolism of phosphocreatine, has been shown to be downregulated in the dorsolateral prefrontal cortex in subjects with BD [25]. Although these results have been inconsistent [26], numerous studies have suggested the involvement of mitochondria-related genes in the etiology and pathophysiology of these mental disorders, at least in part.

3.5 Environmental Factors for Mental Disorders

Environmental stressors triggering activation of the hypothalamic-pituitary-adrenal (HPA) axis cause the brain to be exposed to corticosteroids, affecting neurobehavioral functions with a strong downregulation of hippocampal neurogenesis, and are major risk factors for MDD [27, 28]. In the case of BD, three types of environmental risk factors based on timing and postulated mechanism of action have been identified: neurodevelopment (maternal infection during pregnancy), substances (cannabis, cocaine, opioids, tranquilizers, stimulants, and sedatives), and physical/psychological stress (parental loss, adversities, abuses, and brain injury) [29]. For

SZ, hypoxia, maternal infection, maternal stress, maternal malnutrition, and atypical mother-child interactions have been suggested as perinatal and early childhood risk factors [30, 31]. In addition, drug abuse (especially exposure to cannabis) has been suggested as a later life environmental risk factor for SZ [30, 31]. In addition, evidence suggests that maternal stress can influence central nervous system (CNS) development and consequently exert long-lasting behavioral and cognitive changes in the offspring [32]. Among these environmental factors, stress is a common major risk factor for mental disorders such as MDD, BD, and SZ [33, 34].

3.6 Interactions of Environmental Factors and Mitochondria-Related Genes

Gene-environment interactions have been suggested to play a significant role in the pathogenesis of mental disorders. To study the impact of altered expression levels of mitochondria-related genes on the stress response in vivo, Picard et al. genetically manipulated two mtDNA-encoded respiratory chain subunits, mt-Nd6 (complex I) and mt-Co1 (complex IV), which resulted in selective impairment of mitochondrial respiratory chain function [35]. These mice have shown hyperactivation of the HPA axis and an increased cytokine IL-6 inflammatory response to stress [35]. Intriguingly, the same mice have shown changes in hippocampal gene expression responses to stress [35], consistent with reports that alterations in hippocampal functions have often been reported to occur in individuals with mental disorders [36].

In preclinical studies, Glombik et al. have shown that maternal stress leads to depression-like behaviors in the offspring of rats [37]. Of note, these offspring have shown brain mitochondrial abnormalities, including significant downregulation of Ndufv2 (complex I) [37]. As a possible mechanism of action, Lambertini et al. have shown that maternal stress modifies the mitochondrial gene expression profile in the human placenta and the expression levels of mitochondrial-encoded genes, MT-ND2, MT-ND6, and MT-CO2, have correlated to both maternal stress and infant temperament indices in placental samples from the stress in pregnancy birth cohort study [38]. The placenta plays a role in supporting the fetal allograft throughout gestation, protecting against immune rejection, serving to supply oxygen and nutrients, and removing carbon dioxide and waste products from the developing fetus [39]. From such findings, it has been suggested that dysfunctions at the level of the placenta, the maternal-fetal interface, can contribute to the pathogenesis of mental disorders [40]. These reports suggest that mitochondria respond to maternal psychosocial stress in pregnancy and play an important role in infant brain development.

Taken together, these findings indicate that stress may lead to and augment mitochondrial abnormalities found in individuals with psychiatric conditions.

3.7 Conclusion

Thus far, a single gene responsible for mental disorders has not been identified, and it seems unlikely that any single gene could fully explain the pathogenesis of these complex disorders. Instead, multiple lines of evidence have led to the suggestion that mitochondrial abnormalities, especially those leading to dysfunctions in energy metabolism, often occur in these mental disorders. As a possible pathological mechanism, stressors have been observed to be common environmental risk factors for these disorders, which could lead to altered expression levels of mitochondria-related genes [38]. Of note, the results of animal studies have suggested that mitochondrial abnormalities are augmented by stress [35]. Furthermore, other environmental factors, such as viral infection during pregnancy and hypoxia during labor and delivery, have been also suggested as risk factors for mental disorders [41, 42]. Further investigations into interaction of mitochondrial abnormalities with environmental factors are required to elucidate the pathogenesis of these mental disorders. Such studies could help to identify possible novel drug targets for improved treatment of individuals suffering from these debilitating disorders.

Acknowledgments This work was supported in part by the Japan Society for the Promotion of Science (JSPS) Program for Advancing Strategic International Networks to Accelerate the Circulation of Talented Researchers, Grant No. S2603, and the Japan Foundation for Pediatric Research (to K.I.).

References

1. Uher R (2014) Gene-environment interactions in severe mental illness. Front Psych 5:48. https://doi.org/10.3389/fpsyt.2014.00048
2. Jou SH, Chiu NY, Liu CS (2009) Mitochondrial dysfunction and psychiatric disorders. Chang Gung Med J 32:370–379
3. Uranova NA, Orlovskaia DD, Vikhreva OV, Zimina IS, Rakhmanova VI (2001) Morphometric study of ultrastructural changes in oligodendroglial cells in the postmortem brain in endogenous psychoses. Vestn Ross Akad Med Nauk 7:42–48
4. Kung L, Roberts RC (1999) Mitochondrial pathology in human schizophrenic striatum: a postmortem ultrastructural study. Synapse 31:67–75
5. Shi XF, Carlson PJ, Sung YH, Fiedler KK, Forrest LN, Hellem TL et al (2015) Decreased brain PME/PDE ratio in bipolar disorder: a preliminary (31) P magnetic resonance spectroscopy study. Bipolar Disord 17:743–752
6. Reddy R, Keshavan MS (2003) Phosphorus magnetic resonance spectroscopy: its utility in examining the membrane hypothesis of schizophrenia. Prostaglandins Leukot Essent Fatty Acids 69:401–405
7. Moore CM, Christensen JD, Lafer B, Fava M, Renshaw PF (1997) Lower levels of nucleoside triphosphate in the basal ganglia of depressed subjects: a phosphorous-31 magnetic resonance spectroscopy study. Am J Psychiatry 154:116–118
8. Renshaw PF, Parow AM, Hirashima F, Ke Y, Moore CM, Frederick Bde B et al (2001) Multinuclear magnetic resonance spectroscopy studies of brain purines in major depression. Am J Psychiatry 158:2048–2055

9. Volz HP, Rzanny R, Riehemann S, May S, Hegewald H, Preussler B et al (1998) 31P magnetic resonance spectroscopy in the frontal lobe of major depressed patients. Eur Arch Psychiatry Clin Neurosci 248:289–295
10. Kato T, Takahashi S, Shioiri T, Inubushi T (1992) Brain phosphorous metabolism in depressive disorders detected by phosphorus-31 magnetic resonance spectroscopy. J Affect Disord 26:223–230
11. Iosifescu DV, Bolo NR, Nierenberg AA, Jensen JE, Fava M, Renshaw PF (2008) Brain bioenergetics and response to triiodothyronine augmentation in major depressive disorder. Biol Psychiatry 63:1127–1134
12. Harper DG, Jensen JE, Ravichandran C, Perlis RH, Fava M, Renshaw PF et al (2017) Tissue type-specific bioenergetic abnormalities in adults with major depression. Neuropsychopharmacology 42:876–885
13. Clark JB (1998) N-acetyl aspartate: a marker for neuronal loss or mitochondrial dysfunction. Dev Neurosci 20:271–276
14. Steen RG, Hamer RM, Lieberman JA (2005) Measurement of brain metabolites by 1H magnetic resonance spectroscopy in patients with schizophrenia: a systematic review and meta-analysis. Neuropsychopharmacology 30:1949–1962
15. Yildiz-Yesiloglu A, Ankerst DP (2006) Neurochemical alterations of the brain in bipolar disorder and their implications for pathophysiology: a systematic review of the in vivo proton magnetic resonance spectroscopy findings. Prog Neuropsychopharmacol Biol Psychiatry 30:969–995
16. Caverzasi E, Pichiecchio A, Poloni GU, Calligaro A, Pasin M, Palesi F et al (2012) Magnetic resonance spectroscopy in the evaluation of treatment efficacy in unipolar major depressive disorder: a review of the literature. Funct Neurol 27:13–22
17. Scholl M, Damian A, Engler H (2014) Fluorodeoxyglucose PET in neurology and psychiatry. PET Clin 9:371–390, v. https://doi.org/10.1016/j.cpet.2014.07.005
18. Videbech P (2000) PET measurements of brain glucose metabolism and blood flow in major depressive disorder: a critical review. Acta Psychiatr Scand 101:11–20
19. Su L, Cai Y, Xu Y, Dutt A, Shi S, Bramon E (2014) Cerebral metabolism in major depressive disorder: a voxel-based meta-analysis of positron emission tomography studies. BMC Psychiatry 14:321. https://doi.org/10.1186/s12888-014-0321-9
20. Shao L, Martin MV, Watson SJ, Schatzberg A, Akil H, Myers RM et al (2008) Mitochondrial involvement in psychiatric disorders. Ann Med 40:281–295
21. Karry R, Klein E, Ben Shachar D (2004) Mitochondrial complex I subunits expression is altered in schizophrenia: a postmortem study. Biol Psychiatry 55:676–684
22. Prabakaran S, Swatton JE, Ryan MM, Huffaker SJ, Huang JT, Griffin JL et al (2004) Mitochondrial dysfunction in schizophrenia: evidence for compromised brain metabolism and oxidative stress. Mol Psychiatry 9:684–697, 643
23. Ben-Shachar D, Karry R (2008) Neuroanatomical pattern of mitochondrial complex I pathology varies between schizophrenia, bipolar disorder and major depression. PLoS One 3:e3676. https://doi.org/10.1371/journal.pone.0003676
24. Sun X, Wang JF, Tseng M, Young LT (2006) Downregulation in components of the mitochondrial electron transport chain in the postmortem frontal cortex of subjects with bipolar disorder. J Psychiatry Neurosci 31:189–196
25. MacDonald ML, Naydenov A, Chu M, Matzilevich D, Konradi C (2006) Decrease in creatine kinase messenger RNA expression in the hippocampus and dorsolateral prefrontal cortex in bipolar disorder. Bipolar Disord 8:55–264
26. Anglin RE, Mazurek MF, Tarnopolsky MA, Rosebush PI (2012) The mitochondrial genome and psychiatric illness. Am J Med Genet B Neuropsychiatr Genet 159B:749–759
27. Lopez JF, Akil H, Watson SJ (1999) Neural circuits mediating stress. Biol Psychiatry 46:1461–1471
28. Masi G, Brovedani P (2011) The hippocampus, neurotrophic factors and depression: possible implications for the pharmacotherapy of depression. CNS Drugs 25:913–931

29. Marangoni C, Hernandez M, Faedda GL (2016) The role of environmental exposures as risk factors for bipolar disorder: a systematic review of longitudinal studies. J Affect Disord 193:165–174
30. van Os J, Kapur S (2009) Schizophrenia. Lancet 374:635–645
31. Dean K, Murray RM (2005) Environmental risk factors for psychosis. Dialogues Clin Neurosci 7:69–80
32. Marques AH, Bjorke-Monsen AL, Teixeira AL, Silverman MN (2015) Maternal stress, nutrition and physical activity: impact on immune function, CNS development and psychopathology. Brain Res 1617:28–46
33. Gomes FV, Grace AA (2017) Adolescent stress as a driving factor for schizophrenia development-a basic science perspective. Schizophr Bull 43:486–489
34. Beijers R, Jansen J, Riksen-Walraven M, de Weerth C (2010) Maternal prenatal anxiety and stress predict infant illnesses and health complaints. Pediatrics 126:e401–e409. https://doi.org/10.1542/peds.2009-3226
35. Picard M, McManus MJ, Gray JD, Nasca C, Moffat C, Kopinski PK et al (2015) Mitochondrial functions modulate neuroendocrine, metabolic, inflammatory, and transcriptional responses to acute psychological stress. Proc Natl Acad Sci U S A 112:E6614–E6623
36. Lodge DJ, Grace AA (2011) Developmental pathology, dopamine, stress and schizophrenia. Int J Dev Neurosci 29:207–213
37. Głombik K, Stachowicz A, Ślusarczyk J, Trojan E, Budziszewska B, Suski M et al (2015) Maternal stress predicts altered biogenesis and the profile of mitochondrial proteins in the frontal cortex and hippocampus of adult offspring rats. Psychoneuroendocrinology 60:151–162
38. Lambertini L, Chen J, Nomura Y (2015) Mitochondrial gene expression profiles are associated with maternal psychosocial stress in pregnancy and infant temperament. PLoS One 10:e0138929. https://doi.org/10.1371/journal.pone.0138929
39. Carlin A, Alfirevic Z (2008) Physiological changes of pregnancy and monitoring. Best Pract Res Clin Obstet Gynaecol 22:801–823
40. Hsiao EY, Patterson PH (2012) Placental regulation of maternal-fetal interactions and brain development. Dev Neurobiol 72:1317–1326
41. Wakuda T, Iwata K, Iwata Y, Anitha A, Takahashi T, Yamada K et al (2015) Perinatal asphyxia alters neuregulin-1 and COMT gene expression in the medial prefrontal cortex in rats. Prog Neuropsychopharmacol Biol Psychiatry 56:149–154
42. Paparelli A, Iwata K, Wakuda T, Iyegbe C, Murray RM, Takei N (2017) Perinatal asphyxia in rat alters expression of novel schizophrenia risk genes. Front Mol Neurosci 10:341. https://doi.org/10.3389/fnmol.2017.00341

Chapter 4
Lymphocytes, Platelets, Erythrocytes, and Exosomes as Possible Biomarkers for Alzheimer's Disease Clinical Diagnosis

Ryszard Pluta and Marzena Ułamek-Kozioł

Abstract In the aging world population, Alzheimer's disease accounts for more than 70% of all cases of dementia and is the sixth leading cause of death. The neurodegenerative processes of this disorder can begin 10–20 years before the clinical symptoms develop. Postmortem brain autopsy of Alzheimer's disease cases reveals characteristic hallmarks like extracellular amyloid plaques and intraneuronal neurofibrillary tangles and synaptic and neuronal disintegration with severe brain atrophy. Some studies have reported that platelets contain the amyloid protein precursor and the secretase enzymes required for the amyloidogenic processing of this protein. Thus, platelets can be a good blood cell-based marker to investigate the onset of Alzheimer's disease. Other studies have indicated cellular and molecular alterations in erythrocytes and lymphocytes from Alzheimer's disease subjects, which emphasize the systemic nature of the disorder. In addition, small extracellular vesicles called exosomes appear to be an important factor during the progression of the disease. These vesicles contain disease-associated molecules such as the amyloid protein precursor and tau protein. In this chapter, we will summarize the recent knowledge on the involvement of lymphocytes, erythrocytes, platelets, and exosomes in the development of Alzheimer's disease. The data will be reviewed with a view to applying the above elements as Alzheimer's disease early preclinical and late-stage biomarkers with potential use for clinical diagnosis, prognosis, and monitoring disease progression and treatment responses.

Keywords Alzheimer's disease · Biomarkers · Platelets · Lymphocytes · Erythrocytes · Exosomes · Plasma

---

R. Pluta (✉)
Laboratory of Ischemic and Neurodegenerative Brain Research, Mossakowski Medical Research Centre, Polish Academy of Sciences, Warsaw, Poland
e-mail: pluta@imdik.pan.pl

M. Ułamek-Kozioł
Laboratory of Ischemic and Neurodegenerative Brain Research, Mossakowski Medical Research Centre, Polish Academy of Sciences, Warsaw, Poland

First Department of Neurology, Institute of Psychiatry and Neurology, Warsaw, Poland

P. C. Guest (ed.), *Reviews on Biomarker Studies in Psychiatric and Neurodegenerative Disorders*, Advances in Experimental Medicine and Biology 1118,
https://doi.org/10.1007/978-3-030-05542-4_4

4.1 Introduction

Alzheimer's disease is a progressive and irreversible disorder with diminished cognition, memory, personality, and symptoms that are connected with some psychiatric diseases such as depression [1–5]. The vast majority of patients (>95%) have a sporadic form of the dementia due to Alzheimer's disease. Alzheimer's disease is the leading cause of dementia in the world's aging population (60–80% of cases) and is the sixth leading cause of death [6]. Mortality from Alzheimer's disease has continued to rise, with the number of Alzheimer's disease-related deaths increasing by 123% between 2000 and 2015 [7]. The progressive neuronal death, mainly in vulnerable brain structures such as the hippocampus, and dementia in patients with Alzheimer's disease are related to the presence of extracellular amyloid plaques, intraneuronal neurofibrillary tangles, and brain atrophy. Recent studies have indicated that the number of individuals suffering from Alzheimer's disease accounts for 15–20 million worldwide and 5 million of these occur in the USA [8]. By 2025, the number of people aged 65 and older with dementia due to Alzheimer's disease is estimated to reach 7 million in the USA, and this will increase further to 13.8 million by 2050 [7, 9, 10]. Approximately 3.4 million of Americans over the age of 65 with Alzheimer's disease are women [7]. At the individual level, the risk of developing this disease doubles every 5 years after the age of 65 and approaches 50% by the age of 85.

In 2014, the direct cost of Alzheimer's disease in the USA alone was estimated to be $214 billion [11]. According to the Alzheimer's Association report, it is the most expensive disorder in the USA, costing an estimated $277 billion in 2018, which includes an increase of nearly $20 billion since 2017 [7]. In Great Britain, Alzheimer's disease costs the economy more than cancer and heart disease together [12]. It is forecast that by 2050, the total cost of Alzheimer's disease will have increased to more than $1.1 trillion [7].

The neuropathological alterations associated with Alzheimer's disease begin 10–20 years before the clinical symptoms' development [2, 3]. The etiology of sporadic Alzheimer's disease is unclear, and the mechanisms that lead to neuronal damage and losses especially in the hippocampus during progression of the disease are not completely understood. As a result of the above, antemortem diagnosis or causal therapy for this devastating disorder is not currently possible [13–15]. The search for Alzheimer's disease etiology, early antemortem diagnosis/biomarkers, and causal therapy has been one fraught with a wide range of complications and numerous revisions with a lack of a final solution. Alzheimer's disease is becoming a more common cause of death, and it is the only top ten cause of death that cannot be prevented, slowed, or even cured.

At present Alzheimer's disease is one of the greatest national healthcare system challenges. In December 2013, the G8 stated that dementia, including dementia due to Alzheimer's disease, should be made a global priority. It is their ambition to find a cure or a disease-modifying treatment by the year 2025 [16]. Alzheimer's disease etiology and final antemortem clinical diagnosis have been unresolved issues to date, as can be seen in Alzheimer's Disease Research contest announced by the US Department of Defense at the end of 2017. Recently, President Donald Trump signed a bill to provide a $414 million increase for Alzheimer's disease research and education [7]. With this increase, dementia and dementia due to Alzheimer's disease research funding at the federal US government is $1.8 billion. This bill came in response to the demands from the Alzheimer's Association and others, which recognized that there is much work to be done to alleviate suffering and eliminate this devastating neurodegenerative disorder. The study of the causes of Alzheimer's disease to date has focused on the "amyloid theory" without final conclusion. Hence, other etiological approaches may be necessary [13, 14, 17–25].

The first established molecular biomarkers for Alzheimer's disease which have been accepted and used in Alzheimer's disease clinical diagnosis are β-amyloid peptide 1–42, total tau protein, and hyperphosphorylated tau protein levels in cerebrospinal fluid [26]. Aside from these three cerebrospinal fluid biomarkers, a number of potential candidates have been identified in cerebrospinal fluid and other body fluids such as blood. In order to identify biomarkers for clinical diagnosis, prognosis, early prevention, and response to treatment, multiplex biomarker tests will be required from various platforms and tissue or body fluid sources. This includes proteomic analyses of platelets, lymphocytes, erythrocytes, and exosomes. This chapter highlights the research of platelets, lymphocytes, erythrocytes, and exosomes from Alzheimer's disease patients and their applications and potential use for clinical diagnosis, prognosis, and evaluation of treatment effects and for monitoring disease progression. The discovery of biomarkers for Alzheimer's disease at the early clinical level will mean that fewer deaths will occur from this devastating and debilitating disorder. New antemortem biomarkers are needed for Alzheimer's disease, and this will only be achieved by making it a world healthcare priority and increasing funding for research. It is anticipated that this may lead to earlier clinical diagnosis, causal therapies, and, ultimately, a cure.

4.2 Current Biomarkers in Alzheimer's Disease Clinical Diagnosis

At present clinical diagnosis of Alzheimer's disease is based on cognitive tests, brain imaging, and information collected from the families and medical histories of the patients [26, 27]. Measurements of β-amyloid peptide 1–42 and total and phosphorylated tau protein levels are currently used as biomarkers in the cerebrospinal fluid to establish the diagnosis during the development of Alzheimer's disease. The total tau protein levels reflect the intensification of neuronal cell degeneration, phosphorylated tau protein levels correlate proportionally with tangle generation, and the β-amyloid peptide 1–42 correlate inversely with plaque generation [28]. Currently, clinical diagnosis relies on the decline of cognitive symptoms that are characteristic of the advanced stages of Alzheimer's disease. However, this does not offer a definitive antemortem diagnosis. Actually, the final diagnosis of Alzheimer's disease requires postmortem autopsy of the brain with presentation of amyloid plaques and neurofibrillary tangles. The generally accepted cerebrospinal fluid markers do not translate well to the blood since these show no significant differences between Alzheimer's disease cases and healthy people in plasma or serum samples. Furthermore, considerable variability in the measurement of these proteins is observed across different studies and laboratories [29, 30]. Additionally, one of the main problems in clinical diagnosis of Alzheimer's disease comes from the observation that the neuropathological changes can begin in the brain tissue 20 years before clinical symptoms' development [2, 3]. Moreover, there are no clinical biomarkers associated with the neuropathological progression, especially in preclinical stages of Alzheimer's disease. Given the ease of access compared with cerebrospinal fluid, it would be ideal if such markers could be observed in blood and other peripheral tissues.

4.3 Candidates for New Biomarkers

The search for final, reliable, non-expensive, and noninvasive clinical diagnostic tests based on neuronal pathology in the brain is an important driving force for biomarker-based research. Ideally, a blood-based biomarker would be representative of biochemical and pathological changes in the brain that are diagnostic of a disease and its stage and progression. Because blood circulates through different organs including the brain, it contains a number of molecules that can give insight into pathophysiological changes in the disease [31]. Ideally, biomarkers should identify the molecular mechanisms associated with both the neuropathological alterations and dementia development. In Alzheimer's disease, a number of peripheral cells such as platelets, erythrocytes, and lymphocytes show physiological changes. Platelets are also an important source of the amyloid protein precursor in the blood, and the study of this protein in platelets has been suggested as a possible biomarker in the clinical diagnosis of Alzheimer's disease [12]. Other investigations have noted both molecular and cellular abnormalities in lymphocytes and erythrocytes from Alzheimer's disease patients [32–34]. Furthermore, we discuss new data on the role of exosomes in Alzheimer's disease progression and spreading. We also underline the possible utility of exosome as new clinical diagnostic biomarkers for this disorder [35]. Increasing evidence suggests that during the development of Alzheimer's disease, changes occur in these small brain extracellular vesicles, which comprise disease-associated proteins such as the amyloid protein precursor and tau protein [36]. Exosomes are a likely route that facilitates the spread of β-amyloid peptide and tau protein from their original cells into brain parenchyma.

4.3.1 Platelets as Alzheimer's Disease Biomarkers

Dysfunctional platelets from Alzheimer's disease cases show a reduced amyloid protein precursor ratio, a switch from non-amyloidogenic to amyloidogenic metabolism of the amyloid protein precursor and, as a consequence, an increased generation of β-amyloid peptide [12]. In addition, alterations in the activity of cyclooxygenase 2, cytochrome c oxidase, and monoamine oxidase B enzymes have been documented in platelets isolated from the blood of Alzheimer's disease subjects [37]. The amyloid protein precursor is present in platelet membranes and as a soluble form of different lengths in the platelet α-granules. One study showed that the decreased amyloid protein precursor levels in platelets positively correlated with the development of the disease [12]. This relationship was also noted in the preclinical period of Alzheimer's disease suggesting that this can be used as an early antemortem marker for Alzheimer's disease clinical diagnosis. Other observations showed an increase in the number of platelet-leukocyte complexes and platelet aggregates in the circulation of patients with Alzheimer's disease [12]. This suggests that peripheral platelets from Alzheimer's disease patients present increased activation compared to those from non-Alzheimer's disease patients. In observations made over 1 year, activated platelets showed a significant correlation with the staging of cognitive deficits in Alzheimer's disease patients [38]. Another important finding from this study was an increased number of coated platelets in Alzheimer's disease cases. These coated platelets retained a high level of the amyloid protein precursor on their surface [39] and the platelet levels correlated with Alzheimer's disease stage of development [40]. These observations indicated a possible role of coated platelets as preclinical markers of Alzheimer's disease. Furthermore, the identification of tau protein in peripheral platelets from patients with Alzheimer's disease suggested this as a potential new biomarker for Alzheimer's disease clinical diagnosis [41]. It was noted that platelet tau protein has a significantly higher correlation with Alzheimer's disease progression than the platelet amyloid protein precursor [42] and the platelet tau protein was also found to correlate well with brain atrophy in Alzheimer's disease patients in a separate study [43].

4.3.2 Lymphocytes as Alzheimer's Disease Biomarkers

Changes in Ca^{2+} levels and oxidative stress have been noted to occur in lymphocytes in the early stages of Alzheimer's disease [44, 45]. A greater amount of reactive oxygen species impairment in the function of mitochondria and DNA integrity, along with changed activity of antioxidant enzymes and apoptosis, have been documented in lymphocytes from patients with Alzheimer's disease [45, 46]. The above data suggested the universal nature of oxidative stress in Alzheimer's disease dementia development and supports the premise that oxidative stress is an early hallmark of Alzheimer's disease formation. Cell cycle dysfunction is another early

Alzheimer's disease neuropathological hallmark manifested in peripheral lymphocytes [46, 47]. A number of investigations have now documented the role of T cells in the development of Alzheimer's disease-type dementia in relation with immune system alterations, as shown by a decreased reaction of T cells to some pathogens and increased telomerase activity in lymphocytes that leads to reduction of lymphocyte proliferation and dysfunction of the immune system in Alzheimer's disease patients [48–53]. These findings support the possible input of a dysfunctional peripheral immune system in the development of Alzheimer's disease.

4.3.3 Erythrocytes as Alzheimer's Disease Biomarkers

The existing results indicate that red blood cell β-amyloid peptides may constitute markers for Alzheimer's disease clinical diagnosis [54]. The red blood cell levels of the β-amyloid peptides 1–40 and 1–42 have been found to be 8 and 14 times higher than in serum, respectively [54]. Red blood cell amyloid complexes induce alterations in morphology and adhesion of the cells to the endothelium and thereby influence blood vessel function [33, 55]. Recent data indicate that 98% of erythrocytes from patients with Alzheimer's disease are amyloid binding-positive, but only 38% of erythrocyte binding is found in healthy persons [34]. Moreover, the data indicate that the presence of red blood cell β-amyloid peptides 1–40 and 1–42 may lead to generation of other erythrocyte-based markers for Alzheimer's disease clinical diagnosis. β-Amyloid peptide binding causes reactive oxygen species and oxidative stress generation in red blood cells and induces deposition of phospholipid hydroperoxides, a typical cause of red blood cell damage [56]. Similarly, erythrocytes from patients with Alzheimer's disease presented higher activity of the glycolytic enzymes hexokinase, bisphosphoglycerate mutase, phosphofructokinase, and bisphosphoglycerate phosphatase [34]. Furthermore, red blood cell membrane proteins such as the calpain-1, glucose transporter, band 3 protein, IgG, and Hsp 90 were changed in Alzheimer's disease patients [33].

4.3.4 Exosomes as Alzheimer's Disease Biomarkers

Exosomes can be found in many extracellular human fluids like cerebrospinal fluid, blood, saliva, milk, and urine [57–60]. These vesicles contain a rich source of molecular cargo, including nucleic acids and proteins, and have therefore been proposed to provide a systemic noninvasive source of biomarkers for brain diseases [59]. Exosomes from cerebrospinal fluid contain proteins originating from the brain such as microglial- and neuron-specific markers, apolipoprotein E, and notch homolog protein 3 [61]. A significant increase of both total tau protein and phosphorylated tau protein levels was found in cerebrospinal fluid exosomes in Alzheimer's disease patients compared to healthy control subjects [62]. The amyloid protein precursor has also been found in exosomes from cerebrospinal fluid [63]. In serum, exosomal miRNAs associated with Alzheimer's disease have been identified which show a correlation with neuropsychological and neuroimaging examination results [64]. This suggests the possibility that peripheral exosomal miRNAs and other biomarkers can be used for the identification of early, preclinical Alzheimer's disease cases. However, the diagnostic utility of peripheral exosomes is still not clear. Further molecular investigations are needed to confirm whether the β-amyloid peptides and tau protein are also present in exosomes from experimental models of the disease.

4.4 Exosomes as a Spread Factor in Alzheimer's Disease

The β-amyloid peptide 1–42 aggregates and develops oligomer assemblies, including amyloid plaques in different brain regions. It has been suggested that the spread of these abnormalities is mediated by exosomes [29, 35, 65]. A study showed that the β-amyloid peptide spreads from the hippocampus to septal nuclei over 1 year following the intracerebral injection of this molecule [66]. The confirmation of this spreading neurodegenerative pathology via neuron-to-neuron transfer of β-amyloid peptide has been documented experimentally [67, 68]. These findings indicate that β-amyloid peptide can be relocated from cell to cell and thereby disperses the neuropathology throughout the whole brain in Alzheimer's disease cases. The spread of tau protein by different brain structures has also been shown to occur in tau protein transgenic mice [69, 70]. The occurrence of the cell-to-cell transmission of Alzheimer's disease-associated proteins suggests a potential molecular pathway that could be targeted by novel treatments with the aim of disrupting and/or delaying the spread and progression of the disorder.

4.5 Conclusions

For patients with Alzheimer's disease, growing evidence indicates that pathological changes take place not only in the brain parenchyma but also in blood platelets, lymphocytes, erythrocytes, and exosomes. Platelets, due to shared properties with neuronal cells, can be used for the study of neuronal pathologies [37]. Evidently, all of these blood components are easily available as biomarkers for preclinical and definitive clinical diagnosis and probably for drug screening and treatment monitoring. Evidence suggests that neuroinflammation is an important mechanism in Alzheimer's disease progression in both the early and late disease stages [71]. Among the factors which play a role in Alzheimer's disease inflammatory mechanisms are lymphocytes which migrate through the blood-brain barrier to the brain [32]. For this reason, the inflammatory factors present in the blood, as well as molecular alterations in lymphocytes from patients with Alzheimer's disease, represent a promising source of blood-based markers for clinical diagnosis of both the early and late stages of the disorder. Further blood-based clinical marker progress requires discovery of characteristic immune factors for the preclinical, moderate, and late diagnosis of Alzheimer's disease. This concept may transform diagnostics, shifting the focus away from clinical symptoms and more toward molecular mechanisms. This chapter also describes a number of potentially useful erythrocyte markers for the diagnosis of patients with Alzheimer's disease. These include erythrocyte morphology; their membrane proteins such as the β-amyloid peptide, glucose transporter, IgG, Hsp 90, calpain-1, and band 3 protein; and oxidative stress molecules. Some data suggest that serum β-amyloid peptide binds to red blood cells, implying a pathogenic role of erythrocyte amyloid complexes. In addition, erythrocytes have been found to contain the β-amyloid peptide, as well as alterations in Hsp 90, calpain-1, and band 3 protein. Potentially the most prominent aspect of erythrocytes is their changed morphology in Alzheimer's disease cases [33]. Such a feature can be easily detected and used as a diagnostic marker, even in the early pathogenic stages of the disease. A major advantage of using exosomes as a biomarker source is the fact that these vesicles deliver cargo over distances without degradation or dilution, because the biomolecules are securely transported within their capsules [72–74]. Further studies aimed at increasing our understanding of the physiological functions of exosomes and the factors influencing their switch to a pathological state are important for increasing our knowledge on the spreading and etiology of Alzheimer's disease and therefore identifying a new biomarker source associated with all stages of the disorder. Still, much remains to be explored because we have just scratched the surface of understanding these blood-based vesicles as a source of clinically useful biomarkers. However, prior to drawing any serious conclusions, these data need to be verified at first by various experimental and further clinical studies.

To identify new biomarker candidates that arise from blood, these need to be validated using thousands of samples that are well-matched and come from different geographical locations with regard to Alzheimer's disease research centers. In addition, all validated biomarkers should be translated into inexpensive and clinical useful tests to facilitate early detection of individuals at high risk of developing Alzheimer's disease. This will allow earlier treatment than can be currently achieved using existing methods. This could be more effective than the existing scenario of treatment at a later stage when irreversible changes to neuronal physiology have already occurred. The development of such clinical useful tests will also afford the possibility of monitoring treatment responses for achieving the best possible outcomes for individuals suffering with this devastating neurodegenerative disorder.

Acknowledgments The authors acknowledge the support provided by the Mossakowski Medical Research Centre, Polish Academy of Sciences, Warsaw, Poland (T3).

References

1. Engedal K, Barca ML, Laks J, Selbaek G (2011) Depression in Alzheimer's disease: specificity of depressive symptoms using three different clinical criteria. Int J Geriatr Psychiatry 26:944–951
2. Villemagne VL, Pike KE, Chetelat G, Ellis KA, Mulligan RS, Bourgeat P et al (2011) Longitudinal assessment of Abeta and cognition in aging and Alzheimer disease. Ann Neurol 69:181–192
3. Villemagne VL, Burnham S, Bourgeat P, Brown B, Ellis KA, Salvado O et al (2013) Amyloid beta deposition, neurodegeneration, and cognitive decline in sporadic Alzheimer's disease: a prospective cohort study. Lancet Neurol 12:357–367
4. Rushing NC, Sachs-Ericsson N, Steffens DC (2014) Neuropsychological indicators of preclinical Alzheimer's disease among depressed older adults. Neuropsychol Dev Cogn B Aging Neuropsychol Cogn 21:99–128
5. Zhao QF, Tan L, Wang HF, Jiang T, Tan MS, Tan L et al (2016) The prevalence of neuropsychiatric symptoms in Alzheimer's disease: systematic review and meta-analysis. J Affect Disord 190:264–271
6. Landeiro F, Walsh K, Ghinai I, Mughal S, Nye E, Wace H et al (2018) Measuring quality of life of people with predementia and dementia and their caregivers: a systematic review protocol. BMJ Open 8:e019082. https://doi.org/10.1136/bmjopen-2017-019082
7. www.alz.org. 2018
8. Hu Z, Zeng L, Huang Z, Zhang J, Li T (2007) The study of Golgi apparatus in Alzheimer's disease. Neurochem Res 32:1265–1277
9. Hebert LE, Beckett LA, Scherr PA, Evans DA (2001) Annual incidence of Alzheimer disease in the United States projected to the years 2000 through 2050. Alzheimer Dis Assoc Disord 15:169–173
10. Prince M, Bryce R, Albanese E, Wimo A, Ribeiro W, Ferri CP (2013) The global prevalence of dementia: a systematic review and metaanalysis. Alzheimers Dement 9:63–75.e62. https://doi.org/10.1016/j.jalz.2012.11.007
11. Winblad B, Amouyel P, Andrieu S, Ballard C, Brayne C, Brodaty H et al (2016) Defeating Alzheimer's disease and other dementias: a priority for European science and society. Lancet Neurol 15:455–532
12. Canobbio I, Abubaker AA, Visconte C, Torti M, Pula G (2015) Role of amyloid peptides in vascular dysfunction and platelet dysregulation in Alzheimer's disease. Front Cell Neurosci 9:65. https://doi.org/10.3389/fncel.2015.00065
13. Pluta R, Furmaga-Jabłońska W, Maciejewski R, Ułamek-Kozioł M, Jabłoński M (2013) Brain ischemia activates β- and γ-secretase cleavage of amyloid precursor protein: significance in sporadic Alzheimer's disease. Mol Neurobiol 47:425–434
14. Pluta R, Jabłoński M, Ułamek-Kozioł M, Kocki J, Brzozowska J, Januszewski S et al (2013) Sporadic Alzheimer's disease begins as episodes of brain ischemia and ischemically dysregulated Alzheimer's disease genes. Mol Neurobiol 48:500–515
15. Pluta R, Ułamek-Kozioł M, Januszewski S, Czuczwar SJ (2018) Platelets, lymphocytes and erythrocytes from Alzheimer's disease patients: the quest for blood cell-based biomarkers. Folia Neuropathol 56:14–20
16. Scheltens P, Blennow K, Breteler MM, de Strooper B, Frisoni GB, Salloway S et al (2016) Alzheimer's disease. Lancet 388:505–517
17. Pluta R, Kida E, Lossinsky AS, Golabek AA, Mossakowski MJ, Wisniewski HM (1994) Complete cerebral ischemia with short-term survival in rats induced by cardiac arrest. I. Extracellular accumulation of Alzheimer's β-amyloid protein precursor in the brain. Brain Res 649:323–328
18. Pluta R (2007) Ischemia-reperfusion pathways in Alzheimer's disease. Nova Science Publisher, Inc., New York ISBN-10:1600217443

19. Pluta R, Ułamek M, Jabłoński M (2009) Alzheimer's mechanisms in ischemic brain degeneration. Anat Rec 292:1863–1881
20. Kocki J, Ułamek-Kozioł M, Bogucka-Kocka A, Januszewski S, Jabłoński M, Gil-Kulik P et al (2015) Dysregulation of amyloid precursor protein, β-secretase, presenilin 1 and 2 genes in the rat selectively vulnerable CA1 subfield of hippocampus following transient global brain ischemia. J Alzheimers Dis 47:1047–1056
21. Pluta R, Kocki J, Ułamek-Kozioł M, Bogucka-Kocka A, Gil-Kulik P, Januszewski S et al (2016) Alzheimer-associated presenilin 2 gene is dysregulated in rat medial temporal lobe cortex after complete brain ischemia due to cardiac arrest. Pharmacol Rep 68:155–161
22. Pluta R, Kocki J, Ułamek-Kozioł M, Petniak A, Gil-Kulik P, Januszewski S et al (2016) Discrepancy in expression of β-secretase and amyloid-β protein precursor in Alzheimer-related genes in the rat medial temporal lobe cortex following transient global brain ischemia. J Alzheimers Dis 51:1023–1031
23. Ułamek-Kozioł M, Kocki J, Bogucka-Kocka A, Petniak A, Gil-Kulik P, Januszewski S et al (2016) Dysregulation of autophagy, mitophagy and apoptotic genes in the medial temporal lobe cortex in an ischemic model of Alzheimer's disease. J Alzheimers Dis 54:113–121
24. Ułamek-Kozioł M, Kocki J, Bogucka-Kocka A, Januszewski S, Bogucki J, Czuczwar SJ et al (2017) Autophagy, mitophagy and apoptotic gene changes in the hippocampal CA1 area in a rat ischemic model of Alzheimer's disease. Pharmacol Rep 69:1289–1294
25. Pluta R, Bogucka-Kocka A, Ułamek-Kozioł M, Bogucki J, Januszewski S, Kocki J et al (2018) Ischemic tau protein gene induction as an additional key factor driving development of Alzheimer's phenotype changes in CA1 area of hippocampus in an ischemic model of Alzheimer's disease. Pharmacol 70:881–884
26. Schaffer C, Sarad N, DeCrumpe A, Goswami D, Herrmann S, Morales J et al (2015) Biomarkers in the diagnosis and prognosis of Alzheimer's disease. J Lab Autom 20:589–600
27. McKhann GM, Knopman DS, Chertkow H, Hyman BT, Jack CR Jr, Kawas CH et al (2011) The diagnosis of dementia due to Alzheimer's disease: recommendations from the National Institute on Aging-Alzheimer's Association workgroups on diagnostic guidelines for Alzheimer's disease. Alzheimers Dement 7:263–269
28. Blennow K, Hampel H, Weiner M, Zetterberg H (2010) Cerebrospinal fluid and plasma biomarkers in Alzheimer disease. Nat Rev Neurol 6:131–144
29. Fiandaca MS, Kapogiannis D, Mapstone M, Boxer A, Eitan E, Schwartz JB et al (2014) Identification of preclinical Alzheimer's disease by a profile of pathogenic proteins in neurally derived blood exosomes: a case-control study. Alzheimers Dement 11:600–607
30. Seeburger JL, Holder DJ, Combrinck M, Joachim C, Laterza O, Tanen M et al (2015) Cerebrospinal fluid biomarkers distinguish postmortem-confirmed Alzheimer's disease from other dementias and healthy controls in the optima cohort. J Alzheimers Dis 44:525–539
31. Perneczky R, Guo LH (2016) Plasma proteomics biomarkers in Alzheimer's disease: latest advances and challenges. Methods Mol Biol 1303:521–529
32. Mietelska-Porowska A, Wojda U (2017) T lymphocytes and inflammatory mediators in the interplay between brain and blood in Alzheimer's disease: potential pools of new biomarkers. J Immunol Res 17. https://doi.org/10.1155/2017/4626540
33. Stevenson A, Lopez D, Khoo P, Kalaria RN, Mukaetova-Ladinska EB (2017) Exploring erythrocytes as blood biomarkers for Alzheimer's disease. J Alzheimers Dis 60:845–857
34. Wojsiat J, Laskowska-Kaszub K, Mietelska-Porowska A, Wojda U (2017) Search for Alzheimer's disease biomarkers in blood cells: hypotheses-driven approach. Biomark Med 11:917–931
35. Vella LJ, Hill AF, Cheng L (2016) Focus on extracellular vesicles: exosomes and their role in protein trafficking and biomarker potential in Alzheimer's and Parkinson's disease. Int J Mol Sci 17:173. https://doi.org/10.3390/ijms17020173
36. Yuyama K, Igarashi Y (2017) Exosomes as carriers of Alzheimer's amyloid-β. Front Neurosci 11:229. https://doi.org/10.3389/fnins.2017.00229
37. Veitinger M, Varga B, Guterres SB, Zellner M (2014) Platelets, a reliable source for peripheral Alzheimer's disease biomarkers? Acta Neuropathol Commun 2:65. https://doi.org/10.1186/2051-5960-2-65

38. Sakurai H, Hanyu H, Sato T, Kume K, Hirao K, Kanetaka H et al (2013) Effects of cilostazol on cognition and regional cerebral blood flow in patients with Alzheimer's disease and cerebrovascular disease: a pilot study. Geriatr Gerontol Int 13:90–97
39. Prodan CI, Szasz R, Vincent AS, Ross ED, Dale GL (2006) Coated platelets retain amyloid precursor protein on their surface. Platelets 17:56–60
40. Prodan CI, Ross ED, Vincent AS, Dale GL (2008) Rate of progression in Alzheimer's disease correlates with coated-platelet levels–a longitudinal study. Transl Res 152:99–102
41. Neumann K, Farias G, Slachevsky A, Perez P, Maccioni RB (2011) Human platelet tau: a potential peripheral marker for Alzheimer's disease. J Alzheimers Dis 25:103–109
42. Farias G, Perez P, Slachevsky A, Maccioni RB (2012) Platelet tau pattern correlates with cognitive status in Alzheimer's disease. J Alzheimers Dis 31:65–69
43. Slachevsky A, Guzman-Martınez L, Delgado C, Reyes P, Farıas GA, Munoz-Neira C et al (2017) Tau platelets correlate with regional brain atrophy in patients with Alzheimer's disease. J Alzheimers Dis 55:1595–1603
44. Mota SI, Costa RO, Ferreira IL, Santana I, Caldeira GL, Padovano C et al (2015) Oxidative stress involving changes in Nrf2 and ER stress in early stages of Alzheimer's disease. Biochim Biophys Acta 1852:1428–1441
45. Wojsiat J, Prandelli C, Laskowska-Kaszub K, Martín Requero A, Wojda U (2015) Oxidative stress and aberrant cell cycle in Alzheimer's disease lymphocytes: diagnostic prospects. J Alzheimers Dis 46:329–350
46. Wojda U (2016) Alzheimer's disease lymphocytes: potential for biomarkers. Biomark Med 10:1–4
47. Kuhla A, Ludwig SC, Kuhla B, Münch G, Vollmar B (2015) Advanced glycation end products are mitogenic signals and trigger cell cycle reentry of neurons in Alzheimer's disease brain. Neurobiol Aging 36:753–761
48. Zhang J, Kong Q, Zhang Z, Ge P, Ba D, He W (2003) Telomere dysfunction of lymphocytes in patients with Alzheimer disease. Cogn Behav Neurol 16:170–176
49. Richartz-Salzburger E, Batra A, Stransky E, Laske C, Köhler N, Bartels M et al (2007) Altered lymphocyte distribution in Alzheimer's disease. J Psychiatr Res 41:174–178
50. Da Mesquita SA, Ferreira C, Sousa JC, Correia-Neves M, Sousa N, Marques F (2016) Insights on the pathophysiology of Alzheimer's disease: the crosstalk between amyloid pathology, neuroinflammation and the peripheral immune system. Neurosci Biobehav Rev 68:547–562
51. Licastro F, Porcellini E (2016) Persistent infections, immune-senescence and Alzheimer's disease. Oncoscience 3:135–142
52. Schwartz M, Deczkowska A (2016) Neurological disease as a failure of brain-immune crosstalk: the multiple faces of neuroinflammation. Trends Immunol 37:668–679
53. Deardorff WJ, Grossberg GT (2017) Targeting neuroinflammation in Alzheimer's disease: evidence for NSAIDs and novel therapeutics. Expert Rev Neurother 17:17–32
54. Kiko T, Nakagawa K, Satoh A, Tsuduki T, Furukawa K, Arai H et al (2012) Amyloid beta levels in human red blood cells. PLoS One 7(11):e49620. https://doi.org/10.1371/journal.pone.0049620
55. Nakagawa K, Kiko T, Kuriwada S, Miyazawa T, Kimura F, Miyazawa T (2011) Amyloid β induces adhesion of erythrocytes to endothelial cells and affects endothelial viability and functionality. Biosci Biotechnol Biochem 75:2030–2033
56. Nakagawa K, Kiko T, Miyazawa T, Sookwong P, Tsuduki T, Satoh A et al (2011) Amyloid β-induced erythrocytic damage and its attenuation by carotenoids. FEBS Lett 585:1249–1254
57. Street JM, Barran PE, Mackay CL, Weidt S, Balmforth C, Walsh TS et al (2012) Identification and proteomic profiling of exosomes in human cerebrospinal fluid. J Transl Med 10. https://doi.org/10.1186/1479-5876-10-5
58. Cheng L, Sun X, Scicluna BJ, Coleman BM, Hill AF (2013) Characterization and deep sequencing analysis of exosomal and non-exosomal miRNA in human urine. Kidney Int 86:433–444
59. Cheng L, Sharples RA, Scicluna BJ, Hill AF (2014) Exosomes provide a protective and enriched source of miRNA for biomarker profiling compared to intracellular and cell-free blood. J Extracell Vesicles 3. https://doi.org/10.3402/jev.v3.23743

60. Kalra H, Drummen GPC, Mathivanan S (2016) Focus on extracellular vesicles: introducing the next small big thing. Int J Mol Sci 17:170. https://doi.org/10.3390/ijms17020170
61. Chiasserini D, van Weering JR, Piersma SR, Pham TV, Malekzadeh A, Teunissen CE et al (2014) Proteomic analysis of cerebrospinal fluid extracellular vesicles: a comprehensive dataset. J Proteome 106:191–204
62. Saman S, Kim W, Raya M, Visnick Y, Miro S, Saman S et al (2012) Exosome-associated tau is secreted in tauopathy models and is selectively phosphorylated in cerebrospinal fluid in early Alzheimer disease. J Biol Chem 287:3842–3849
63. Lopez-Font I, Cuchillo-Ibanez I, Sogorb-Esteve A, Garcia-Ayllon MS, Saez-Valero J (2015) Transmembrane amyloid-related proteins in CSF as potential biomarkers for Alzheimer's disease. Front Neurol 6:125. https://doi.org/10.3389/fneur.2015.00125
64. Cheng L, Doecke JD, Sharples RA, Villemagne VL, Fowler CJ, Rembach A et al (2014) Prognostic serum miRNA biomarkers associated with Alzheimer's disease shows concordance with neuropsychological and neuroimaging assessment. Mol Psychiatry 20:1188–1196
65. Vingtdeux V, Sergeant N, Buee L (2012) Potential contribution of exosomes to the prion-like propagation of lesions in Alzheimer's disease. Front Physiol 3:229. https://doi.org/10.3389/fphys.2012.00229
66. Hamaguchi T, Eisele YS, Varvel NH, Lamb BT, Walker LC, Jucker M (2012) The presence of Aβ seeds, and not age per se, is critical to the initiation of Aβ deposition in the brain. Acta Neuropathol 123:31–37
67. Nath S, Agholme L, Kurudenkandy FR, Granseth B, Marcusson J, Hallbeck M (2012) Spreading of neurodegenerative pathology via neuron-to-neuron transmission of β-amyloid. J Neurosci 32:8767–8777
68. Domert J, Rao SB, Agholme L, Brorsson AC, Marcusson J, Hallbeck M et al (2014) Spreading of amyloid-β peptides via neuritic cell-to-cell transfer is dependent on insufficient cellular clearance. Neurobiol Dis 65:82–92
69. Clavaguera F, Bolmont T, Crowther RA, Abramowski D, Frank S, Probst A et al (2009) Transmission and spreading of tauopathy in transgenic mouse brain. Nat Cell Biol 11:909–913
70. Liu L, Drouet V, Wu JW, Witter MP, Small SA, Clelland C et al (2012) Trans-synaptic spread of tau pathology in vivo. PLoS One 7:e31302. doi: 10.1371/journal.pone.0031302
71. Heneka TM, Carson MJ, Khoury JE, Landreth GE, Brosseron F, Feinstein DL et al (2015) Neuroinflammation in Alzheimer's disease. Lancet Neurol 14:388–405
72. Ghidoni R, Squitti R, Siotto M, Benussi L (2018) Innovative biomarkers for Alzheimer's disease: focus on the hidden disease biomarkers. J Alzheimers Dis 62:1507–1518
73. Polanco JC, Li C, Durisic N, Sullivan R, Götz J (2018) Exosomes taken up by neurons hijack the endosomal pathway to spread to interconnected neurons. Acta Neuropathol Commun 6(1):10. https://doi.org/10.1186/s40478-018-0514-4
74. Zheng T, Wu X, Wei X, Wang M, Zhang B (2018) The release and transmission of amyloid precursor protein via exosomes. Neurochem Int 114:18–253

Chapter 5
Genetic Risk Factors for Alzheimer Disease: Emerging Roles of Microglia in Disease Pathomechanisms

Sho Takatori, Wenbo Wang, Akihiro Iguchi, and Taisuke Tomita

Abstract The accumulation of aggregated amyloid β (Aβ) peptides in the brain is deeply involved in Alzheimer disease (AD) pathogenesis. Mutations in APP and presenilins play major roles in Aβ pathology in rare autosomal-dominant forms of AD, whereas pathomechanisms of sporadic AD, accounting for the majority of cases, remain unknown. In this chapter, we review current knowledge on genetic risk factors of AD, clarified by recent advances in genome analysis technology. Interestingly, TREM2 and many genes associated with disease risk are predominantly expressed in microglia, suggesting that these risk factors are involved in pathogenicity through common mechanisms involving microglia. Therefore, we focus on factors closely associated with microglia and discuss their possible roles in pathomechanisms of AD. Furthermore, we review current views on the pathological roles of microglia and emphasize the importance of microglial changes in response to Aβ deposition and mechanisms underlying the phenotypic changes. Importantly, functional outcomes of microglial activation can be both protective and deleterious to neurons. We further describe the involvement of microglia in tau pathology and the activation of other glial cells. Through these topics, we shed light on microglia as a promising target for drug development for AD and other neurological disorders.

Keywords Alzheimer disease (AD) · Amyloid β (Aβ) · Tau · Amyloid hypothesis · Sporadic AD · Genetic risk factor · Genome-wide association study (GWAS) · Single nucleotide polymorphism (SNP) · Rare variant · Microglia · Triggering receptor expressed on myeloid cells 2 (TREM2) · Neuroinflammation · Neurodegenerative disease

---

S. Takatori · W. Wang · A. Iguchi · T. Tomita (✉)
Laboratory of Neuropathology and Neuroscience, Graduate School of Pharmaceutical Sciences, The University of Tokyo, Tokyo, Japan
e-mail: taisuke@mol.f.u-tokyo.ac.jp

P. C. Guest (ed.), *Reviews on Biomarker Studies in Psychiatric and Neurodegenerative Disorders*, Advances in Experimental Medicine and Biology 1118,
https://doi.org/10.1007/978-3-030-05542-4_5

5.1 Introduction

Alzheimer disease (AD) is the most common neurodegenerative disease, which causes progressive neuronal loss in multiple regions of the brain, including the cerebral cortex [1]. Its core clinical symptoms include memory impairment, cognitive dysfunction, and various psychological symptoms, all of which severely limit the social activities of patients and their family members. The greatest risk factor for AD is aging, and the number of patients is increasing due to the increasing elderly population around the world. Therefore, AD has become one of the greatest social problems in recent years.

AD is pathologically characterized by senile plaques and neurofibrillary tangles observed in broad brain regions [2]. Senile plaques are composed of amyloid β (Aβ) peptides that aggregate and are extracellularly deposited, whereas neurofibrillary tangles are composed of tau, a cytoskeleton-associated protein, which aggregates and accumulates inside neurons. Aggregated tau is also known to be hyperphosphorylated at multiple residues.

In recent years, positron-emission tomography (PET) probes for aggregated Aβ and tau have been developed, and their deposition in the brain can be visualized in real time [3]. Changes in the levels of cerebrospinal fluid (CSF) Aβ and tau were also found to correlate well with the results of diagnostic imaging [4] (https://www.alzforum.org/alzbiomarker). These novel methodologies are now replacing diagnostic methods based on clinical manifestations and are being established as a new "definition" of AD [5–8].

The majority of AD patients have the so-called sporadic form that does not show a family history, although there is also another type of the disease showing an autosomal-dominant form of inheritance. Our current knowledge about the pathomechanisms of AD owes much to the identification and subsequent analyses of the causative genes for autosomal-dominant AD. Through these studies, Aβ peptides constituting senile plaques were found to play an active role in the pathogenesis of AD.

Aβ is a partial peptide produced by the stepwise cleavage of a type I transmembrane protein called Aβ precursor protein (APP). APP first undergoes cleavage by β-secretase in the extracellular region in the vicinity of the membrane. As a result, a C-terminal portion of 99 amino acids remains on the membrane, called C-terminal fragment or C99. C99 is further subjected to a second cleavage by γ-secretase in the transmembrane domain, releasing the Aβ peptide. γ-Secretase is a complex-type protease composed of four transmembrane proteins named presenilin, nicastrin, anterior pharynx defective 1, and presenilin enhancer-2 [9]. As several variations of the γ-secretase cleavage site are known, Aβ species differing in length from 37 to 43 amino acids are produced. Of these, Aβ40 is the most abundant. On the other hand, Aβ42 is present at approximately one-tenth of the levels of Aβ40, but it is known to have a higher tendency to aggregate.

Interestingly, all of the causative mutations of autosomal-dominant AD were identified in the genes for APP, PSEN1, or PSEN2, the latter two of which encode presenilin isoforms. Genetic and biochemical studies have shown that presenilin mutants alter the activity of γ-secretase and increase the production ratio of Aβ42. On the other hand, APP mutations promoted Aβ accumulation via one of the following mechanisms: (1) increasing Aβ production, (2) increasing the production ratio of Aβ42, or (3) enhancing Aβ aggregation. Taken together, the "amyloid hypothesis," in which abnormalities of Aβ production or aggregation are thought to be the key to AD pathogenesis, has been widely accepted [10]. However, it remains unclear whether this pathomechanism is also true for sporadic AD cases. Therefore, in this chapter we review the recent advances in our understanding of the genetic risk factors for AD and highlight the emerging roles of non-neuronal cells called microglia in the pathomechanisms of AD.

5.2 Sporadic Alzheimer Disease and Its Genetic Risk Factors

The aberrant accumulation of aggregated Aβ is a general hallmark of AD and is not restricted to autosomal-dominant cases. This suggests the importance of the amyloid hypothesis as a universal pathogenic mechanism, but the pathomechanisms of AD in patients without mutations in APP or PSEN1/PSEN2 remain unknown.

Many clinical trials have so far been unsuccessful based on strategies that interfere with Aβ production (e.g., γ-secretase inhibitors), and some researchers regard it as important evidence disproving the amyloid hypothesis. However, it has been demonstrated that Aβ accumulation in the brain begins 10–20 years before disease onset and that there is already a substantial amount of Aβ accumulation by the time clinical symptoms appear [11]. Therefore, it is likely that intervention at earlier stages, called preclinical or mild cognitive impairment, might be necessary for therapeutic strategies targeting Aβ to be effective [12]. The finding that a rare coding variant of APP (p.A673T) reduces AD risk is in good accordance with this idea. This APP mutant was less susceptible to β-cleavage and therefore produced less Aβ [13], suggesting that Aβ plays a central role in the pathomechanisms of AD. Furthermore, the treatment of an antibody against aggregated forms of Aβ reduced Aβ burden in the brains of AD patients as well as delayed their disease progression [14]. These data corroborated the idea that strategies based on the amyloid hypothesis are still promising as AD therapeutics.

As the proportion of AD patients with mutations in either APP or PSEN1/PSEN2 is small, if Aβ accumulation is crucial to the development of AD, it is important to know why it occurs in patients. To explain this, several hypotheses have been proposed: (1) Aβ production is enhanced; (2) the rate of degradation or metabolism of Aβ is decreased in patients; or (3) responsiveness and sensitivity of neurons and glial cells to Aβ are enhanced, irrespective of the degree of Aβ accumulation. Associated with this topic, AD patients reportedly showed a decreased clearance rate of Aβ from the CSF [15], suggesting that abnormalities in the extracerebral efflux of Aβ underlie the pathogenesis of AD. This may somehow be associated with the fact that AD patients show lower levels of Aβ42 in their CSF, as well as plasma [16, 17]. It is likely that there are genetic factors that underlie such defects in Aβ metabolism. Importantly, such genetic factors are likely to be different from mutations in APP and PSEN1/PSEN2 in familial AD but are factors causing imperfect penetrance, and that increases or decreases the risk of developing AD. In this section, we will describe such genetic "risk factors" in detail.

Genetic polymorphisms within the APOE locus have long been known as strong risk factors of AD. APOE encodes an apolipoprotein that constitutes plasma lipoprotein particles. Whereas ApoE functions in peripheral tissues, it is also known to be produced from astrocytes in the brain and plays an important role in supplying cholesterol to neurons. Three common variants of the APOE gene, namely, ε2, ε3, and ε4, are known, and these isoforms encode the same polypeptide of 299 amino acids except for differences in amino acids number 112 and 158. Whereas the ε3 isoform has a cysteine and an arginine at positions 112 and 158, respectively, both residues are substituted to cysteines in ε2 and to arginines in ε4. To date, it has been shown that the ε4 allele raises the risk of developing AD, whereas the ε2 allele has the opposite effect [18, 19]. It has also been shown that AD patients with an ε4 allele develop AD about 8 years earlier on the average per allele [20]. These facts have suggested that ApoE plays a crucial role in the pathogenesis of AD. In fact, ApoE is known to accumulate in senile plaques together with Aβ, and the ε4 isoform has been shown to increase Aβ aggregation in animal models, as well as in some in vitro experiments [21]. Furthermore, there are multiple lines of evidence that ApoE is also involved in Aβ clearance. Collectively, ApoE is thought to contribute to AD pathogenesis by modifying the aggregation or clearance of Aβ. However, because the proportion of late-onset AD cases associated with the ε4 allele is estimated as only 27.3% [22], it has been assumed that there are yet unidentified genetic risk factors other than APOE.

Several large-scale genome-wide association studies (GWAS) have been performed for late-onset AD, and more than 20 loci have been identified to date [22–27]. The functions of many of these genes remain unknown and are expected to provide new clues to elucidate the pathomechanisms of AD. On the other hand, most susceptibility loci identified in the GWASs showed small effect sizes, which were in contrast to the moderate effect sizes of the risk/protective alleles of APOE. From this point of view, some people view the GWAS for AD as unsuccessful. However, these multiple weak risk factors may work additively, and the aggregation of such factors can reduce the age of onset by up to 10 years [28].

Geneticists are still searching for other genetic risk factors of AD. Owing to their predicted small effect sizes, some researchers are attempting to increase the sample size by recruiting not only AD patients but also healthy individuals with a family history of AD (called "AD-by-proxy"). Such an attempt, called GWAS by proxy, or GWAX, is being performed at present for several common diseases [29] and has successfully identified the association between AD and the ADAM10 gene [30], which is involved in APP metabolism and was previously identified as a possible causative gene for late-onset AD [31].

In principle, GWAS is a method to analyze disease associations among common variants or single nucleotide polymorphisms (SNPs), which exist comparatively abundantly in the population. As for AD, the fact that almost every factor showed only a weak association suggests that rare variants that were not included in GWAS analyses have stronger effects on disease risk. Of course, there have been attempts to identify such factors so far. However, since most previous studies have targeted a small number of candidate genes, even if these studies identified a factor that is "significantly" associated with cases in case-control comparisons, the association is not always true in the genome-wide context. In recent years, however, the rapid progress of genome analysis technology has enabled the identification of risk factors with genome-wide or exome-wide significance.

To identify such rare genetic factors efficiently, many researchers are taking advantage of the following two methods: (1) whole-exome sequencing (WES) or whole-genome sequencing (WGS) and (2) chip-array technologies similar to the SNP array used in GWAS analyses. For example, in the identification of the protective allele of the APP gene described above, data obtained from WGS conducted on nearly 2000 Icelanders were utilized. Identified coding variants in APP were integrated with the chip-genotyped data using an "imputation" technique and analyzed for their association with AD [13]. In addition, rare variants found in the WGS were further investigated for their association with AD on a genome-wide scale, which led to the discovery of the novel AD risk factor TREM2 [32]. The disease association of this coding variant of TREM2 (p.R47H) has also been identified in an independent study [33] and has been repeatedly replicated. A meta-analysis of these studies estimated that the TREM2 mutation increased the risk of developing AD by about threefold, which is almost the same effect as that of APOE-ε4 [34]. More importantly, TREM2 was found to be expressed in microglia but not in neurons in the central nervous system (CNS). This unequivocally indicates that non-neuronal cells also play an important role in AD pathogenesis, in contrast to a conventional view that neuronal cells have only been regarded as important in the context of the amyloid hypothesis.

In addition, an extremely large-scale association analysis of about 85,000 people was recently reported, utilizing a novel genotyping chip equipped with a large number of probes for rare variants with a minor allele frequency of less than 0.5% [35]. This study also replicated the association between TREM2 mutations and AD and furthermore demonstrated novel disease associations of PLCG2 and ABI3, which are also specifically expressed in microglia. These facts again emphasize the importance of microglia in AD pathology.

Several reports have utilized WGS and WES technologies to identify the disease-susceptible genes in AD families with no mutations in APP or PSEN1/PSEN2 and successfully identified interesting genes, such as SORL1, PLD3, AKAP9, and UNC5C [36–39]. Unlike mutations of APP and PSEN1/PSEN2, these alleles are not always penetrant, and hence it may be more accurate to express these as risk factors rather than pathogenic genes. Nevertheless, the SORL1 locus was found to have multiple loss-of-function mutations in AD families, which showed high penetrance, and these mutations were rare in healthy subjects in population-based studies, suggesting that SORL1 mutations have a strong effect similar to that of APP or PSEN1/PSEN2. In addition, a recent WES analysis conducted on more than 20,000 people identified multiple, ultrarare, loss-of-function variants of SORL1 in some AD patients and found that SORL1 was the only gene in which variants were significantly associated with AD in the genome-wide context [40]. Similarly, the association between AD and rare variants in the ABCA7 gene, one of the GWAS hits (see below), has increasingly been identified. These suggest that additional polymorphisms with a larger effect size may be discovered in the genes that were previously identified by GWAS.

Besides these examples, various searches for genetic factors using novel approaches are being performed. The most unique one is a search for genetic factors in people who have reached old age without developing various age-associated disorders, such as AD [41]. Similar to this, another group searched for a "resilience factor" from elderly non-AD people with at least one copy of the APOE ε4allele [42]. These approaches may offer additional therapeutic strategies against AD.

To summarize, the identification of genetic risk factors for AD has been accelerated by recent advances in genome analysis technology. We summarized the major genetic risk factors in (Table 5.1). Almost all of these factors were found to have low to moderate effect sizes, indicating that none of these factors alone can accurately predict disease, unlike autosomal-dominant AD cases. Instead, multiple factors may cumulatively affect disease susceptibility. In this regard, it is noteworthy that molecular functions of these genes roughly cluster within several pathways, including membrane traffic (SORL1, BIN1, PICALM, and CD2AP), lipid metabolism (APOE, CLU, and ABCA7), and inflammatory process (TREM2, CD33, CR1, PLCG2, and INPP5D), and therefore it is possible that risk factors contribute to pathogenicity possibly through a few such common pathways.

5.3 Genetic Risk Factors Associated with Microglia

The finding that rare variants of TREM2 showed an effect size comparable to that of APOE prompted us to reconsider the pathological roles of microglia, which had not been considered as a central player in the amyloid hypothesis. In addition, many of the genes associated with AD risk were reportedly expressed specifically or highly in microglia [43], suggesting that their common functions are relevant to the pathomechanisms of AD. To elucidate such mechanisms, it is important to understand these gene functions and the effects of the risk/protective variants on AD pathology. Therefore, in this section, we review our current knowledge of individual risk factors, particularly by focusing on factors that are linked with microglial function in previous reports. We particularly emphasize (1) how their disease associations were identified, (2) what their molecular/cellular functions are, and (3) their possible roles in AD pathology.

Table 5.1 Summary of genetic risk factors for AD

Genes	Cellular functions	Possible roles in AD pathology	Cell types[a]	References
ABCA7	Lipid metabolism [98], phagocytosis [99]	Aβ clearance [104], APP processing [105]	Neuron, microglia	[26, 27]
ABI3	Actin cytoskeletal organization [111, 112]	Unknown	Microglia	[35]
APOE	Lipid metabolism [165]	Aβ clearance [166, 167], Aβ aggregation [168], tau pathology, and neurotoxicity [155]	Astrocyte, microglia	[18, 19]
BIN1	Endocytosis [169]	APP processing [169, 170], tau pathology [171]	Microglia, oligodendrocyte, neuron	[25]
CASS4	Cell adhesion and migration [172]	Tau pathology [173]	Microglia	[22]
CD2AP	Cell adhesion [174], endocytosis [175]	APP processing [169], tau pathology [176]	Microglia	[26, 27]
CD33	Phagocytosis [69]	Aβ clearance [69]	Microglia	[26, 27]
CELF1	Posttranscriptional regulation [177]	Tau pathology [176]	Astrocyte, neuron, microglia	[22]
CLU	Lipid metabolism [178], chaperone-like activity [179, 180]	Aβ clearance [181]	Astrocyte	[23, 24]
CR1	Complement system [72, 73]	Aβ clearance [74]	Microglia	[24]
DSG2	Cell adhesion [182]	Unknown	Endothelium, neuron, microglia	[22]
EPHA1	Cell adhesion and migration [183], immune response [184]	Tau pathology [173]	Oligodendrocyte, microglia	[26, 27]
FERMT2	Cell adhesion [185]	APP processing [186], tau pathology [176]	Astrocyte	[22]
GRN	Phagocytosis [86], complement system [87]	Aβ clearance [86], tau pathology [85]	Microglia	[81]
Human leukocyte antigen gene cluster	Antigen presentation [187]	Unknown	Microglia	[22]
IL1RAP	Immune response [188]	Aβ clearance [88]	Microglia, oligodendrocyte	[88]
INPP5D	Immune response [189], PI3K signaling [120]	Unknown	Microglia	[22]

(continued)

Table 5.1 (continued)

Genes	Cellular functions	Possible roles in AD pathology	Cell types[a]	References
MEF2C	Transcription factor [190], immune response [127]	Unknown	Neuron, microglia	[22]
MS4A gene cluster	Immune response [123], lipid sensing [122]	Unknown	Microglia	[26, 27, 121]
NME8	Primary cilia function [191]	Unknown	Oligodendrocyte, microglia	[22]
PICALM	Endocytosis [192]	APP processing [193, 194], Aβ clearance [195], tau pathology [196]	Microglia, oligodendrocyte, neuron	[23]
PLCG2	Immune response [107], Ca^{2+} signaling [197]	Unknown	Microglia	[35]
PLD3	Lysosomal function [198]	APP processing [37], but not replicated in [198]	Microglia, astrocyte	[37]
PTK2B	Cell adhesion [199], mitogen-activated protein kinase signaling [200]	Tau pathology [173]	Microglia, neuron	[22]
SLC24A4-RIN3	*RIN3*: endocytosis [201] *SLC24A4*: ion transport [202]	Unknown	*RIN3*: microglia *SLC24A4*: neuron	[22]
SORL1	Endocytic receptor [203], intracellular trafficking [204]	APP processing [205], Aβ clearance [206]	Astrocyte, microglia	[22, 36]
SPI1	Transcription factor [118], phagocytosis [117]	Aβ clearance [117]	Microglia	[22, 117]
TREM2	Phagocytosis [207], cell differentiation and proliferation [50, 51]	Aβ clearance [48], tau pathology [154], neurotoxicity [61, 153]	Microglia	[32, 33]
ZCWPW1	Epigenetic regulation [208]	Unknown	Oligodendrocyte, microglia	[22]

[a]Brain cell types in which the gene is highly expressed in mouse and human are shown based on the brain RNA-seq database (http://www.brainrnaseq.org/) [209, 210]. When results are inconsistent between mouse and human or when expression levels are similar in different cell types, all the cell types are described together

5.3.1 TREM2

Triggering receptor expressed on myeloid cells 2 (TREM2) was originally identified as a causative gene for polycystic lipomembranous osteodysplasia with sclerosing leukoencephalopathy (PLOSL, also known as Nasu-Hakola disease), which is characterized by multiple bone cysts and early-onset dementia. Homozygous mutations, including nonsense and missense mutations (Y38C, T66M, etc.), cause PLOSL, but the same mutations also cause frontotemporal dementia without bone symptoms [44]. Several TREM2 variants are associated with AD risk as mentioned above. Interestingly, these variants are almost distinct from those found in PLOSL and contribute to disease risk in heterozygotes [32, 33].

TREM2 is a single-pass transmembrane receptor expressed almost exclusively in immune cells, including dendritic cells and macrophages, as well as microglia in the CNS [45]. Whereas TREM2 has only a short cytoplasmic tail without any functional domain, it forms a complex with DNAX-activation protein 12 (DAP12), which is a transmembrane protein with an immunoreceptor tyrosine-based activation motif (ITAM) in the cytoplasm, and therefore it can transmit signals through DAP12 upon recognition of extracellular ligands. TYROBP, the gene encoding DAP12, was also identified as a causative gene for PLOSL, and its rare variants were identified in early-onset AD patients [46].

Many examples of TREM2 ligands have been identified so far. These include various negatively charged lipids [47]. TREM2 also binds to Aβ [48], as well as to apolipoproteins ApoE and clusterin (encoded by the CLU gene), both of which are associated with AD risk [49]. Upon ligand binding, multivalent interactions of TREM2 with its ligands lead to the clustering of DAP12 in the membrane. This results in the phosphorylation of DAP12 at tyrosine residues in the ITAM in a SRC family kinase-dependent manner and, thereby, the membrane recruitment of spleen tyrosine kinase SYK. SYK activates several branches of the signal network, including phosphatidylinositol-3 kinase (PI3K) and mitogen-activated protein kinase pathways, through which TREM2 can stimulate the nuclear translocation of transcription factors, such as nuclear factor of activated T-cells (NFAT) and β-catenin [50, 51]. The functional outcomes of these signals include cellular proliferation, regulation of phagocytosis, and anti-inflammatory responses [52].

Of note, although phosphorylation of ITAM typically causes the activation of downstream signaling, partial phosphorylation of ITAM tyrosines can mediate inhibitory signals by recruiting a different set of effectors, including Src homology domain 2 (SH2)-containing protein-tyrosine phosphatase (SHP)-1/2, as well as phosphoinositide phosphatase inositol polyphosphate-5-phosphatase D (INPP5D), which is also a GWAS hit for AD [22, 53]. Accordingly, TREM2/DAP12 can transduce both anti-inflammatory and inflammatory signals, and which pathway is chosen depends on the type of ligand or the strength of the interaction [52, 54].

Some of PLOSL-associated mutations of TREM2 were found to affect its anterograde trafficking and thereby impair the cell surface function of TREM2 [55, 56]. On the other hand, AD-associated variants, such as R47H and R62H, only showed a weak effect on surface expression, suggesting that the mutations affect another functional aspect of TREM2. Indeed, the crystal structure of TREM2 R47H showed that the amino-acid substitution caused a structural change in the region required for binding to a negatively charged lipid phosphatidylserine [57]. These suggest that the loss of ligand-binding activity of TREM2 is relevant to AD pathogenesis.

To investigate the in vivo role of TREM2, Trem2-deficient mice were crossed with AD model mice showing Aβ deposition. These mice showed a marked reduction in the number of microglia around Aβ plaques, making a clear contrast to control mice that showed plaques that were often completely surrounded by multiple microglia [47, 58–60]. In addition, Trem2-knockout mice were characteristic in their plaque morphology [61]. These plaques were less compact, had spikelike fibrils extending radially, and lacked a dense core, which was often observed in the plaques of control mice. Similar phenotypes were also observed in Tyrobp-deficient mice, as well as in AD patients with the R47H mutation. Importantly, dystrophic neurites, which are swollen or distorted neurites that are found in close vicinity of Aβ plaques, were exacerbated around these "filamentous" plaques, suggesting the hypothesis that microglia can mitigate neurotoxicity by sequestering Aβ fibrils [61, 62]. We will discuss this and other functional aspects of plaque-associated microglia in Sect. 5.4.

Besides functioning as a receptor, TREM2 undergoes extracellular proteolysis and releases a soluble fragment called sTREM2. sTREM2 functions in stimulating microglial survival and the production of inflammatory cytokines [54]. Interestingly, sTREM2 can be detected in human CSF [55, 63], and changes in CSF levels of sTREM2 were documented in several neurological disorders, including AD [63, 64]. Since the change likely reflects an alteration in the activation status of microglia, CSF sTREM2 levels may be useful as a potential biomarker of neuroinflammation [65].

5.3.2 CD33

Polymorphisms in the CD33 locus (rs3865444, rs3826656, and rs114282264) were associated with late-onset AD risk [26, 27, 66, 67]. CD33 encodes a type I transmembrane receptor expressed in immune cells in peripheral tissues, as well as in microglia in the brain. CD33, also known as sialic acid-binding immunoglobulin-like lectin (SIGLEC)-3, is a member of the SIGLEC family of lectins and recognizes sialic acid in its extracellular immunoglobulin-like fold. It also has multiple immunoreceptor tyrosine-based inhibition motifs in its cytoplasmic tail and thereby can inhibit cell signaling through recruiting the inhibitory phosphatases SHP-1/2 [68].

Previous studies showed that CD33 expression was upregulated in AD patients' brains and that the number of CD33-positive microglia was positively correlated with the amount of Aβ42 accumulation [69]. On the other hand, the protective rs3865444 allele was associated with decreases in both the expression level of CD33 and the amount of insoluble Aβ42 in the brains of AD patients, suggesting that reduced expression of CD33 has a beneficial role in suppressing Aβ pathology. In line with this notion, Cd33 deficiency reduced Aβ burden in AD model mice [69]. Of note is that Aβ uptake activity was increased in microglia derived from knockout mice, suggesting that CD33 inhibits the Aβ clearing mechanism by microglia. We will discuss the microglial involvement in Aβ metabolism in Sect. 5.4.

Furthermore, the risk allele of CD33 was reportedly associated with an increased cell surface expression level of TREM2 [70]. Suppression of CD33 signaling with an antibody to CD33 also reduced the surface level of TREM2. Therefore, CD33 may play pleiotropic roles in AD pathology by modulating the function of TREM2.

5.3.3 CR1

Polymorphisms of CR1 were originally found to be associated with AD in a GWAS [24], and this disease association has been replicated in subsequent studies [71].

CR1 encodes a receptor for the complement factors C1q, C3b, and C4b [72, 73]. A copy number variation of CR1 increasing the number of C3b-/C4b-binding sites was found to increase the risk of AD [74], suggesting that complement-associated functions of CR1 are involved in AD pathomechanisms.

In peripheral tissues, the best-characterized function of complement factors is opsonization, by which complement factors stimulate the phagocytic clearance of marked pathogens by macrophages. CR1 is a complement receptor expressed on macrophages, as well as on microglia in the brain. In the CNS, it has been demonstrated that microglial phagocytosis plays an important role in the removal of unnecessary synapses, where complement factors function in opsonizing the synapses [75, 76]. This activity, called synaptic pruning, is suggested to be dysregulated in several diseases, including AD. Indeed, decreases in synaptic density were observed in the brains of AD patients as well as AD model mice [77]. Moreover, complement factor deficiency, such as of C1q and C3, ameliorated both synaptic loss and cognitive decline in AD model mice [43]. In contrast to these beneficial effects, Aβ accumulation was exacerbated in C3-deficient mice. In addition, AD risk polymorphisms of CR1 (rs646817G and rs12034383G) were associated with increased levels of CSF Aβ [74], which likely reflect reduced Aβ accumulation in the brain parenchyma. These results imply that the complement system plays an additional role in the efflux or clearance of Aβ from the brain. Collectively, these data suggest that the aberrant activation of synaptic pruning has a greater effect on the cognitive deficits of AD model mice than the accumulation of Aβ itself.

5.3.4 GRN

Progranulin is a growth factor that is expressed in neurons and microglia [78]. Haploinsufficiency of GRN, the gene encoding the precursor of progranulin, causes frontotemporal lobar dementia with accumulation of TAR DNA/RNA binding protein 43, or FTLD-TDP43 [79, 80]. Recently, GRN polymorphisms, including rs4792939, rs850713, and rs5848, were implicated as susceptibility loci for AD [81]. Among them, the rs5848T variant reportedly reduced the expression level of GRN and increased the risk of AD [82–84], suggesting a protective role of GRN against AD. Interestingly, the rs5848T allele showed no significant effects on either PET imaging of Aβ or the level of CSF Aβ, but instead it was associated with an increased level of CSF tau [85], suggesting the association between GRN and tau.

In line with their human evidence, Takahashi and colleagues found that Grn deficiency had no exacerbating effect on Aβ pathology in APP/PS1 mice [85]. On the contrary, Grn-deficient mice showed reduced growth of Aβ plaques, as well as amelioration of axonal dystrophy and memory deficit, although these data are inconsistent with a previous observation that microglia-specific deletion of Grn increased Aβ burden in a different AD model (J20) [86]. Although the reason for this discrepancy remains unknown, Takahashi and colleagues also reported that Grn deficiency in a tau overexpression model increased the amount of phosphorylated tau [85], further supporting their notion that GRN contributes to tau pathology.

Grn-deficient mice demonstrated other characteristic abnormalities, including repetitive self-grooming. This is reminiscent of human obsessive-compulsive disorder, which is a clinical feature of frontotemporal dementia. Mechanistically, the phenotype observed in mice was caused by hyperactivity of the thalamocortical circuit [87]. In detail, Grn deficiency caused lysosomal dysfunction and increased the production of complement factors, such as C1q and C3 in an age-dependent manner, resulting in enhanced phagocytic elimination of inhibitory synapses in the thalamus. This aberrant microglial activation may underlie the synaptic loss that is also observed in AD pathology.

5.3.5 IL1RAP

A polymorphism of the IL1RAP gene (rs12053868G) was associated with accelerated cognitive decline as well as a higher rate of Aβ accumulation, based on longitudinal studies in AD patients [88]. The carriers of this polymorphism also showed lower signals of a marker for microglial activation, suggesting that microglial dysfunction led to the increased Aβ burden, possibly through affecting the Aβ clearance mechanism of microglia.

IL1RAP encodes accessory protein of type I interleukin (IL)-1 receptor or IL1RI, and IL1RI and IL1RAP constitute a functional receptor for proinflammatory cytokines IL-1α and IL-1β. Of these, IL-1β was reportedly upregulated in AD brains [89], although its functional significance remains controversial. For example, the overexpression of IL-1β in AD model mice increased the number of microglia around plaques and reduced Aβ burden [90, 91], suggesting the role of IL-1β in Aβ clearance. However, reduced levels of Aβ accumulation were also observed in Nlrp3-deficient mice, which lack functional IL-1β [92]. Therefore, further studies are required to clarify this discrepancy as well as the precise roles of IL1RAP in AD pathogenesis.

5.3.6 ABCA7

GWASs of late-onset AD and their meta-analyses identified several variants in the ABCA7 gene that were associated with disease risk, including rs3764650, rs3752246, and rs115550680 [26, 27, 67]. Among them, the risk allele rs3764650G was found to reduce the expression level of ABCA7 [93]. In addition, several loss-of-function variants of ABCA7, including nonsense mutations, were associated with an increased risk of AD [94], suggesting that the normal function of ABCA7 plays a preventive role against AD. Furthermore, many coding variants of ABCA7 with an unknown functional outcome have been identified in AD patients [26]. Among them, it is noteworthy that one mutant (p.G215S) was associated with a reduced risk of AD [95]. These studies strongly suggest that functional alterations of ABCA7 have a large effect on disease susceptibility.

ABCA7 encodes a 12-pass transmembrane protein belonging to the ATP-binding cassette transporter family. ABCA1, the closest homolog to ABCA7 of the family members, is well characterized to encode a lipid transporter that is essential for the formation of high-density lipoprotein complexes [96]. Specifically, ABCA1 mobilizes lipids from the cytoplasmic leaflet of the membrane to the outer leaflet and then onto apolipoproteins. This lipid transport activity is conserved in ABCA7. However, the lipoprotein complex formed by ABCA7 had lower cholesterol content and higher lysophosphatidylcholine-to-phosphatidylcholine ratio, suggesting a difference in their biochemical properties [97, 98]. Of note, ABCA7 has been shown to have lipid-transporting activity for ApoE, implying the possible involvement of this activity in the pathomechanisms of AD.

Besides lipid transport, ABCA1 and ABCA7 are involved in the regulation of phagocytosis. ABCA7 deficiency affected the uptake of apoptotic cells and microbes by macrophages [99, 100]. Similarly, ABCA1 was found to be involved in the phagocytic clearance of neuronal debris by astrocytes [101]. These functions have been evolutionarily conserved, as CED-7, a nematode homolog of ABCA1 and ABCA7, has long been known as an essential gene for the clearance of apoptotic cells [102, 103]. However, it remains unknown as to how and whether lipid transport activity is involved in these processes.

Several reports have addressed the in vivo roles of ABCA7 in AD model mice. These consistently demonstrated that Abca7 deficiency exacerbated Aβ accumulation [104–106]. A reduction of phagocytic activity against Aβ was observed in microglia and macrophages derived from the knockout mice [104], suggesting that the reduced clearance underlies the accumulation of Aβ. On the other hand, Kanekiyo and colleagues provided evidence that Abca7 deficiency did not alter the clearance rate of Aβ based on in vivo microdialysis and instead proposed that the loss of Abca7 is responsible for the increased generation of Aβ from neurons [105]. Future analyses on the cell-specific roles of ABCA7 will be important for understanding their involvement in the pathomechanisms of AD.

5.3.7 Other Factors

PLCG2. A rare coding variant of PLCG2 was found as a protective allele for AD (rs72824905; p.P522R) [35]. PLCG2 encodes the γ2 isoform of phospholipase C, which converts the membrane lipid phosphatidylinositol 4,5-bisphosphate into diacylglycerol and inositol 1,4,5-trisphosphate (IP3). PLCγ2 is specifically expressed in immune cells, including B cells and osteoclasts, as well as microglia in the CNS. PLCγ2 activity is regulated by phosphorylation, which is triggered by upstream receptors, such as the B-cell receptor. Activated PLCγ2 mobilizes calcium ions from the endoplasmic reticulum into the cytoplasm by opening IP3 receptors, causing the Ca^{2+}-sensitive nuclear translocation of NFAT. Of note, PLCγ2 was reportedly activated downstream of DAP12 in osteoclasts [107]. PLCγ2 is a protein with multiple domains, including a pleckstrin homology (PH) domain, two EF-hand motifs, two SH2 domains, a Src homology 3 (SH3) domain, and a C2 domain [108]. Additionally, PLCγ1/2 has a "split" PH domain with a PH-like fold, which is formed by the assembly of its N-terminal and C-terminal halves that are encoded by two spatially separated sequences in the protein. The split PH domain in PLCγ2 is responsible for its interaction with the small GTPase RAC2 [109, 110]. Proline 522, which was linked with AD risk, is positioned within the inserted sequence of the split PH domain, and therefore the function of this domain may be relevant to the pathomechanism of AD.

ABI3. A rare coding variant of ABI3 was associated with an increased risk of AD (rs616338; p.S209F) [35]. The ABI3 protein is a member of the ABL-interactor (ABI) family, together with ABI1 and ABI2. Whereas the expression and function of ABI3 were observed in neurons [111, 112], an immunohistochemical analysis on human brains demonstrated the specific expression of ABI3 in microglia [113]. The ABI3 protein is a multidomain scaffold protein with a Wiskott-Aldrich syndrome protein family verprolin-homologous protein (WAVE)-binding domain at the N-terminus, several proline-rich regions, and an SH3 domain at the C-terminus [114]. ABI1 is best characterized as a component of the WAVE regulatory complex, which is a regulator of actin polymerization. ABI3 reportedly constitutes a similar complex, but the function of the complex was suggested to be distinct from the ABI1-containing complex [115]. Furthermore, ABI3 function was reported to be regulated by the PI3K pathway [116].

SPI1. Cis-expression quantitative trait locus analysis showed that the CELF1 gene variant rs1057233G, previously identified as a protective allele for AD, was associated with the reduced expression of the SPI1 gene [22, 117]. Moreover, carriers of this protective allele showed an increased level of CSF Aβ42, which likely reflects decreased Aβ accumulation in the brain [117]. SPI1 encodes a transcription factor called PU.1, which is highly expressed in myeloid and B-lymphoid cells, as well as in microglia in the CNS [118]. Interestingly, chromatin-immunoprecipitation sequencing demonstrated that PU.1 binds to cis-regulatory elements of many AD-associated genes, including ABCA7, CD33, MS4A4A, MS4A6A, TREM2, TREML2, and TYROBP, and alteration of PU.1 levels affected the expression of these genes [117]. Therefore, it is likely that PU.1 is involved in multiple aspects of microglial pathology in AD through regulating the expression of these genes. As one such involvement, effects on phagocytosis have been suggested. The knockdown of PU.1 reduced the phagocytic activity of microglia, whereas the overexpression of PU.1 increased phagocytic activity [117]. Similarly, PU.1 knockdown in primary human microglia was found to reduce the phagocytic activity of Aβ [119]. These observations suggest that the protective allele of SPI1 likely reduces the phagocytic activity of microglia, although it is inconsistent with the hypothesis that microglia mitigate Aβ pathology through the phagocytic clearance of Aβ. Therefore, further studies on the functional outcomes of the reduced expression of SPI1 are required to understand its pathological significance.

INPP5D. A common variant of the INPP5D gene was identified in a meta-analysis of GWASs for late-onset AD [22]. INPP5D, also known as SH2-containing phosphatidylinositol 3,4,5-trisphosphate 5′-phosphatase or SHIP1, encodes a phosphatase that converts the lipid second messenger phosphatidylinositol 3,4,5-trisphosphate, or $PI(3,4,5)P_3$, into phosphatidylinositol 3,4-bisphosphate and thereby inhibits PI3K signaling, which is activated downstream of many receptors [120]. Of note, INPP5D has an inhibitory role in TREM2/DAP12 signaling of osteoclasts through a mechanism that depends on $PI(3,4,5)P_3$ dephosphorylation activities as well as via competition with the DAP12-SYK interaction [50].

Membrane-spanning 4-domain family, subfamily A (MS4A) gene cluster. Several polymorphisms in the MS4A gene cluster were identified in GWASs of AD. These are located near the MS4A4A, MS4A4E, MS4A6A, and MS4A6E genes [26, 27, 121]. MS4A family proteins constitute a newly identified class of odor receptors, expressed in necklace sensory neurons [122]. Each MS4A gene encodes a four-pass transmembrane protein that recognizes a distinct set of pheromones or fatty acids and triggers calcium mobilization through unknown mechanisms. MS4A genes are also expressed in macrophages and microglia, and the expression of MS4A4A was found to be differentially regulated in different polarization states of macrophages [123]. Given that TREM2 also senses various lipid molecules, microglia may recognize a pathophysiological environmental change using MS4As as chemosensors in combination with TREM2.

MEF2C. An SNP in the MEF2C gene was identified in a meta-analysis of GWAS for AD [22]. Haploinsufficiency of MEF2C is thought to contribute to a neurodevelopmental disorder with intellectual disability and epilepsy, in which patients have point mutations and microdeletions of this gene [124]. The MEF2C protein belongs to the MCM1, Agamous, Deficiens, and SRF-box family of transcription factors. MEF2C is essential for cardiovascular development and also participates in the development of various tissues, such as the bones, neurons, and hematopoietic cells [125]. In the brain, the expression of MEF2C was found in neurons and microglia. Of note, MEF2-binding motifs were found to be enriched in the enhancer regions of microglia-specific genes [126], suggesting an important role of MEF2C in microglia-specific functions. Indeed, Mef2c-deficient microglia showed a reduced expression of CX3C chemokine receptor 1, a microglia-specific chemokine receptor, and responded to immune stimuli, such as tumor necrosis factor and lipopolysaccharides, more strongly than control microglia [127]. In addition, the level of Mef2c in microglia was downregulated in aged mice through a type I interferon-dependent mechanism [127], suggesting that age-dependent alterations of microglial phenotypes can be partially explained by the reduced expression levels of MEF2C.

5.4 Pathological Roles of Microglia in Alzheimer Disease

In this section, we will describe the possible involvements of microglia in several aspects of AD pathology (Fig. 5.1). It has long been known that microglia congregate around Aβ plaques and that the plaque-associated microglia demonstrate

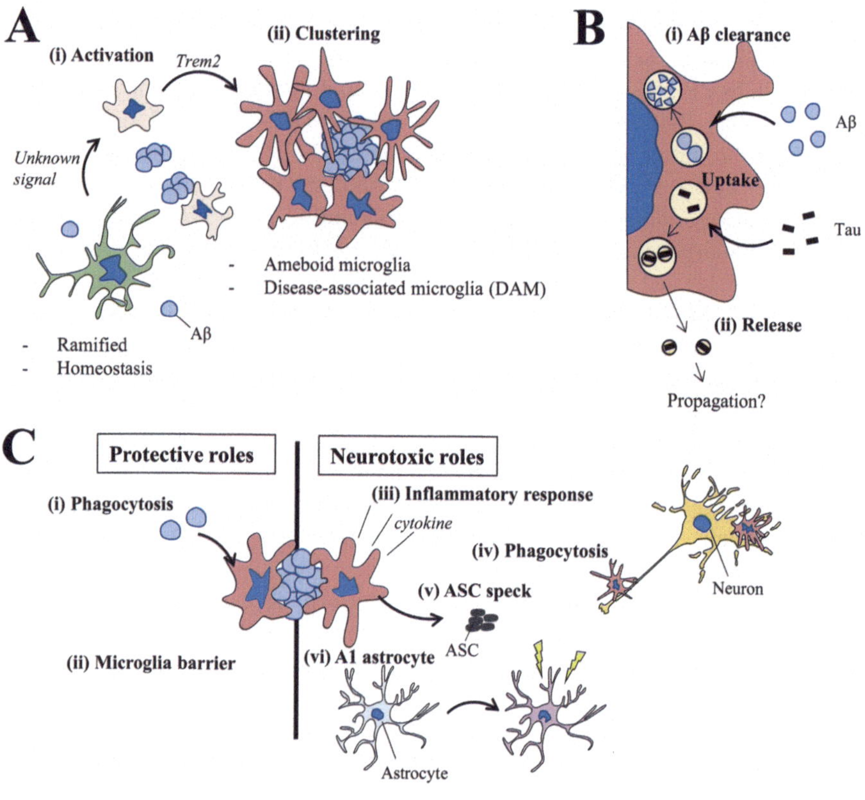

Fig. 5.1 Schematic illustrations of proposed functions of microglia in AD pathology. (**a**) Microglia are activated in response to Aβ accumulations and change their shapes from ramified to ameboid forms. The activated microglia cluster around plaques and alter their gene expression patterns in Trem2-dependent manners. (**b**) Microglia internalize Aβ and tau. (i) Aβ uptake facilitates its degradation. (ii) A proportion of internalized tau was shown to be secreted together with exosomes and to participate in propagation of tau pathology. (**c**) Activated microglia exert both protective and detrimental effects on neurons. Although microglia can mitigate Aβ toxicity by (i) phagocytosing and (ii) sequestering Aβ aggregates, they also participate in (iii) inflammatory responses, (iv) aberrant phagocytosis of neuronal cell bodies and synapses, (v) promotion of Aβ aggregation through secreting ASC specks, and (vi) conversion of astrocytes into a neurotoxic A1 phenotype

different properties from steady-state cells, but it is only recently that the molecular mechanisms behind such a phenotypic change have been addressed. Thus, we will summarize our current knowledge of such molecular mechanisms and then discuss the idea that plaque-associated microglia have both protective and toxic roles against neuronal cells. Furthermore, we will also discuss the roles of microglia in tau pathology, as well as in the activation of other glial cells.

5.4.1 Microglial Changes Around Aβ Plaques

In AD brains, the density of microglia in the vicinity of senile plaques is increased [128]. These microglia show several characteristic features. For example, they are altered from a ramified form having long branched protrusions to an ameboid form in which protrusions are retracted, which is considered to be an activated state. As tau pathology is evident in the brain, dystrophic microglia with fragmented protrusions also appear [129]. Electron microscopic analysis demonstrated that the microglia around plaques look darker, that is, more electron-dense than usual, and they were hence named "dark microglia" [130]. Alterations in electrophysiological properties were also reported in the plaque-associated microglia [131]. PET studies also indicated the alteration of microglial activity in the brains of AD patients [132].

Gene expression changes that are the cause of such phenotypic changes are beginning to be uncovered. Genes that characterize steady-state or homeostatic microglia were demonstrated by cell-type-specific transcriptomic analyses [133, 134]. Among these, transforming growth factor-β (TGF-β) receptor was crucial for maintaining the homeostatic state of microglia, whereas the expression of such homeostatic genes was attenuated in mice lacking TGF-β receptor. Interestingly, the absence of the receptor induced the upregulation of a characteristic set of genes, including ApoE [133].

Single-cell RNA-seq analysis of microglia from AD model mice demonstrated a novel subtype of microglia, called disease-associated microglia (DAM), which specifically appeared in AD model mice [135]. This report suggested that microglia first undergo a change from a homeostatic state to intermediate DAM (stage 1) and then to complete DAM (stage 2). Importantly, Trem2 was necessary for the change to stage 2, and Trem2 deficiency caused an arrest at stage 1. Considering that the loss-of-function of TREM2 increases the risk of AD, it is likely that stage 2 cells have an indispensable role in neuroprotection and/or that stage 1 cells have deleterious effects on neuronal survivability.

The clustering of microglia around senile plaques is thought to result from the proliferation and chemotactic attraction of microglia around plaques. Of these, the proliferation of microglia is commonly observed in various pathological conditions, which is in clear contrast with the fact that steady-state microglia are long-lived and proliferate slowly [136]. The decreased association of both Trem2-deficient and Tyrobp-deficient microglia with senile plaques is associated with their reduced proliferation and survival. Mechanistically, TREM2 was found to be important for the responsiveness of microglia to macrophage colony-stimulating factor, which is essential for the survival of microglia, and apoptosis was enhanced in Trem2-deficient mice [47]. Furthermore, metabolic abnormalities were found in Trem2-deficient microglia, which were associated with a compensatory increase in the autophagic activity of the microglia [137]. In addition, TREM2 is involved in chemotactic activity [138], which may also contribute to the clustering of microglia around the plaques.

Although it is not clear what induces the phenotypic changes of microglia in response to Aβ plaques, it is suggested that microglia recognize and phagocytose dead or dying cells around the plaques, which in turn triggers the phenotypic changes [139]. Interestingly, both TREM2 and ApoE are suggested to function in this pathway, although the exact role of ApoE in microglia remains unknown.

Taken together, it is becoming clear that the microglia around plaques have different characteristics to those of steady-state microglia and even non-plaque-associated microglia. However, the pathological roles of such microglial clustering remain unclear.

5.4.2 Protective Roles of Microglia

Considering their phagocytic nature, it has been assumed that microglia play a role in Aβ clearance. Indeed, in vitro studies have demonstrated that microglia phagocytose Aβ. However, it is controversial whether or not microglia contribute to Aβ clearance in vivo. Several reports have stated that the number and size of amyloid plaques in AD model mice did not change even after microglia were pharmacologically depleted from their brains [140–143]. However, these results do not necessarily exclude the involvement of microglia in Aβ clearance. Rather, there are many instances in which Aβ burden in the brain was decreased or increased upon pharmacological/genetic interventions that affect microglial function, suggesting that clearance activity can be modulated by various conditions. These examples are listed in (Table 5.2). It should be noted that such pharmacological depletion experiments only analyze the effects on Aβ burden in a short time range, from weeks to months. Considering that humans usually develop AD in their later stages of life, it is possible that a slight decline in Aβ clearance has a major effect if it continues for decades.

Besides phagocytosis, microglia may play another role in neuroprotection. By sequestering Aβ plaques, microglia serve as a barrier to protect neurons from Aβ toxicity [62]. In addition, microglia mask the surface of plaques and suppress their growth by preventing Aβ monomers from being incorporated into the plaques. It has also been suggested that microglia affect the morphology of Aβ fibrils; Aβ plaques surrounded by microglia appeared more compact than plaques not surrounded by microglia. Such a barrier function of microglia was impaired in Trem2-knockout mice, as well as in AD patients with the TREM2 R47H mutation [61], suggesting the crucial role of TREM2 in this process.

Table 5.2 Genetic and pharmacological manipulations of microglia that affected Aβ accumulation in vivo

Manipulations	Aβ accumulation	References
Il10 knockout	Decreased	[211]
Cxcr3 knockout	Decreased	[212]
Cd33 knockout	Decreased	[69]
Nlrp3, *Casp1* knockout	Decreased	[92]
Cyclooxygenase-1 inhibitor (SC-560)	Decreased	[213]
IL-12/IL-23 knockouts and neutralizing antibody	Decreased	[214]
Peroxisome proliferator-activated receptor γ and retinoid X receptor agonist (DSP-8658)	Decreased	[215]
Tumor necrosis factor α adeno-associated virus (AAV)-mediated overexpression	Decreased	[216]
Interferon-γ AAV-mediated overexpression	Decreased	[217]
IL-6 AAV-mediated overexpression	Decreased	[218]
IL-4 AAV-mediated overexpression	Increased/decreased	[219, 220]
Cx3cr1 knockout	Decreased	[148, 221, 222]
Systemic administration of macrophage colony-stimulating factor	Decreased	[223]
Dominant-negative TGF-β receptor expressed from CD11c promoter	Decreased	[224]
IL-1β transgenic	Decreased	[90]
TGF-β1 transgenic	Decreased in parenchyma but increased in vasculature	[225, 226]
TREM1 lentivirus-mediated overexpression	Decreased	[227]
IL-10 AAV-mediated overexpression	Increased	[228]
Grn knockout	Increased/decreased	[85, 86]
Msr1 knockout	Unchanged/increased	[229, 230]
Becn1 knockout	Increased	[231]
Ccr2 knockout	Increased	[232]
C3 knockout	Increased	[233, 234]
Overexpression of loss-of-function mutant of *TLR4*	Increased	[235]
Ptprc knockout	Increased	[236]

5.4.3 Neurotoxic Roles of Microglia

Pharmacological depletion of microglia was shown to restore the learning deficits of AD model mice, despite its small effect on Aβ clearance, as mentioned before [141–144]. This means that microglial activation around the plaques can have deleterious effects on neuronal cell function, which are likely to be through multiple mechanisms, such as inflammatory responses [145, 146]. In addition, phagocytic activity may also contribute to the neurotoxicity. Pioneering work by Brown and Neher demonstrated that microglia can promote neuronal death by phagocytosing live cells [147]. There are also several lines of evidence that microglia are involved in the elimination of nerve cells in vivo [148, 149]. In addition to engulfing whole cell bodies, microglia also perform selective phagocytosis of synapses in both physiological and pathological conditions. In AD model mice, complement factors were found to accumulate at synapses and trigger the phagocytosis by microglia [77].

Furthermore, it has also been hypothesized that microglia are involved in Aβ aggregation. In activated microglia, inflammation-dependent formation of the NOD-like receptor family, pyrin domain-containing-3 protein complex induces the conversion of apoptosis-associated speck-like protein containing A caspase recruitment domain (ASC) into its fibrillar form, called ASC specks [150]. ASC specks were found to leak out of activated microglia and demonstrate prion-like seeding activity, which in turn promotes Aβ aggregation [151].

5.4.4 Roles of Microglia in Tau Pathology

Microglia are also associated with tau pathology. Mice overexpressing the highly aggregation-prone mutant tau (P301S) showed hyperactivation of microglia, as well as substantial neuronal loss, which was suppressed by the administration of an immunosuppressant [152]. Interestingly, Trem2 deficiency in this model also suppressed neuronal death [153], suggesting that TREM2 can exert neurotoxic functions, which is in clear contrast to its protective role suggested in Aβ pathology. However, it should be noted that there is also a conflicting report demonstrating that the lack of Trem2 increases the phosphorylation and aggregation of tau in another mouse model of AD [154].

It is also noteworthy that neuronal cell death by tau overexpression was exacerbated in APOE-ε4 knock-in mice [155]. Considering that microglia are highly activated in this model and that ApoE expression is induced in activated microglia [139], it is likely that ApoE plays some active roles in microglia and the ε4 isoform is deleterious for their function.

Finally, microglia have some roles in the "propagation" of tau pathology in the brain. Although the details will not be mentioned here, multiple lines of evidence suggest that tau aggregation in neurons spreads along a neural circuit through a mechanism in which aggregation-prone "seeds" are physically transferred from one cell to intact neighboring cells [156]. Interestingly, Ikezu and colleagues found that tau propagation along a circuit was suppressed upon pharmacological depletion of microglia [157], suggesting that microglia participated in the progression of tau pathology. Mechanistically, it was assumed that microglia phagocytosed tau molecules and released them together with exosomes, thereby contributing to the spread of tau molecules with aggregation-prone properties.

5.4.5 *Association with Astrocytes*

Microglia are not the sole type of glial cells to be activated in AD pathogenesis. Astrocytes are also known to be activated in the vicinity of Aβ plaques, where they express several characteristic activation markers, such as glial fibrillary acidic protein [158].

Interestingly, microglia play an active role in the phenotypic changes of astrocytes. Systemic inflammation by means of the peripheral administration of lipopolysaccharides induced a characteristic gene expression profile in astrocytes, and the change to this state, termed A1, was dependent on microglia [159]. Mechanistically, activated microglia secrete three factors (IL-1α, tumor necrosis factor α, and C1q) that are necessary and sufficient for the induction of A1. A1 astrocytes lacked several functions that are mediated by normal astrocytes and that are important for neuronal functions. Importantly, A1 astrocytes were found in human brains of patients with several neurological disorders, such as AD and Parkinson disease. It was also induced in tau-transgenic mice expressing human APOE-ε4 [155], as well as in normal aged mice [160]. Furthermore, A1 astrocytes were found to be crucially involved in the neuronal death occurring in Parkinson disease model mice [161]. Therefore, it is possible that microglia contribute to neurotoxicity in the pathogenesis of AD not only directly but also indirectly via astrocytes.

5.5 Concluding Remarks

In this article, we described that many genetic risk factors for AD have been identified as a result of the recent advances in genome analysis technologies and that microglia-associated genes are enriched in the AD risk genes. Because previous studies had focused on the association between neurons and Aβ or tau, glial pathology had often been regarded as a secondary effect that was just a response to the pathological changes. However, with the accumulation of many lines of genetic evidence, it is almost certain that microglia are not just a bystander but are an active player in the pathogenesis of AD. In this regard, recent studies have clarified that glial cells other than microglia, such as astrocytes and vascular endothelial cells, also play important roles in the pathogenesis of AD [162]. Therefore, understanding the association between neurons and these non-neuronal cells will be increasingly important in the future.

The fact that the pharmacological depletion of microglia restored memory deficits in AD model mice suggests that the complete suppression of microglia may be effective as a treatment for AD. However, it is likely that such a treatment has severe adverse effects by also suppressing the physiological functions of microglia. Furthermore, as mentioned in Sect. 5.4, microglial involvement in AD pathology is highly complicated. Some functions are deleterious to neurons, whereas others are neuroprotective. Therefore, it is important for future therapeutics to precisely target a specific function of microglia. Many pharmaceutical companies are currently working on such projects, and indeed several candidate drugs are or will soon be in clinical trials [163]. To achieve this goal, we need to understand the roles of microglia more precisely. In particular, it is important to clarify how much each of these functions contribute to disease progression and how they change with aging. It is also promising that detailed analyses of genetic risk factors will provide us with hitherto unknown roles of microglia in AD pathology.

Given that the activation of microglia may precede or occur simultaneously with disease onset, a biomolecule with levels correlating with microglial activity might be useful as a potential biomarker for diagnosing AD. From this point of view, it may be promising to focus on sTREM2, because its levels are reportedly altered with the progression of AD [65].

Finally, microglia are thought to play important roles in other neurological disorders, including Parkinson disease, amyotrophic lateral sclerosis, and schizophrenia [164]. Although the detailed roles of microglia in these diseases remain largely unknown, a revolutionary drug that can be applied to multiple diseases may be possible by targeting microglia. Therefore, it will also be important to understand the roles of microglia in the pathogenesis of these other diseases.

Acknowledgment This work was supported in part by Grant-in-Aid for Scientific Research (A) [15H02492 to T.T.] and Grant-in-Aid for Young Scientists (B) [17K15446 to S.T.] from the Japan Society for the Promotion of Science (JSPS), by the Brain Mapping by Integrated Neurotechnologies for Disease Studies (Brain/MINDS) from the Japan Agency for Medical Research and Development (AMED) [JP18dm0207014h to T.T.], by the Sunbor Grant from the Suntory Foundation for Life Sciences [to S.T.], by the Uehara Memorial Foundation [to T.T.], and by the Mitsubishi Foundation [to T.T.].

References

1. Masters CL, Bateman R, Blennow K, Rowe CC, Sperling RA, Cummings JL (2015) Alzheimer's disease. Nat Rev Dis Prim 1:15056. https://doi.org/10.1038/nrdp.2015.56
2. Serrano-Pozo A, Frosch MP, Masliah E, Hyman BT (2011) Neuropathological alterations in Alzheimer disease. Cold Spring Harb Perspect Med 1:a006189. https://doi.org/10.1101/cshperspect.a006189
3. Nordberg A, Rinne JO, Kadir A, Långström B (2010) The use of PET in Alzheimer disease. Nat Rev Neurol 6(2):78-87
4. Olsson B, Lautner R, Andreasson U, Öhrfelt A, Portelius E, Bjerke M et al (2016) CSF and blood biomarkers for the diagnosis of Alzheimer's disease: a systematic review and meta-analysis. Lancet Neurol 15:673–684
5. Khachaturian AS, Hayden KM, Mielke MM, Tang Y, Lutz MW, Gustafson DR et al (2018) Future prospects and challenges for Alzheimer's disease drug development in the era of the NIA-AA Research Framework. Alzheimers Dement 14:532–534
6. Jack CR, Bennett DA, Blennow K, Carrillo MC, Dunn B, Haeberlein SB et al (2018) NIA-AA Research Framework: toward a biological definition of Alzheimer's disease. Alzheimers Dement 14:535–562

7. Knopman DS, Haeberlein SB, Carrillo MC, Hendrix JA, Kerchner G, Margolin R et al (2018) The National Institute on Aging and the Alzheimer's Association Research Framework for Alzheimer's disease: perspectives from the research roundtable. Alzheimers Dement 14:563–575
8. Silverberg N, Elliott C, Ryan L, Masliah E, Hodes R (2018) NIA commentary on the NIA-AA Research Framework: towards a biological definition of Alzheimer's disease. Alzheimers Dement 14:576–578
9. Wolfe MS (2006) The γ-secretase complex: membrane-embedded proteolytic ensemble. Biochemistry 45:7931–7939
10. Selkoe DJ, Hardy J (2016) The amyloid hypothesis of Alzheimer's disease at 25 years. EMBO Mol Med 8:595–608
11. Bateman RJ, Xiong C, Benzinger TLS, Fagan AM, Goate A, Fox NC et al (2012) Clinical and biomarker changes in dominantly inherited Alzheimer's disease. N Engl J Med 367:795–804
12. Reiman EM, Langbaum JB, Tariot PN, Lopera F, Bateman RJ, Morris JC et al (2016) CAP—advancing the evaluation of preclinical Alzheimer disease treatments. Nat Rev Neurol 12:56–61
13. Jonsson T, Atwal JK, Steinberg S, Snaedal J, Jonsson PV, Bjornsson S et al (2012) A mutation in APP protects against Alzheimer's disease and age-related cognitive decline. Nature 488:96–99
14. Sevigny J, Chiao P, Bussière T, Weinreb PH, Williams L, Maier M et al (2016) The antibody aducanumab reduces Aβ plaques in Alzheimer's disease. Nature 537:50–56
15. Mawuenyega KG, Sigurdson W, Ovod V, Munsell L, Kasten T, Morris JC et al (2010) Decreased clearance of CNS-amyloid in Alzheimer's disease. Science 330(6012):1774. https://doi.org/10.1126/science.1197623
16. Ovod V, Ramsey KN, Mawuenyega KG, Bollinger JG, Hicks T, Schneider T et al (2017) Amyloid β concentrations and stable isotope labeling kinetics of human plasma specific to central nervous system amyloidosis. Alzheimers Dement 13:841–849
17. Nakamura A, Kaneko N, Villemagne VL, Kato T, Doecke J, Doré V et al (2018) High performance plasma amyloid-β biomarkers for Alzheimer's disease. Nature 554:249–254
18. Corder EH, Saunders AM, Strittmatter WJ, Schmechel DE, Gaskell PC, Small GW et al (1993) Gene dose of apolipoprotein E type 4 allele and the risk of Alzheimer's disease in late onset families. Science 261:921–923
19. Corder EH, Saunders AM, Risch NJ, Strittmatter WJ, Schmechel DE, Gaskell PC et al (1994) Protective effect of apolipoprotein E type 2 allele for late onset Alzheimer disease. Nat Genet 7:180–184
20. Liu C-C, Liu C-C, Kanekiyo T, Xu H, Bu G (2013) Apolipoprotein E and Alzheimer disease: risk, mechanisms and therapy. Nat Rev Neurol 9:106–118
21. Kim J, Basak JM, Holtzman DM (2009) The role of apolipoprotein E in Alzheimer's disease. Neuron 63:287–303
22. Lambert JC, Ibrahim-Verbaas CA, Harold D, Naj AC, Sims R, Bellenguez C et al (2013) Meta-analysis of 74,046 individuals identifies 11 new susceptibility loci for Alzheimer's disease. Nat Genet 45:1452–1458
23. Harold D, Abraham R, Hollingworth P, Sims R, Gerrish A, Hamshere ML et al (2009) Genome-wide association study identifies variants at CLU and PICALM associated with Alzheimer's disease. Nat Genet 41:1088–1093
24. Lambert J-C, Heath S, Even G, Campion D, Sleegers K, Hiltunen M et al (2009) Genome-wide association study identifies variants at CLU and CR1 associated with Alzheimer's disease. Nat Genet 41:1094–1099
25. Seshadri S, Fitzpatrick AL, Ikram MA, DeStefano AL, Gudnason V, Boada M et al (2010) Genome-wide analysis of genetic loci associated with Alzheimer disease. JAMA 303:1832. https://doi.org/10.1001/jama.2010.574
26. Naj AC, Jun G, Beecham GW, Wang L-S, Vardarajan BN, Buros J et al (2011) Common variants at MS4A4/MS4A6E, CD2AP, CD33 and EPHA1 are associated with late-onset Alzheimer's disease. Nat Genet 43:436–441

27. Hollingworth P, Harold D, Sims R, Gerrish A, Lambert J-C, Carrasquillo MM et al (2011) Common variants at ABCA7, MS4A6A/MS4A4E, EPHA1, CD33 and CD2AP are associated with Alzheimer's disease. Nat Genet 43:429–435
28. Desikan RS, Fan CC, Wang Y, Schork AJ, Cabral HJ, Cupples LA et al (2017) Genetic assessment of age-associated Alzheimer disease risk: development and validation of a polygenic hazard score. PLoS Med 14:e1002258. https://doi.org/10.1371/journal.pmed.1002258
29. Liu JZ, Erlich Y, Pickrell JK (2017) Case–control association mapping by proxy using family history of disease. Nat Genet 49:325–331
30. Marioni RE, Harris SE, Zhang Q, McRae AF, Hagenaars SP, Hill WD et al (2018) GWAS on family history of Alzheimer's disease. Transl Psychiatry 8:99. https://doi.org/10.1038/s41398-018-0150-6
31. Suh J, Choi SH, Romano DM, Gannon MA, Lesinski AN, Kim DY et al (2013) ADAM10 missense mutations potentiate β-amyloid accumulation by impairing prodomain chaperone function. Neuron 80:385–401
32. Jonsson T, Stefansson H, Steinberg S, Jonsdottir I, Jonsson PV, Snaedal J et al (2013) Variant of TREM2 associated with the risk of Alzheimer's disease. N Engl J Med 368:107–116
33. Guerreiro R, Wojtas A, Bras J, Carrasquillo M, Rogaeva E, Majounie E et al (2013) TREM2 variants in Alzheimer's disease. N Engl J Med 368:117–127
34. Condello C, Yuan P, Grutzendler J (2018) Microglia-mediated neuroprotection, TREM2, and Alzheimer's disease: evidence from optical imaging. Biol Psychiatry 83:377–387
35. Sims R, van der Lee SJ, Naj AC, Bellenguez C, Badarinarayan N, Jakobsdottir J et al (2017) Rare coding variants in PLCG2, ABI3, and TREM2 implicate microglial-mediated innate immunity in Alzheimer's disease. Nat Genet 49:1373–1384
36. Pottier C, Hannequin D, Coutant S, Rovelet-Lecrux A, Wallon D, Rousseau S et al (2012) High frequency of potentially pathogenic SORL1 mutations in autosomal dominant early-onset Alzheimer disease. Mol Psychiatry 17:875–879
37. Cruchaga C, Karch CM, Jin SC, Benitez BA, Cai Y, Guerreiro R et al (2014) Rare coding variants in the phospholipase D3 gene confer risk for Alzheimer's disease. Nature 505:550–554
38. Logue MW, Schu M, Vardarajan BN, Farrell J, Bennett DA, Buxbaum JD et al (2014) Two rare AKAP9 variants are associated with Alzheimer's disease in African Americans. Alzheimers Dement 10:609–618.e11. https://doi.org/10.1016/j.jalz.2014.06.010
39. Wetzel-Smith MK, Hunkapiller J, Bhangale TR, Srinivasan K, Maloney JA, Atwal JK et al (2014) A rare mutation in UNC5C predisposes to late-onset Alzheimer's disease and increases neuronal cell death. Nat Med 20:1452–1457
40. Raghavan NS, Brickman AM, Andrews H, Manly JJ, Schupf N, Lantigua R et al (2018) Whole-exome sequencing in 20,197 persons for rare variants in Alzheimer's disease. Ann Clin Transl Neurol 5:832–842
41. Erikson GA, Bodian DL, Rueda M, Molparia B, Scott ER, Scott-Van Zeeland AA et al (2016) Whole-genome sequencing of a healthy aging cohort. Cell 165:1002–1011
42. Ridge PG, Karch CM, Hsu S, Arano I, Teerlink CC, Ebbert MTW et al (2017) Linkage, whole genome sequence, and biological data implicate variants in RAB10 in Alzheimer's disease resilience. Genome Med 9:100. https://doi.org/10.1186/s13073-017-0486-1
43. Hansen DV, Hanson JE, Sheng M (2018) Microglia in Alzheimer's disease. J Cell Biol 217:459–472
44. Guerreiro RJ, Lohmann E, Brás JM, Gibbs JR, Rohrer JD, Gurunlian N et al (2013) Using exome sequencing to reveal mutations in TREM2 presenting as a frontotemporal dementia-like syndrome without bone involvement. JAMA Neurol 70:78. https://doi.org/10.1001/jamaneurol.2013.579
45. Paradowska-Gorycka A, Jurkowska M (2013) Structure, expression pattern and biological activity of molecular complex TREM-2/DAP12. Hum Immunol 74:730–737
46. Pottier C, Ravenscroft TA, Brown PH, Finch NA, Baker M, Parsons M et al (2016) TYROBP genetic variants in early-onset Alzheimer's disease. Neurobiol Aging 48:222.e9–222.e15. https://doi.org/10.1016/j.neurobiolaging.2016.07.028

47. Wang Y, Cella M, Mallinson K, Ulrich JD, Young KL, Robinette ML et al (2015) TREM2 lipid sensing sustains the microglial response in an Alzheimer's disease model. Cell 160:1061–1071
48. Zhao Y, Wu X, Li X, Jiang L-L, Gui X, Liu Y et al (2018) TREM2 is a receptor for β-amyloid that mediates microglial function. Neuron 97:1023–1031.e7. https://doi.org/10.1016/j.neuron.2018.01.031
49. Yeh FL, Wang Y, Tom I, Gonzalez LC, Sheng M (2016) TREM2 binds to apolipoproteins, including APOE and CLU/APOJ, and thereby facilitates uptake of amyloid-beta by microglia. Neuron 91:328–340
50. Peng Q, Malhotra S, Torchia JA, Kerr WG, Coggeshall KM, Humphrey MB (2010) TREM2- and DAP12-dependent activation of PI3K requires DAP10 and is inhibited by SHIP1. Sci Signal 3:ra38. https://doi.org/10.1126/scisignal.2000500
51. Otero K, Turnbull IR, Poliani PL, Vermi W, Cerutti E, Aoshi T et al (2009) Macrophage colony-stimulating factor induces the proliferation and survival of macrophages via a pathway involving DAP12 and β-catenin. Nat Immunol 10:734–743
52. Jay TR, von Saucken VE, Landreth GE (2017) TREM2 in neurodegenerative diseases. Mol Neurodegener 12:56. https://doi.org/10.1186/s13024-017-0197-5
53. Barrow AD, Trowsdale J (2006) You say ITAM and I say ITIM, let's call the whole thing off: the ambiguity of immunoreceptor signalling. Eur J Immunol 36:1646–1653
54. Zhong L, Chen X-F, Wang T, Wang Z, Liao C, Wang Z et al (2017) Soluble TREM2 induces inflammatory responses and enhances microglial survival. J Exp Med 214:597–607
55. Kleinberger G, Yamanishi Y, Suarez-Calvet M, Czirr E, Lohmann E, Cuyvers E et al (2014) TREM2 mutations implicated in neurodegeneration impair cell surface transport and phagocytosis. Sci Transl Med 6:243ra86. https://doi.org/10.1126/scitranslmed.3009093
56. Kober DL, Alexander-Brett JM, Karch CM, Cruchaga C, Colonna M, Holtzman MJ et al (2016) Neurodegenerative disease mutations in TREM2 reveal a functional surface and distinct loss-of-function mechanisms. Elife 5. https://doi.org/10.7554/eLife.20391
57. Sudom A, Talreja S, Danao J, Bragg E, Kegel R, Min X et al (2018) Molecular basis for the loss-of-function effects of the Alzheimer's disease-associated R47H variant of the immune receptor TREM2. J Biol Chem 293:12634–12646
58. Wang Y, Ulland TK, Ulrich JD, Song W, Tzaferis JA, Hole JT et al (2016) TREM2-mediated early microglial response limits diffusion and toxicity of amyloid plaques. J Exp Med 213:667–675
59. Jay TR, Miller CM, Cheng PJ, Graham LC, Bemiller S, Broihier ML et al (2015) TREM2 deficiency eliminates TREM2+ inflammatory macrophages and ameliorates pathology in Alzheimer's disease mouse models. J Exp Med 212:287–295
60. Jay TR, Hirsch AM, Broihier ML, Miller CM, Neilson LE, Ransohoff RM et al (2017) Disease progression-dependent effects of TREM2 deficiency in a mouse model of Alzheimer's disease. J Neurosci 37:637–647
61. Yuan P, Condello C, Keene CD, Wang Y, Bird TD, Paul SM et al (2016) TREM2 haplodeficiency in mice and humans impairs the microglia barrier function leading to decreased amyloid compaction and severe axonal dystrophy. Neuron 92:252–264
62. Condello C, Yuan P, Schain A, Grutzendler J (2015) Microglia constitute a barrier that prevents neurotoxic protofibrillar Aβ42 hotspots around plaques. Nat Commun 6:6176. https://doi.org/10.1038/ncomms7176
63. Piccio L, Buonsanti C, Cella M, Tassi I, Schmidt RE, Fenoglio C et al (2008) Identification of soluble TREM-2 in the cerebrospinal fluid and its association with multiple sclerosis and CNS inflammation. Brain 131:3081–3091
64. Piccio L, Deming Y, Del-Águila JL, Ghezzi L, Holtzman DM, Fagan AM et al (2016) Cerebrospinal fluid soluble TREM2 is higher in Alzheimer disease and associated with mutation status. Acta Neuropathol 131:925–933
65. Suárez-Calvet M, Kleinberger G, Araque Caballero MÁ, Brendel M, Rominger A, Alcolea D et al (2016) sTREM2 cerebrospinal fluid levels are a potential biomarker for microglia activ-

ity in early-stage Alzheimer's disease and associate with neuronal injury markers. EMBO Mol Med 8:466–476
66. Bertram L, Lange C, Mullin K, Parkinson M, Hsiao M, Hogan MF et al (2008) Genome-wide association analysis reveals putative Alzheimer's disease susceptibility loci in addition to APOE. Am J Hum Genet 83:623–632
67. Reitz C, Jun G, Naj A, Rajbhandary R, Vardarajan BN, Wang L-S et al (2013) Variants in the ATP-binding cassette transporter (ABCA7), apolipoprotein E ϵ4, and the risk of late-onset Alzheimer disease in African Americans. JAMA 309:1483. https://doi.org/10.1001/jama.2013.2973
68. Ravetch JV, Lanier LL (2000) Immune inhibitory receptors. Science 290:84–89
69. Griciuc A, Serrano-Pozo A, Parrado AR, Lesinski AN, Asselin CN, Mullin K et al (2013) Alzheimer's disease risk gene CD33 inhibits microglial uptake of amyloid beta. Neuron 78:631–643
70. Chan G, White CC, Winn PA, Cimpean M, Replogle JM, Glick LR et al (2015) CD33 modulates TREM2: convergence of Alzheimer loci. Nat Neurosci 18:1556–1558
71. Zhu X-C, Yu J-T, Jiang T, Wang P, Cao L, Tan L (2015) CR1 in Alzheimer's disease. Mol Neurobiol 51:753–765
72. Klickstein LB, Bartow TJ, Miletic V, Rabson LD, Smith JA, Fearon DT (1988) Identification of distinct C3b and C4b recognition sites in the human C3b/C4b receptor (CR1, CD35) by deletion mutagenesis. J Exp Med 168:1699–1717
73. Klickstein LB, Barbashov SF, Liu T, Jack RM, Nicholson-Weller A (1997) Complement receptor type 1 (CR1, CD35) is a receptor for C1q. Immunity 7:345–355
74. Brouwers N, Van Cauwenberghe C, Engelborghs S, Lambert JC, Bettens K, Le Bastard N et al (2012) Alzheimer risk associated with a copy number variation in the complement receptor 1 increasing C3b/C4b binding sites. Mol Psychiatry 17:223–233
75. Stevens B, Allen NJ, Vazquez LE, Howell GR, Christopherson KS, Nouri N et al (2007) The classical complement cascade mediates CNS synapse elimination. Cell 131:1164–1178
76. Schafer DP, Lehrman EK, Kautzman AG, Koyama R, Mardinly AR, Yamasaki R et al (2012) Microglia sculpt postnatal neural circuits in an activity and complement-dependent manner. Neuron 74:691–705
77. Hong S, Beja-Glasser VF, Nfonoyim BM, Frouin A, Li S, Ramakrishnan S et al (2016) Complement and microglia mediate early synapse loss in Alzheimer mouse models. Science 352:712–716
78. Kao AW, McKay A, Singh PP, Brunet A, Huang EJ (2017) Progranulin, lysosomal regulation and neurodegenerative disease. Nat Rev Neurosci 18:325–333
79. Baker M, Mackenzie IR, Pickering-Brown SM, Gass J, Rademakers R, Lindholm C et al (2006) Mutations in progranulin cause tau-negative frontotemporal dementia linked to chromosome 17. Nature 442:916–919
80. Cruts M, Gijselinck I, van der Zee J, Engelborghs S, Wils H, Pirici D et al (2006) Null mutations in progranulin cause ubiquitin-positive frontotemporal dementia linked to chromosome 17q21. Nature 442:920–924
81. Viswanathan J, Mäkinen P, Helisalmi S, Haapasalo A, Soininen H, Hiltunen M (2009) An association study between granulin gene polymorphisms and Alzheimer's disease in Finnish population. Am J Med Genet B Neuropsychiatr Genet 150B:747–750
82. Sheng J, Su L, Xu Z, Chen G (2014) Progranulin polymorphism rs5848 is associated with increased risk of Alzheimer's disease. Gene 542:141–145
83. Kamalainen A, Viswanathan J, Natunen T, Helisalmi S, Kauppinen T, Pikkarainen M et al (2013) GRN variant rs5848 reduces plasma and brain levels of granulin in Alzheimer's disease patients. J Alzheimers Dis 33:23–27
84. Fenoglio C, Galimberti D, Cortini F, Kauwe JSK, Cruchaga C, Venturelli E et al (2009) Rs5848 variant influences GRN mRNA levels in brain and peripheral mononuclear cells in patients with Alzheimer's disease. J Alzheimers Dis 18:603–612

85. Takahashi H, Klein ZA, Bhagat SM, Kaufman AC, Kostylev MA, Ikezu T et al (2017) Opposing effects of progranulin deficiency on amyloid and tau pathologies via microglial TYROBP network. Acta Neuropathol 133:785–807
86. Minami SS, Min SW, Krabbe G, Wang C, Zhou Y, Asgarov R et al (2014) Progranulin protects against amyloid β deposition and toxicity in Alzheimer's disease mouse models. Nat Med 20:1157–1164
87. Lui H, Zhang J, Makinson SR, Cahill MK, Kelley KW, Huang H-Y et al (2016) Progranulin deficiency promotes circuit-specific synaptic pruning by microglia via complement activation. Cell 165:921–935
88. Ramanan VK, Risacher SL, Nho K, Kim S, Shen L, McDonald BC et al (2015) GWAS of longitudinal amyloid accumulation on 18F-florbetapir PET in Alzheimer's disease implicates microglial activation gene IL1RAP. Brain 138:3076–3088
89. Griffin WS, Stanley LC, Ling C, White L, MacLeod V, Perrot LJ et al (1989) Brain interleukin 1 and S-100 immunoreactivity are elevated in Down syndrome and Alzheimer disease. Proc Natl Acad Sci U S A 86:7611–7615
90. Shaftel SS, Kyrkanides S, Olschowka JA, Miller JH, Johnson RE, O'Banion MK (2007) Sustained hippocampal IL-1 beta overexpression mediates chronic neuroinflammation and ameliorates Alzheimer plaque pathology. J Clin Invest 117:1595–1604
91. Ghosh S, Wu MD, Shaftel SS, Kyrkanides S, LaFerla FM, Olschowka JA et al (2013) Sustained interleukin-1 overexpression exacerbates tau pathology despite reduced amyloid burden in an Alzheimer's mouse model. J Neurosci 33:5053–5064
92. Heneka MT, Kummer MP, Stutz A, Delekate A, Schwartz S, Vieira-Saecker A et al (2012) NLRP3 is activated in Alzheimer's disease and contributes to pathology in APP/PS1 mice. Nature 493:674–678
93. Vasquez JB, Fardo DW, Estus S (2013) ABCA7 expression is associated with Alzheimer's disease polymorphism and disease status. Neurosci Lett 556:58–62
94. Steinberg S, Stefansson H, Jonsson T, Johannsdottir H, Ingason A, Helgason H et al (2015) Loss-of-function variants in ABCA7 confer risk of Alzheimer's disease. Nat Genet 47:445–447
95. Sassi C, Nalls MA, Ridge PG, Gibbs JR, Ding J, Lupton MK et al (2016) ABCA7 p.G215S as potential protective factor for Alzheimer's disease. Neurobiol Aging 46:235.e1–235.e9. https://doi.org/10.1016/j.neurobiolaging.2016.04.004
96. Kaminski WE, Orsó E, Diederich W, Klucken J, Drobnik W, Schmitz G (2000) Identification of a novel human sterol-sensitive ATP-binding cassette transporter (ABCA7). Biochem Biophys Res Commun 273:532–538
97. Hayashi M, Abe-Dohmae S, Okazaki M, Ueda K, Yokoyama S (2005) Heterogeneity of high density lipoprotein generated by ABCA1 and ABCA7. J Lipid Res 46:1703–1711
98. Wang N, Lan D, Gerbod-Giannone M, Linsel-Nitschke P, Jehle AW, Chen W et al (2003) ATP-binding cassette transporter A7 (ABCA7) binds apolipoprotein A-I and mediates cellular phospholipid but not cholesterol efflux. J Biol Chem 278:42906–42912
99. Jehle AW, Gardai SJ, Li S, Linsel-Nitschke P, Morimoto K, Janssen WJ et al (2006) ATP-binding cassette transporter A7 enhances phagocytosis of apoptotic cells and associated ERK signaling in macrophages. J Cell Biol 174:547–556
100. Tanaka N, Abe-Dohmae S, Iwamoto N, Fitzgerald ML, Yokoyama S (2010) Helical apolipoproteins of high-density lipoprotein enhance phagocytosis by stabilizing ATP-binding cassette transporter A7. J Lipid Res 51:2591–2599
101. Morizawa YM, Hirayama Y, Ohno N, Shibata S, Shigetomi E, Sui Y et al (2017) Reactive astrocytes function as phagocytes after brain ischemia via ABCA1-mediated pathway. Nat Commun 8:28. https://doi.org/10.1038/s41467-017-00037-1
102. Hedgecock EM, Sulston JE, Thomson JN (1983) Mutations affecting programmed cell deaths in the nematode Caenorhabditis elegans. Science 220:1277–1279
103. Ellis RE, Jacobson DM, Horvitz HR (1991) Genes required for the engulfment of cell corpses during programmed cell death in Caenorhabditis elegans. Genetics 129:79–94

104. Kim WS, Li H, Ruberu K, Chan S, Elliott DA, Low JK et al (2013) Deletion of Abca7 increases cerebral amyloid-β accumulation in the J20 mouse model of Alzheimer's disease. J Neurosci 33:4387–4394
105. Sakae N, Liu C-C, Shinohara M, Frisch-Daiello J, Ma L, Yamazaki Y et al (2016) ABCA7 deficiency accelerates amyloid-beta generation and Alzheimer's neuronal pathology. J Neurosci 36:3848–3859
106. Satoh K, Abe-Dohmae S, Yokoyama S, St George-Hyslop P, Fraser PE (2015) ATP-binding cassette transporter A7 (ABCA7) loss of function alters Alzheimer amyloid processing. J Biol Chem 290:24152–24165
107. Mao D, Epple H, Uthgenannt B, Novack DV, Faccio R (2006) PLCγ2 regulates osteoclastogenesis via its interaction with ITAM proteins and GAB2. J Clin Invest 116:2869–2879
108. Bunney TD, Katan M (2011) PLC regulation: emerging pictures for molecular mechanisms. Trends Biochem Sci 36:88–96
109. Walliser C, Retlich M, Harris R, Everett KL, Josephs MB, Vatter P et al (2008) rac regulates its effector phospholipase Cgamma2 through interaction with a split pleckstrin homology domain. J Biol Chem 283:30351–30362
110. Bunney TD, Opaleye O, Roe SM, Vatter P, Baxendale RW, Walliser C et al (2009) Structural insights into formation of an active signaling complex between Rac and phospholipase C Gamma 2. Mol Cell 34:223–233
111. Bae J, Sung BH, Cho IH, Song WK (2012) F-actin-dependent regulation of NESH dynamics in rat hippocampal neurons. PLoS One 7:e34514. https://doi.org/10.1371/journal.pone.0034514
112. Bae J, Sung BH, Cho IH, Kim S-M, Song WK (2012) NESH regulates dendritic spine morphology and synapse formation. PLoS One 7:e34677. https://doi.org/10.1371/journal.pone.0034677
113. Satoh J-I, Kino Y, Yanaizu M, Tosaki Y, Sakai K, Ishida T et al (2017) Microglia express ABI3 in the brains of Alzheimer's disease and Nasu-Hakola disease. Intractable Rare Dis Res 6:262–268
114. Ichigotani Y, Fujii K, Hamaguchi M, Matsuda S (2002) In search of a function for the E3B1/Abi2/Argbp1/NESH family (Review). Int J Mol Med 9:591–595
115. Sekino S, Kashiwagi Y, Kanazawa H, Takada K, Baba T, Sato S et al (2015) The NESH/Abi-3-based WAVE2 complex is functionally distinct from the Abi-1-based WAVE2 complex. Cell Commun Signal 13:41. https://doi.org/10.1186/s12964-015-0119-5
116. Moraes L, Zanchin NIT, Cerutti JM (2017) ABI3, a component of the WAVE2 complex, is potentially regulated by PI3K/AKT pathway. Oncotarget 8:67769–67781
117. Huang KL, Marcora E, Pimenova AA, Di Narzo AF, Kapoor M, Jin SC et al (2017) A common haplotype lowers PU.1 expression in myeloid cells and delays onset of Alzheimer's disease. Nat Neurosci 20:1052–1061
118. Fisher RC, Scott EW (1998) Role of PU.1 in hematopoiesis. Stem Cells 16:25–37
119. Smith AM, Gibbons HM, Oldfield RL, Bergin PM, Mee EW, Faull RLM et al (2013) The transcription factor PU.1 is critical for viability and function of human brain microglia. Glia 61:929–942
120. Pauls SD, Marshall AJ (2017) Regulation of immune cell signaling by SHIP1: a phosphatase, scaffold protein, and potential therapeutic target. Eur J Immunol 47:932–945
121. Antunez C, Boada M, Gonzalez-Perez A, Gayan J, Ramirez-Lorca R, Marin J et al (2011) The membrane-spanning 4-domains, subfamily A (MS4A) gene cluster contains a common variant associated with Alzheimer's disease. Genome Med 3:33. https://doi.org/10.1186/gm249
122. Greer PL, Bear DM, Lassance J-M, Bloom ML, Tsukahara T, Pashkovski SL et al (2016) A family of non-GPCR chemosensors defines an alternative logic for mammalian olfaction. Cell 165:1734–1748
123. Sanyal R, Polyak MJ, Zuccolo J, Puri M, Deng L, Roberts L et al (2017) MS4A4A: a novel cell surface marker for M2 macrophages and plasma cells. Immunol Cell Biol 95:611–619

124. Rocha H, Sampaio M, Rocha R, Fernandes S, Leão M (2016) MEF2C haploinsufficiency syndrome: report of a new MEF2C mutation and review. Eur J Med Genet 59:478–482
125. Canté-Barrett K, Pieters R, Meijerink JPP (2014) Myocyte enhancer factor 2C in hematopoiesis and leukemia. Oncogene 33:403–410
126. Lavin Y, Winter D, Blecher-Gonen R, David E, Keren-Shaul H, Merad M et al (2014) Tissue-resident macrophage enhancer landscapes are shaped by the local microenvironment. Cell 159:1312–1326
127. Deczkowska A, Matcovitch-Natan O, Tsitsou-Kampeli A, Ben-Hamo S, Dvir-Szternfeld R, Spinrad A et al (2017) Mef2C restrains microglial inflammatory response and is lost in brain ageing in an IFN-I-dependent manner. Nat Commun 8:717. https://doi.org/10.1038/s41467-017-00769-0
128. Akiyama H (1994) Inflammatory response in Alzheimer's disease. Tohoku J Exp Med 174:295–303
129. Streit WJ, Braak H, Xue Q-S, Bechmann I (2009) Dystrophic (senescent) rather than activated microglial cells are associated with tau pathology and likely precede neurodegeneration in Alzheimer's disease. Acta Neuropathol 118:475–485
130. Bisht K, Sharma KP, Lecours C, Sánchez MG, El Hajj H, Milior G et al (2016) Dark microglia: a new phenotype predominantly associated with pathological states. Glia 64:826–839
131. Plescher M, Seifert G, Hansen JN, Bedner P, Steinhäuser C, Halle A (2018) Plaque-dependent morphological and electrophysiological heterogeneity of microglia in an Alzheimer's disease mouse model. Glia 66:1464–1480
132. Yasuno F, Ota M, Kosaka J, Ito H, Higuchi M, Doronbekov TK et al (2008) Increased binding of peripheral benzodiazepine receptor in Alzheimer's disease measured by positron emission tomography with [11C]DAA1106. Biol Psychiatry 64:835–841
133. Butovsky O, Jedrychowski MP, Moore CS, Cialic R, Lanser AJ, Gabriely G et al (2014) Identification of a unique TGF-β-dependent molecular and functional signature in microglia. Nat Neurosci 17:131–143
134. Bennett ML, Bennett FC, Liddelow SA, Ajami B, Zamanian JL, Fernhoff NB et al (2016) New tools for studying microglia in the mouse and human CNS. Proc Natl Acad Sci U S A 113:E1738–E1746. https://doi.org/10.1073/pnas.1525528113
135. Keren-Shaul H, Spinrad A, Weiner A, Matcovitch-Natan O, Dvir-Szternfeld R, Ulland TK et al (2017) A unique microglia type associated with restricting development of Alzheimer's disease. Cell 169:1276–1290.e17. https://doi.org/10.1016/j.cell.2017.05.018
136. Füger P, Hefendehl JK, Veeraraghavalu K, Wendeln A-C, Schlosser C, Obermüller U et al (2017) Microglia turnover with aging and in an Alzheimer's model via long-term in vivo single-cell imaging. Nat Neurosci 20:1371–1376
137. Ulland TK, Song WM, Huang SC-C, Ulrich JD, Sergushichev A, Beatty WL, et al (2017) TREM2 Maintains Microglial Metabolic Fitness in Alzheimer's Disease. Cell 170:649–663.e13. https://doi.org/10.1016/j.cell.2017.07.023
138. Mazaheri F, Snaidero N, Kleinberger G, Madore C, Daria A, Werner G et al (2017) TREM2 deficiency impairs chemotaxis and microglial responses to neuronal injury. EMBO Rep 18:1186–1198
139. Krasemann S, Madore C, Cialic R, Baufeld C, Calcagno N, El Fatimy R et al (2017) The TREM2-APOE pathway drives the transcriptional phenotype of dysfunctional microglia in neurodegenerative diseases. Immunity 47:566–581.e9. https://doi.org/10.1016/j.immuni.2017.08.008
140. Grathwohl SA, Kälin RE, Bolmont T, Prokop S, Winkelmann G, Kaeser SA et al (2009) Formation and maintenance of Alzheimer's disease β-amyloid plaques in the absence of microglia. Nat Neurosci 12:1361–1363
141. Dagher NN, Najafi AR, Kayala KMN, Elmore MRP, White TE, Medeiros R et al (2015) Colony-stimulating factor 1 receptor inhibition prevents microglial plaque association and improves cognition in 3xTg-AD mice. J Neuroinflammation 12:139. https://doi.org/10.1186/s12974-015-0366-9

142. Olmos-Alonso A, Schetters STT, Sri S, Askew K, Mancuso R, Vargas-Caballero M et al (2016) Pharmacological targeting of CSF1R inhibits microglial proliferation and prevents the progression of Alzheimer's-like pathology. Brain 139:891–907
143. Spangenberg EE, Lee RJ, Najafi AR, Rice RA, Elmore MRP, Blurton-Jones M et al (2016) Eliminating microglia in Alzheimer's mice prevents neuronal loss without modulating amyloid-β pathology. Brain 139:1265–1281
144. Sosna J, Philipp S, Albay R, Reyes-Ruiz JM, Baglietto-Vargas D, LaFerla FM et al (2018) Early long-term administration of the CSF1R inhibitor PLX3397 ablates microglia and reduces accumulation of intraneuronal amyloid, neuritic plaque deposition and pre-fibrillar oligomers in 5XFAD mouse model of Alzheimer's disease. Mol Neurodegener 13:11. https://doi.org/10.1186/s13024-018-0244-x
145. Glass CK, Saijo K, Winner B, Marchetto MC, Gage FH (2010) Mechanisms underlying inflammation in neurodegeneration. Cell 140:918–934
146. Wyss-Coray T, Rogers J (2012) Inflammation in Alzheimer disease-a brief review of the basic science and clinical literature. Cold Spring Harb Perspect Med 2:a006346. https://doi.org/10.1101/cshperspect.a006346
147. Brown GC, Neher JJ (2014) Microglial phagocytosis of live neurons. Nat Rev Neurosci 15:209–216
148. Fuhrmann M, Bittner T, Jung CKE, Burgold S, Page RM, Mitteregger G et al (2010) Microglial Cx3cr1 knockout prevents neuron loss in a mouse model of Alzheimer's disease. Nat Neurosci 13:411–413
149. Fourgeaud L, Través PG, Tufail Y, Leal-Bailey H, Lew ED, Burrola PG et al (2016) TAM receptors regulate multiple features of microglial physiology. Nature 532:240–244
150. Lu A, Magupalli VG, Ruan J, Yin Q, Atianand MK, Vos MR et al (2014) Unified polymerization mechanism for the assembly of ASC-dependent inflammasomes. Cell 156:1193–1206
151. Venegas C, Kumar S, Franklin BS, Dierkes T, Brinkschulte R, Tejera D et al (2017) Microglia-derived ASC specks cross-seed amyloid-β in Alzheimer's disease. Nature 552:355–361
152. Yoshiyama Y, Higuchi M, Zhang B, Huang S-M, Iwata N, Saido TC et al (2007) Synapse loss and microglial activation precede tangles in a P301S tauopathy mouse model. Neuron 53:337–351
153. Leyns CEG, Ulrich JD, Finn MB, Stewart FR, Koscal LJ, Remolina Serrano J et al (2017) TREM2 deficiency attenuates neuroinflammation and protects against neurodegeneration in a mouse model of tauopathy. Proc Natl Acad Sci U S A 114:11524–11529
154. Bemiller SM, McCray TJ, Allan K, Formica SV, Xu G, Wilson G et al (2017) TREM2 deficiency exacerbates tau pathology through dysregulated kinase signaling in a mouse model of tauopathy. Mol Neurodegener 12:74. https://doi.org/10.1186/s13024-017-0216-6
155. Shi Y, Yamada K, Liddelow SA, Smith ST, Zhao L, Luo W et al (2017) ApoE4 markedly exacerbates tau-mediated neurodegeneration in a mouse model of tauopathy. Nature 549:523–527
156. Goedert M, Eisenberg DS, Crowther RA (2017) Propagation of Tau aggregates and neurodegeneration. Annu Rev Neurosci 40:189–210
157. Asai H, Ikezu S, Tsunoda S, Medalla M, Luebke J, Haydar T et al (2015) Depletion of microglia and inhibition of exosome synthesis halt tau propagation. Nat Neurosci 18:1584–1593
158. Maragakis NJ, Rothstein JD (2006) Mechanisms of disease: astrocytes in neurodegenerative disease. Nat Clin Pract Neurol 2:679–689
159. Liddelow SA, Guttenplan KA, Clarke LE, Bennett FC, Bohlen CJ, Schirmer L et al (2017) Neurotoxic reactive astrocytes are induced by activated microglia. Nature 541:481–487
160. Clarke LE, Liddelow SA, Chakraborty C, Münch AE, Heiman M, Barres BA (2018) Normal aging induces A1-like astrocyte reactivity. Proc Natl Acad Sci U S A 115:E1896–E1905. https://doi.org/10.1073/pnas.1800165115
161. Yun SP, Kam T-I, Panicker N, Kim S, Oh Y, Park J-S et al (2018) Block of A1 astrocyte conversion by microglia is neuroprotective in models of Parkinson's disease. Nat Med 24:931–938

162. De Strooper B, Karran E (2016) The cellular phase of Alzheimer's disease. Cell 164:603–615
163. Mullard A (2018) Microglia-targeted candidates push the Alzheimer drug envelope. Nat Rev Drug Discov 17:303–305
164. Salter MW, Stevens B (2017) Microglia emerge as central players in brain disease. Nat Med 23:1018–1027
165. Mahley RW (1988) Apolipoprotein E: cholesterol transport protein with expanding role in cell biology. Science 240:622–630
166. Castellano JM, Kim J, Stewart FR, Jiang H, DeMattos RB, Patterson BW et al (2011) Human apoE isoforms differentially regulate brain amyloid-β peptide clearance. Sci Transl Med 3:89ra57. https://doi.org/10.1126/scitranslmed.3002156
167. Deane R, Sagare A, Hamm K, Parisi M, Lane S, Finn MB et al (2008) apoE isoform-specific disruption of amyloid beta peptide clearance from mouse brain. J Clin Invest 118:4002–4013
168. Hashimoto T, Serrano-Pozo A, Hori Y, Adams KW, Takeda S, Banerji AO et al (2012) Apolipoprotein E, especially apolipoprotein E4, increases the oligomerization of amyloid beta peptide. J Neurosci 32:15181–15192
169. Ubelmann F, Burrinha T, Salavessa L, Gomes R, Ferreira C, Moreno N et al (2017) Bin1 and CD2AP polarise the endocytic generation of beta-amyloid. EMBO Rep 18:102–122
170. Miyagawa T, Ebinuma I, Morohashi Y, Hori Y, Young Chang M, Hattori H et al (2016) BIN1 regulates BACE1 intracellular trafficking and amyloid-beta production. Hum Mol Genet 25:2948–2958
171. Calafate S, Flavin W, Verstreken P, Moechars D (2016) Loss of Bin1 promotes the propagation of Tau pathology. Cell Rep 17:931–940
172. Singh MK, Dadke D, Nicolas E, Serebriiskii IG, Apostolou S, Canutescu A et al (2008) A novel Cas family member, HEPL, regulates FAK and cell spreading. Mol Biol Cell 19:1627–1636
173. Dourlen P, Fernandez-Gomez FJ, Dupont C, Grenier-Boley B, Bellenguez C, Obriot H et al (2017) Functional screening of Alzheimer risk loci identifies PTK2B as an in vivo modulator and early marker of Tau pathology. Mol Psychiatry 22:874–883
174. Schaefer A, van Duijn TJ, Majolee J, Burridge K, Hordijk PL (2017) Endothelial CD2AP binds the receptor ICAM-1 to control mechanosignaling, leukocyte adhesion, and the route of leukocyte diapedesis in vitro. J Immunol 198:4823–4836
175. Kobayashi S, Sawano A, Nojima Y, Shibuya M, Maru Y (2004) The c-Cbl/CD2AP complex regulates VEGF-induced endocytosis and degradation of Flt-1 (VEGFR-1). FASEB J 18:929–931
176. Shulman JM, Imboywa S, Giagtzoglou N, Powers MP, Hu Y, Devenport D et al (2014) Functional screening in Drosophila identifies Alzheimer's disease susceptibility genes and implicates Tau-mediated mechanisms. Hum Mol Genet 23:870–877
177. Kress C, Gautier-Courteille C, Osborne HB, Babinet C, Paillard L (2007) Inactivation of CUG-BP1/CELF1 causes growth, viability, and spermatogenesis defects in mice. Mol Cell Biol 27:1146–1157
178. Calero M, Tokuda T, Rostagno A, Kumar A, Zlokovic B, Frangione B et al (1999) Functional and structural properties of lipid-associated apolipoprotein J (clusterin). Biochem J 344(Pt 2):375–383
179. Bartl MM, Luckenbach T, Bergner O, Ullrich O, Koch-Brandt C (2001) Multiple receptors mediate apoJ-dependent clearance of cellular debris into nonprofessional phagocytes. Exp Cell Res 271:130–141
180. Lakins JN, Poon S, Easterbrook-Smith SB, Carver JA, Tenniswood MPR, Wilson MR (2002) Evidence that clusterin has discrete chaperone and ligand binding sites. Biochemistry 41:282–291
181. Bell RD, Sagare AP, Friedman AE, Bedi GS, Holtzman DM, Deane R et al (2007) Transport pathways for clearance of human Alzheimer's amyloid beta-peptide and apolipoproteins E and J in the mouse central nervous system. J Cereb Blood Flow Metab 27:909–918
182. Ramani VC, Hennings L, Haun RS (2008) Desmoglein 2 is a substrate of kallikrein 7 in pancreatic cancer. BMC Cancer 8:373. https://doi.org/10.1186/1471-2407-8-373

183. Yamazaki T, Masuda J, Omori T, Usui R, Akiyama H, Maru Y (2009) EphA1 interacts with integrin-linked kinase and regulates cell morphology and motility. J Cell Sci 122:243–255
184. Ivanov AI, Romanovsky AA (2006) Putative dual role of ephrin-Eph receptor interactions in inflammation. IUBMB Life 58:389–394
185. Yasuda-Yamahara M, Rogg M, Frimmel J, Trachte P, Helmstaedter M, Schroder P et al (2018) FERMT2 links cortical actin structures, plasma membrane tension and focal adhesion function to stabilize podocyte morphology. Matrix Biol 68–69:263–279
186. Chapuis J, Flaig A, Grenier-Boley B, Eysert F, Pottiez V, Deloison G et al (2017) Genome-wide, high-content siRNA screening identifies the Alzheimer's genetic risk factor FERMT2 as a major modulator of APP metabolism. Acta Neuropathol 133:955–966
187. Pinet V, Vergelli M, Martin R, Bakke O, Long EO (1995) Antigen presentation mediated by recycling of surface HLA-DR molecules. Nature 375:603–606
188. Wesche H, Korherr C, Kracht M, Falk W, Resch K, Martin MU (1997) The interleukin-1 receptor accessory protein (IL-1RAcP) is essential for IL-1-induced activation of interleukin-1 receptor-associated kinase (IRAK) and stress-activated protein kinases (SAP kinases). J Biol Chem 272:7727–7731
189. Chen Z, Shojaee S, Buchner M, Geng H, Lee JW, Klemm L et al (2015) Signalling thresholds and negative B-cell selection in acute lymphoblastic leukaemia. Nature 521:357–361
190. Gossett LA, Kelvin DJ, Sternberg EA, Olson EN (1989) A new myocyte-specific enhancer-binding factor that recognizes a conserved element associated with multiple muscle-specific genes. Mol Cell Biol 9:5022–5033
191. Duriez B, Duquesnoy P, Escudier E, Bridoux A-M, Escalier D, Rayet I et al (2007) A common variant in combination with a nonsense mutation in a member of the thioredoxin family causes primary ciliary dyskinesia. Proc Natl Acad Sci U S A 104:3336–3341
192. Takatori S, Tomita T (2018) AP180 N-terminal homology (ANTH) and epsin N-terminal homology (ENTH) domains: physiological functions and involvement in disease. Adv Exp Med Biol May 18. https://doi.org/10.1007/5584_2018_218. [Epub ahead of print]
193. Kanatsu K, Morohashi Y, Suzuki M, Kuroda H, Watanabe T, Tomita T et al (2014) Decreased CALM expression reduces Aβ42 to total Aβ ratio through clathrin-mediated endocytosis of γ-secretase. Nat Commun 5:3386. https://doi.org/10.1038/ncomms4386
194. Kanatsu K, Hori Y, Takatori S, Watanabe T, Iwatsubo T, Tomita T (2016) Partial loss of CALM function reduces Aβ42 production and amyloid deposition in vivo. Hum Mol Genet 25:3988–3997
195. Zhao Z, Sagare AP, Ma Q, Halliday MR, Kong P, Kisler K et al (2015) Central role for PICALM in amyloid-β blood-brain barrier transcytosis and clearance. Nat Neurosci 18:978–987
196. Moreau K, Fleming A, Imarisio S, Lopez Ramirez A, Mercer JL, Jimenez-Sanchez M et al (2014) PICALM modulates autophagy activity and tau accumulation. Nat Commun 5:4998. https://doi.org/10.1038/ncomms5998
197. Watanabe D, Hashimoto S, Ishiai M, Matsushita M, Baba Y, Kishimoto T et al (2001) Four tyrosine residues in phospholipase C-gamma 2, identified as Btk-dependent phosphorylation sites, are required for B cell antigen receptor-coupled calcium signaling. J Biol Chem 276:38595–38601
198. Fazzari P, Horre K, Arranz AM, Frigerio CS, Saito T, Saido TC et al (2017) PLD3 gene and processing of APP. Nature 541:E1–E2. https://doi.org/10.1038/nature21030
199. Soni D, Regmi SC, Wang D-M, DebRoy A, Zhao Y-Y, Vogel SM et al (2017) Pyk2 phosphorylation of VE-PTP downstream of STIM1-induced Ca(2+) entry regulates disassembly of adherens junctions. Am J Physiol Lung Cell Mol Physiol 312:L1003–L1017. https://doi.org/10.1152/ajplung.00008.2017
200. Dikic I, Tokiwa G, Lev S, Courtneidge SA, Schlessinger J (1996) A role for Pyk2 and Src in linking G-protein-coupled receptors with MAP kinase activation. Nature 383:547–550
201. Kajiho H, Saito K, Tsujita K, Kontani K, Araki Y, Kurosu H et al (2003) RIN3: a novel Rab5 GEF interacting with amphiphysin II involved in the early endocytic pathway. J Cell Sci 116:4159–4168

202. Li X-F, Kraev AS, Lytton J (2002) Molecular cloning of a fourth member of the potassium-dependent sodium-calcium exchanger gene family, NCKX4. J Biol Chem 277:48410–48417
203. Taira K, Bujo H, Hirayama S, Yamazaki H, Kanaki T, Takahashi K et al (2001) LR11, a mosaic LDL receptor family member, mediates the uptake of ApoE-rich lipoproteins in vitro. Arterioscler Thromb Vasc Biol 21:1501–1506
204. Klinger SC, Hojland A, Jain S, Kjolby M, Madsen P, Svendsen AD et al (2016) Polarized trafficking of the sorting receptor SorLA in neurons and MDCK cells. FEBS J 283:2476–2493
205. Andersen OM, Reiche J, Schmidt V, Gotthardt M, Spoelgen R, Behlke J et al (2005) Neuronal sorting protein-related receptor sorLA/LR11 regulates processing of the amyloid precursor protein. Proc Natl Acad Sci U S A 102:13461–13466
206. Caglayan S, Takagi-Niidome S, Liao F, Carlo A-S, Schmidt V, Burgert T et al (2014) Lysosomal sorting of amyloid - by the SORLA receptor is impaired by a familial Alzheimer's disease mutation. Sci Transl Med 6:223ra20. https://doi.org/10.1126/scitranslmed.3007747
207. Takahashi K, Rochford CDP, Neumann H (2005) Clearance of apoptotic neurons without inflammation by microglial triggering receptor expressed on myeloid cells-2. J Exp Med 201:647–657
208. He F, Umehara T, Saito K, Harada T, Watanabe S, Yabuki T et al (2010) Structural insight into the zinc finger CW domain as a histone modification reader. Structure 18:1127–1139
209. Zhang Y, Chen K, Sloan SA, Bennett ML, Scholze AR, O'Keeffe S et al (2014) An RNA-sequencing transcriptome and splicing database of glia, neurons, and vascular cells of the cerebral cortex. J Neurosci 34:11929–11947
210. Zhang Y, Sloan SA, Clarke LE, Caneda C, Plaza CA, Blumenthal PD et al (2016) Purification and characterization of progenitor and mature human astrocytes reveals transcriptional and functional differences with mouse. Neuron 89:37–53
211. Guillot-Sestier M-V, Doty KR, Gate D, Rodriguez J, Leung BP, Rezai-Zadeh K et al (2015) Il10 deficiency rebalances innate immunity to mitigate Alzheimer-like pathology. Neuron 85:534–548
212. Krauthausen M, Kummer MP, Zimmermann J, Reyes-Irisarri E, Terwel D, Bulic B et al (2015) CXCR3 promotes plaque formation and behavioral deficits in an Alzheimer's disease model. J Clin Invest 125:365–378
213. Choi S-H, Aid S, Caracciolo L, Sakura Minami S, Niikura T, Matsuoka Y et al (2013) Cyclooxygenase-1 inhibition reduces amyloid pathology and improves memory deficits in a mouse model of Alzheimer's disease. J Neurochem 124:59–68
214. vom Berg J, Prokop S, Miller KR, Obst J, Kälin RE, Lopategui-Cabezas I et al (2012) Inhibition of IL-12/IL-23 signaling reduces Alzheimer's disease–like pathology and cognitive decline. Nat Med 18:1812–1819
215. Yamanaka M, Ishikawa T, Griep A, Axt D, Kummer MP, Heneka MT (2012) PPARγ/RXRα-induced and CD36-mediated microglial amyloid-β phagocytosis results in cognitive improvement in amyloid precursor protein/presenilin 1 mice. J Neurosci 32:17321–17331
216. Chakrabarty P, Herring A, Ceballos-Diaz C, Das P, Golde TE (2011) Hippocampal expression of murine TNFα results in attenuation of amyloid deposition in vivo. Mol Neurodegener 6:16. https://doi.org/10.1186/1750-1326-6-16
217. Chakrabarty P, Ceballos-Diaz C, Beccard A, Janus C, Dickson D, Golde TE et al (2010) IFN-gamma promotes complement expression and attenuates amyloid plaque deposition in amyloid beta precursor protein transgenic mice. J Immunol 184:5333–5343
218. Chakrabarty P, Jansen-West K, Beccard A, Ceballos-Diaz C, Levites Y, Verbeeck C et al (2010) Massive gliosis induced by interleukin-6 suppresses Abeta deposition in vivo: evidence against inflammation as a driving force for amyloid deposition. FASEB J 24:548–559
219. Kiyota T, Okuyama S, Swan RJ, Jacobsen MT, Gendelman HE, Ikezu T (2010) CNS expression of anti-inflammatory cytokine interleukin-4 attenuates Alzheimer's disease-like pathogenesis in APP+PS1 bigenic mice. FASEB J 24:3093–3102
220. Chakrabarty P, Tianbai L, Herring A, Ceballos-Diaz C, Das P, Golde TE (2012) Hippocampal expression of murine IL-4 results in exacerbation of amyloid deposition. Mol Neurodegener 7:36. https://doi.org/10.1186/1750-1326-7-36

221. Lee S, Varvel NH, Konerth ME, Xu G, Cardona AE, Ransohoff RM et al (2010) CX3CR1 deficiency alters microglial activation and reduces beta-amyloid deposition in two Alzheimer's disease mouse models. Am J Pathol 177:2549–2562
222. Liu Z, Condello C, Schain A, Harb R, Grutzendler J (2010) CX3CR1 in microglia regulates brain amyloid deposition through selective protofibrillar amyloid- phagocytosis. J Neurosci 30:17091–17101
223. Boissonneault V, Filali M, Lessard M, Relton J, Wong G, Rivest S (2008) Powerful beneficial effects of macrophage colony-stimulating factor on -amyloid deposition and cognitive impairment in Alzheimer's disease. Brain 132:1078–1092
224. Town T, Laouar Y, Pittenger C, Mori T, Szekely CA, Tan J et al (2008) Blocking TGF-β–Smad2/3 innate immune signaling mitigates Alzheimer-like pathology. Nat Med 14:681–687
225. Wyss-Coray T, Masliah E, Mallory M, McConlogue L, Johnson-Wood K, Lin C et al (1997) Amyloidogenic role of cytokine TGF-beta1 in transgenic mice and in Alzheimer's disease. Nature 389:603–606
226. Wyss-Coray T, Lin C, Yan F, Yu G-Q, Rohde M, McConlogue L et al (2001) TGF-β1 promotes microglial amyloid-β clearance and reduces plaque burden in transgenic mice. Nat Med 7:612–618
227. Jiang T, Zhang Y-D, Gao Q, Zhou J-S, Zhu X-C, Lu H et al (2016) TREM1 facilitates microglial phagocytosis of amyloid beta. Acta Neuropathol 132:667–683
228. Chakrabarty P, Li A, Ceballos-Diaz C, Eddy JA, Funk CC, Moore B et al (2015) IL-10 alters immunoproteostasis in APP mice, increasing plaque burden and worsening cognitive behavior. Neuron 85:519–533
229. Huang F, Buttini M, Wyss-Coray T, McConlogue L, Kodama T, Pitas RE et al (1999) Elimination of the class A scavenger receptor does not affect amyloid plaque formation or neurodegeneration in transgenic mice expressing human amyloid protein precursors. Am J Pathol 155:1741–1747
230. Frenkel D, Wilkinson K, Zhao L, Hickman SE, Means TK, Puckett L et al (2013) Scara1 deficiency impairs clearance of soluble amyloid-β by mononuclear phagocytes and accelerates Alzheimer's-like disease progression. Nat Commun 4:2030. https://doi.org/10.1038/ncomms3030
231. Lucin KM, O'Brien CE, Bieri G, Czirr E, Mosher KI, Abbey RJ et al (2013) Microglial Beclin 1 regulates retromer trafficking and phagocytosis and is impaired in Alzheimer's disease. Neuron 79:873–886
232. Naert G, Rivest S (2011) CC chemokine receptor 2 deficiency aggravates cognitive impairments and amyloid pathology in a transgenic mouse model of Alzheimer's disease. J Neurosci 31:6208–6220
233. Maier M, Peng Y, Jiang L, Seabrook TJ, Carroll MC, Lemere CA (2008) Complement C3 deficiency leads to accelerated amyloid plaque deposition and neurodegeneration and modulation of the microglia/macrophage phenotype in amyloid precursor protein transgenic mice. J Neurosci 28:6333–6341
234. Shi Q, Chowdhury S, Ma R, Le KX, Hong S, Caldarone BJ et al (2017) Complement C3 deficiency protects against neurodegeneration in aged plaque-rich APP/PS1 mice. Sci Transl Med 9(392):eaaf6295. https://doi.org/10.1126/scitranslmed.aaf6295
235. Song M, Jin J, Lim J-E, Kou J, Pattanayak A, Rehman JA et al (2011) TLR4 mutation reduces microglial activation, increases Aβ deposits and exacerbates cognitive deficits in a mouse model of Alzheimer's disease. J Neuroinflammation 8:92. https://doi.org/10.1186/1742-2094-8-92
236. Zhu Y, Hou H, Rezai-Zadeh K, Giunta B, Ruscin A, Gemma C et al (2011) CD45 deficiency drives amyloid-β peptide oligomers and neuronal loss in Alzheimer's disease mice. J Neurosci 31:1355–1365

Chapter 6
Neuroimaging Studies of Cognitive Function in Schizophrenia

Rafael Penadés, Nicolas Franck, Laura González-Vallespí, and Marie Dekerle

Abstract Persons suffering from schizophrenia present cognitive impairments that have a major functional impact on their lives. Particularly, executive functions and episodic memory are consistently found to be impaired. Neuroimaging allows the investigation of affected areas of the brain associated with these impairments and, moreover, the detection of brain functioning improvements after cognitive remediation interventions. For instance, executive function impairments have been associated with prefrontal cortex volume and thickness; cognitive control impairments are correlated with an increased activation in the anterior cingulate cortex, and episodic memory impairments are linked to hippocampal reduction. Some findings suggest the presence of brain compensatory mechanisms in schizophrenia, e.g. recruiting broader cortical areas to perform identical tasks. Similarly, neuroimaging studies of cognitive remediation in schizophrenia focus differentially on structural, functional and connectivity changes. Cognitive remediation improvements have been reported in two main areas: the prefrontal and thalamic regions. It has been suggested that those changes imply a functional reorganisation of neural networks, and cognitive remediation interventions might have a neuroprotective effect. Future studies should use multimodal neuroimaging procedures and more complex theoretical models to identify, confirm and clarify these and newer outcomes. This chapter highlights neuroimaging findings in anatomical and functional brain correlates of schizophrenia, as well as its application and potential use for identifying brain changes after cognitive remediation.

R. Penadés (✉)
Hospital Clínic Barcelona, IDIBAPS, CIBERSAM, University of Barcelona, Barcelona, Spain
e-mail: rpenades@clinic.ub.es

N. Franck
Centre Ressource de Réhabilitation Psychosociale et Remédiation Cognitive, UMR 5229 CNRS, Centre Hospitalier Le Vinatier, University of Lyon, Lyon, France

L. González-Vallespí
Consorci Sanitari del Maresme, Mataró, Spain

M. Dekerle
Centre Ressource de Réhabilitation Psychosociale et Remédiation Cognitive, Lyon, France

P. C. Guest (ed.), *Reviews on Biomarker Studies in Psychiatric and Neurodegenerative Disorders*, Advances in Experimental Medicine and Biology 1118,
https://doi.org/10.1007/978-3-030-05542-4_6

Keywords Schizophrenia · Neuroimaging · Functional brain correlates · Cognitive remediation · Hippocampus

6.1 Introduction

It has been widely emphasised that people suffering from schizophrenia show cognitive impairments in multiple areas [1–3] although those impairments might, according to some authors, be overestimated [4]. More precisely, executive functions, episodic memory and social cognition are consistently found to be impaired and are the favoured target for cognitive remediation [5, 6].

Interestingly, these impairments are found in all disease stages, including the schizophrenia prodromal phase [7] and first episodes [8]. They are even found in children or adolescents who later will develop schizophrenia [9]. These impairments also are present to a certain extent in patient's relatives, and some of them are thought to constitute an endophenotype of the disorder [10, 11]. However, they usually do not increase with disease duration [7, 12].

Cognitive impairments have an important functional impact on the daily life of patients, and antipsychotics drugs show little effect on them [13]. Consequently, a substantial body of literature has been developed on the neural correlates of these impairments, both from an anatomical and a functional point of view.

Furthermore, schizophrenia is characterised by multiple brain anomalies at many levels, structural and functional, and in terms of both activity and connectivity [14–16]. The majority of studies are based on differences in activation patterns although an increasing number of studies presenting data of cerebral connectivity are being published (see Canu et al. [17] for a review on connectivity).

The first part of this chapter is organised around cognitive functions frequently found to be impaired in patients suffering from schizophrenia, i.e. executive functions (working memory and cognitive control) and episodic memory. The latest neuroimaging studies about the anatomical and functional correlates of these impairments are presented. The second part of the chapter reviews the neuroimaging evidence of structural, functional and connectivity changes found in patients with schizophrenia after cognitive remediation psychotherapies.

6.2 Executive Functions

Executive functions are an umbrella expression that designate a group of abilities aiming at organising and controlling cognitive functions, behaviours and emotions. The cognitive part of executive functions refers to an ensemble of top-down processing recruited to perform efficiently a demanding and/or new task. Executive functions are mostly developed in human beings and enable us to efficiently face new situations [18]. They are composed of a large number of cognitive functions, such as working memory, planning, inhibition, flexibility and reasoning [19, 20]. These functions are difficult to disentangle as high-level executive functions involve many other lower-level executive functions, e.g. planning involves working memory, inhibition and flexibility.

Despite this clinical description, working memory is often studied apart from other executive functions, as it is also strongly linked to episodic memory [21].

6.2.1 Anatomical Data

Executive functions are known to be dependent on the prefrontal area [18, 22], and prefrontal volume and thickness are associated to executive performances in healthy participants [22].

In schizophrenia, performances at executive functioning tasks were found to be related to the volume of some part of the prefrontal cortex. Indeed, Bonilha et al. [23] found a significant correlation between flexibility measured by tests and volume of left dorsolateral prefrontal cortex (DLPFC). The relationship between volume and cognitive control abilities seems to be limited, although a study conducted by Frascarelli et al. [24] evidenced a decrease of volume with duration of illness in medial frontal gyrus but failed to link it to performance of flexibility tasks. Another frontal area, the left orbital inferior cortex, has been found to be linked to performances on the Stroop task [25]. Links between cognitive control and prefrontal cortex volume are limited to several regions. Working memory performances have also been linked to smaller hippocampal grey matter volume [25].

6.2.2 Functional Data

6.2.2.1 Working Memory

Working memory is one of the core cognitive impairments in schizophrenia. It frequently is found to be associated with inefficient engagement of the DLPFC [26]. This hypoactivation is correlated with performance, with a greater hypoactivation leading to poorer performances. However, some studies did not replicate these results and found no differences of activation or even hyperactivation of DLPFC [27]. This apparent lack of consistency could be explained by the variation of task difficulty. Indeed, it has been hypothesised that the activation follows an inverted U-shape function. Therefore, a small activation is needed for an easy task, but it increases with a high task demand. Nonetheless, when the task is too difficult, activation decreases. It is argued that patients suffering from schizophrenia show a shift of this function to the left. Therefore, when the task is relatively easy, they show a greater activation than healthy controls, whereas when the task is difficult, they show a hypoactivation as they have already reached their limit [21, 27, 28]. Kraguljac et al. [21], in a review of the literature, suggested that these findings could reflect a frontal-based, top-down, cognitive control impairment. This impairment would lead to compensatory strategies, and patients suffering from schizophrenia would show a wider pattern of activation, particularly in regions accounting for attention.

6.2.2.2 Cognitive Control

As cognitive control comprises several cognitive functions, it can be studied using different tasks. We first will focus on studies using the Wisconsin Card Sorting Test (WCST), one of the most used tasks to evaluate cognitive control. Patients must classify cards according to a criterion that they have to find by themselves. After a certain amount of correct trials, the criterion changes, and patients must find the new one. Using event-related, functional magnetic resonance imaging (fMRI), authors can compare cortical response after a positive feedback and a negative feedback, therefore aiming to study the effort of changing strategy (i.e. set-shifting). In healthy participants, the paradigm reveals a cortical-subcortical loop comprising the prefrontal and parietal cortices and basal ganglia [29]. When comparing healthy participants to patients with schizophrenia, Wilmsmeier et al. [30] found a more extensive activation network in patients compared to controls, although in their study, both groups performed equally. More specifically, they evidenced an increased response in both dorsal and rostral anterior cingulate cortex (ACC) in patients. According to the authors, the increased activation in dorsal ACC reflects a stronger cognitive effort to shift, whereas rostral ACC overactivation reflects a stronger emotional response to negative feedback. Other studies, however, evidenced that rostral overactivation in patients with schizophrenia in these conditions was associated with better performances and learning abilities, suggesting that rostral ACC might be involved in cognitive control [31].

Wilmsmeier et al. [30] found an overactivation in the insula, known to be related to unsuccessful inhibition of negative state. In addition, the inferior frontal gyrus (IFG), associated with set-shifting, and the bilateral caudate nucleus, which has been shown to play an important role in executive processing [29], were both overactivated in patients. Therefore, a negative feedback would generate a stronger and more distributed activation network in patients, reflecting enhanced cognitive effort to change strategy and obtain the same performances as healthy participants. In addition, it can also be speculated that disappointment might be more difficult to handle for patients.

Interestingly, no overactivation was found in healthy controls when compared to patients. These results suggest that patients show more activation both in terms of intensity and number of activated areas. That could suggest compensatory mechanisms at work, recruiting a broader cortical area to perform the same task.

As explained earlier, executive functioning is recruited by many tasks, and, although they all refer to the same concept, they might rely on slightly different processing and therefore different secondary neuronal correlates [32, 33]. To avoid this pitfall, Minzenberg et al. [34] performed a meta-analysis, comprising 41 studies, to investigate neural correlates of executive functions in schizophrenia using different paradigms (including go/no-go, mental arithmetic, n-back, Stroop test, WCST). Within-group results showed that healthy controls and patients with schizophrenia activated a similarly distributed cortical subcortical network. That network comprises DLPFC, ACC, ventrolateral prefrontal cortex (VLPFC), premotor cortex, lateral temporal cortical areas, parietal areas, cerebellum and thalamus. Significant co-occurrence of activation across studies was found among the DLPFC, ACC and mediodorsal thalamus. When comparing both groups, hypoactivation in patients was found in the DLPFC, VLPFC, dorsal ACC, occipital and parietal cortices and thalamus.

These results are interpreted in the context of models of cognitive control [18], suggesting that lateral PFC provides top-down control on other brain areas. ACC is known to monitor performance [32] and would modulate the engagement of DLPFC. Therefore, dysfunction of DLPFC in schizophrenia would lead to poor engagement of other brain regions related to a given task and explain poor executive performances. The results are in line with the Wilmsmeier et al. [30] study specifying that overactivation of the ACC could reflect a compensatory mechanism to increase the engagement of the DLPFC. That mechanism of compensation would help to control functioning in other brain regions and eventually to obtain similar performances as healthy controls.

6.3 Episodic Memory

Memory relies on three basic processes: encoding, storage and retrieval. Verbal memory evaluations therefore usually comprise an immediate recall accounting for encoding and a delayed recall accounting for storage and retrieval abilities. Verbal memory is frequently found to be impaired in schizophrenia, and the encoding phase is particularly difficult for patients. As this verbal declarative memory impairment is consistently found in patients and their relatives, it is sometimes considered as an endophenotype of schizophrenia [11].

6.3.1 Anatomical Data

In healthy populations, episodic memory is known to be highly dependent on hippocampus integrity [35]. Studies consistently find a reduction of hippocampus volume in patients suffering from schizophrenia [36, 37] and in their unaffected healthy relatives [25], compared to healthy controls. This reduction does not worsen during illness [36], and a large sample study investigating subcortical brain volumes in 15 centres across the world found that hippocampal atrophy was more important in a sample scaled with a proportion of unmedicated patients [37].

Links between cortical thickness and memory impairment were evaluated by Guimond et al. [38]. They investigated cortical thickness in regions known to be involved in episodic memory like the parahippocampal gyrus, frontal cortex and hippocampus in a group of patients showing low to mild memory impairment and in a group of patients showing moderate to severe memory impairment. The results showed a greater cortical thickness in the latter group of patients. Particularly, they were found to show a thinning in the left inferior frontal gyrus, left middle frontal gyrus and orbitofrontal cortex (OFC). The left parahippocampal gyrus was also thinner than in healthy control participants. They failed, however, to find any hippocampal atrophy.

A recent meta-analysis by Antoniades et al. [39] investigated the links between hippocampal volume reduction and functionality. Interestingly, verbal memory performances were found to be correlated with hippocampal volume. The authors showed that verbal learning and performances at delayed recall were associated with both right and left hippocampal volume in patients suffering from schizophrenia, whereas this correlation was not found in healthy controls. The volume of other subcortical structures, such as the amygdala and putamen, also was found to be linked to verbal memory performances although they were less systematically investigated.

6.3.2 Functional Data

Episodic memory functioning is considered to rely mainly on the medial temporal lobe (MTL) with contribution of the prefrontal cortex. In the prefrontal cortex, three regions are principally recruited: ACC, which is thought to support adjustment in cognitive controls; DLPFC, which is considered to process links between items; and VLPFC, supposed to be responsible for semantic processing of the item [40].

6.3.2.1 Prefrontal Implications

In patients suffering from schizophrenia, episodic memory impairment is usually accompanied by a lesser activation of the prefrontal cortex. Indeed, in a meta-analysis, Ragland et al. [40] evidenced that during encoding, patients suffering from schizophrenia showed a hypoactivation of the left frontopolar cortex, VLPFC and DLPFC. This suggests that the observed differences in performance between patients and healthy controls rely more on information monitoring than on pure memory processes. Indeed, these regions are involved in working memory processes, particularly for linking information with its context and with each other [41].

A second analysis was performed including only studies in which patients were given explicit strategies to improve encoding. The authors found a similar pattern of results, except for VLPFC hypoactivation which was not found. This suggests that when strategies are given to patients, the activation of the ventrolateral prefrontal cortex reaches the level of healthy controls, suggesting that patients are able, when initiated, to semantically process and link stimulus with context.

Ragland et al. [40] also performed an analysis accounting for performance differences in the retrieval phase. As for encoding, the differences in the pattern of activation in patients compared to control participants reflected more difficulty in task managing than in memory per se. It revealed hypoactivation in the DLPFC, right ACC, thalamus and cerebellum. The authors hypothesised—given that hypoactivation was found in structures involved in error monitoring, working memory, attention and mental flexibility—that people suffering from schizophrenia show difficulties in monitoring their responses, detecting their errors and adjusting their behaviour. Other evidences in favour of this hypothesis were found recently, suggesting that the VLPFC can be recruited by participants suffering from schizophrenia when explicit encoding strategies are recommended to them [42].

Surprisingly, Ragland et al. [40] found many differences in the prefrontal cortex and not in the MTL, as one could have hypothesised. That has one exception: the hyperactivation found in the right parahippocampal gyrus that might reflect a compensation mechanism. Differences in MTL activation are not systematic in the literature, and it seems that they can be determined by the method of data analysis used in each study. Indeed, studies using region of interest (ROI) analysis seem to find more often significant differences in activation between patients and control groups in the MTL [40, 43].

6.3.2.2 Medial Temporal Implications

Using this type of analysis, several studies found different significant patterns of activation of the hippocampal and parahippocampal areas in patients with schizophrenia during declarative memory tasks as compared to healthy controls [43, 44]. Indeed, when comparing successful encoding with fixation condition, it seems that the successful encoding of an item is characterised by a hyperactivation of the anterior hippocampus in patients as compared to healthy controls and unaffected relatives [44]. This overactivation might reflect an improved effort for patients to successfully encode a stimulus. This hypothesis was supported by the finding that the parietal superior areas also were overactivated. These regions are known to be linked to the hippocampus and involved in integrated perception processing and conscious organisation during encoding. Therefore, this activation pattern suggests that successful encoding of information is more effortful in patients and/or that information processing is longer.

As declarative memory impairment in schizophrenia might be an endophenotype of the disorder, the pattern of activation in the memory task in unaffected relatives was also investigated. Interestingly some anterior left hippocampal hypoactivation was found in both patients and relatives, when comparing successful encoding to unsuccessful encoding, suggesting that some part of irregular cerebral activity might be linked to genetic liability factors [44].

To conclude this section, it seems that verbal episodic memory impairment in schizophrenia relies on abnormal hippocampal volume and prefrontal hypoactivation. Prefrontal hypoactivation suggests that impairment of patients in monitoring information generates difficulties in encoding and retrieval. These difficulties might also be enhanced by a more effortful processing of stimuli.

6.4 Cognitive Remediation Therapies

Cognitive remediation is an evidence-based psychotherapy for schizophrenia aiming to remediate cognitive impairments. There are different programmes and formats, but all of them have in common some principles and methods. The Cognitive Remediation Expert Working group has proposed a definition of cognitive remediation, considering those principles common to all cognitive remediation approaches: a behavioural training intervention targeting cognitive impairment, including attention, memory, executive functions or metacognition, using scientific principles of learning with the ultimate goal of improving functional outcomes. Several studies have tested the outcomes of cognitive remediation in schizophrenia, and two meta-analyses have established its efficacy [45, 46]. Cognitive remediation is helpful for improving cognition as well as daily functioning, but the underlying neural mechanisms of this treatment are not fully understood.

To demonstrate the existence of brain functional changes, researchers have followed the simple strategy of scanning participants before and after the cognitive intervention. In both moments, the participants should be performing a cognitive task that facilitates the activation of the targeted brain areas. Initially, different studies followed a single-case or case-series methodology [47–49]. Those studies helped to demonstrate the possibility of detecting changes in brain functioning after cognitive remediation. The design of the studies was theoretically driven and based on the selection of a few regions of interest, mainly prefrontal areas. All of the studies showed positive effects after cognitive remediation in terms of frontal activation improvement, but some studies showed that some patients could not present any improvement [50]. Despite the obvious limitations of the single-case methodology, these sorts of studies could still be helpful in understanding intersubject variability. There are an important number of studies focusing on the effects of cognitive remediation with neuroimaging procedures (Table 6.1), and the majority of these are randomised and controlled trials.

6.4.1 Findings in Frontal Lobe

Mostly, the main preference of researchers has been testing the activation of the prefrontal areas. Consequently, these have relied on cognitive paradigms that are dependent on the activation of those areas such as the n-back, verbal fluency or relational learning tasks [51–59].

Table 6.1 Neuroimaging studies of cognitive remediation in schizophrenia

Reference	Participants EG/CG	Treatment/ control condition	Treatment duration (weeks)	Imaging method	Experimental task	Changes and affected areas of the brain
Wykes et al. [47]	2	CRT	12	SPECT	Verbal fluency	↑ Bilateral frontal, temporal, parietal and occipital activity
Penadés et al. [50]	2	CRT	12	SPECT	Tower of London	↑ Prefrontal activity
Wexler et al. [48]	8	CRT	10	fMRI	Auditory verbal memory	↑ L inferior frontal activity
Penadés et al. [49]	8	CRT	12	SPECT	Tower of London	↑ Prefrontal activity
Wykes et al. [53]	12/6 (hc)	CRT/OT	12	fMRI	N-back	↑ R inferior frontal gyrus and bilateral occipital activity
Eack et al. [63]	30/23	CET/EST	52	MRI	–	↑ Pr GM in L hippocampus, L parahippocampal gyrus and L fusiform gyrus/↑ GM in L amygdala
Haut et al. [54]	18/9 (hc)	CRT/CBSST	6	fMRI	N-back, lexical task	↑ L DLPFC, ACC, bilateral frontopolar cortex activity
Rowland et al. [51]	17/17 (hc)	CRT	<1	fMRI, VBM	N-back, lexical decision	↑ L amygdala and bilateral inferior parietal regions activity
Edwards et al. [55]	22/14 (hc)	CRT	22	fMRI	Continuous performance task	↑↓ R middle frontal/R superior parietal cortex/R inferior frontal junction/R visual cortex/cerebellum activity
Bor et al. [56]	20/15 (hc)	CRT	8	fMRI	N-back	↑ L inferior middle frontal gyrus, cingulate gyrus and inferior parietal lobule activity
Subramaniam et al. [52]	31/16 (hc)	AT	13	fMRI	Word generation and recognition	↑ Medial PFC activity

(continued)

Table 6.1 (continued)

Reference	Participants EG/CG	Treatment/ control condition	Treatment duration (weeks)	Imaging method	Experimental task	Changes and affected areas of the brain
Penadés et al. [60]	30/15 (hc)	CRT /SST	15	fMRI, DTI	N-back	↑ L superior parietal lobule and bilateral middle frontal gyri activity/↓ DMN activity in L precuneus and middle frontal gyrus/↑ corpus callosum and R posterior thalamic radiations
Vianin et al. [59]	8/8	CRT/TAU	8	fMRI	Verbal fluency	↑ Inferior parietal lobule, precentral gyrus, Broca's area, middle occipital cortex, middle CC and superior parietal lobule
Subramaniam et al. [58]	30/15 (hc)	AT	15	fMRI	N-back	↑ Middle frontal and inferior frontal gyri activity
Keshavan et al. [61]	25/16	CET/EST	104	fMRI	Cognitive control task	↑ R DLPFC activity/↓ DLPFC—ACC connectivity
Eack et al. [62]	25/16	CET/EST	104	fMRI	Resting	↑ Pr L DLPFC—RSN connectivity/↑ R insula—RSN connectivity
Ramsay et al. [57]	15/12	CRT/CST	16	fMRI	N-back	↑ L DLPFC; L middle frontal gyrus and L precentral gyrus
Morimoto et al. [64]	16/15	CRT/TAU	12	MRI	–	↑ R hippocampal GM volume

ACC anterior cingulate cortex, *AT* auditory-based cognitive training, *CBSST* cognitive behavioural social skills training, *CC* cingulate cortex, *CET* cognitive enhancement therapy, *CG* control group, *CRT* cognitive remediation therapy, *CST* computer skill training, *DLPFC* dorsolateral prefrontal cortex, *DMN* default mode network, *DTI* diffusion tensor imaging, *EG* experimental group, *EST* enriched supportive therapy, *FA* fractional anisotropy, *fMRI* functional magnetic resonance imaging, *GM* grey matter, *hc* healthy controls, *L* left, *MRI* magnetic resonance imaging, *OT* occupational therapy, *Pr* preservation, *PFC* prefrontal cortex, *R* right, *RSN* resting state network, *SPECT* single photon emission computed tomography, *SST* social skills training, *TAU* treatment as usual, *VBM* voxel-based morphometry

Thus, with the n-back task, some interesting results have been described. Wykes et al. [53] were the first authors to demonstrate the potentiality of cognitive remediation to improve brain functioning in a randomised and controlled trial. The improvements found in different areas of cognition were significantly related with an increase of activation of the inferior frontal gyrus. A number of researchers then conducted similar studies to replicate those findings. Haut et al. [54] found increases in the activation of the left DLPFC and the ACC. Edwards et al. [55] implicated more areas of improved activation, including the frontal, parietal, inferior frontal junction and visual cortex. Similar results were found by Bor et al. [56] involving the frontal gyrus, cingulate gyrus and inferior parietal cortex. Recently, Ramsay et al. [57] also found increases in left lateral prefrontal activation during an n-back task. Subramanian et al. [58] used a different cognitive paradigm for scanning. They used a word generation and recognition task. After cognitive remediation, patients showed a significant improvement of the medial prefrontal activation pattern. It is interesting to highlight a tendency toward the normalisation of the prefrontal activation pattern although patients still showed poorer activation than healthy controls. Finally, Vianin et al. [59] reported increased activation after treatment in different areas including the parietal lobule, precentral gyrus, occipital cortex, middle cingulate cortex and superior parietal lobule, using a verbal fluency paradigm. Strikingly, the authors reported increased activation in Broca's area, claiming that verbalisation might be the main factor underlying these brain changes.

In general, it seems to be proven that patients showed some increases in the activity of some parts of the frontal lobe after the cognitive remediation. These areas were mainly the left DLPFC, left DLPFC, anterior cingulate and right and left prefrontal cortex.

6.4.2 Findings in Connectivity and Brain Functioning

It has been suggested, however, that high levels of activation should not necessarily be considered an indicator of better brain functioning. A decrease in activation in some brain areas could be correlated with better cognitive performance in healthy people [27]. Penadés et al. [60] used the nback task and showed that two different networks, the central executive network and the default mode network, were overactive when compared to the healthy participants despite the finding that their task performances are similar. Interestingly, decreased activation was found in both networks after treatment. In healthy people, the default-mode network has been proven to be anti-correlated with the other networks, but in schizophrenia, this always remains overactive. Particularly interesting is the decreased activity in the default mode network after treatment. This could mean a better synchronisation in the activation of one network, the central executive network, and deactivation of the other, the default mode network. On the whole, the activation patterns after treatment were more similar to the patterns observed in healthy controls. Those findings could be interpreted as an improvement in the efficiency of both networks.

Recently, a number of researchers have begun focusing on new targets that come from new theoretical frameworks like brain networks theory. The use of the whole-brain approach has allowed focusing on the totality of brain regions and not only on the frontal lobes. Connectivity between different regions of the brain also is becoming the focus of new research [60–62]. Additionally, new methods of analysis of neuroimaging data enable us to test changes not only on taskrelated performance but also on restrelated brain functioning. Those innovative aspects are potentially constructing a more accurate picture of the effects that cognitive remediation may produce on brain functioning.

In this context, Penadés et al. [60] conducted a trial using a wholebrain approach that combined fMRI and diffusion tensor imaging (DTI). They investigated the effect of cognitive remediation on brain functioning, trying to test the potential changes on white matter fibres. The analysis of white matter on DTI showed an increase in the fractional anisotropy index after treatment for participants who received cognitive remediation. That increment was found in the anterior part of the genu of the corpus callosum, and it was correlated with cognitive gains. Although these findings need to be replicated, the authors speculated about the possibility that cognitive remediation could be improving connectivity between the two prefrontal hemispheres. The prefrontal cortex in both hemispheres is connected via the anterior part of the corpus callosum, and the whole system is an important element of the working memory system.

Eack et al. [62] conducted a pioneering research, testing the potential effects of cognitive remediation over brain connectivity. A longitudinal, randomised and controlled trial was carried out with 45 participants. These individuals were treated for 2 years with cognitive enhancement therapy (CET), which combines cognitive computer exercises on an individual basis and social cognitive exercises in group format. Functional MRI data was collected annually, and a resting-state functional analysis was performed. To test potential effects of cognitive remediation on frontotemporal connectivity, they used a seed-to-voxel general linear model. Statistical analyses showed that only participants who underwent CET showed two sorts of changes in connectivity. Firstly, connectivity loss between the resting network and the DLPFC was inferior for patients who followed the cognitive intervention. Secondly, connectivity with the insular cortex extending to the superior temporal gyrus significantly increased. Both networks seem to be related with problem-solving and emotional processing. The authors concluded that CET may improve brain connectivity, particularly between frontal and temporal regions.

To sum up, the findings of these studies suggest that cognitive remediation could be acting in different areas of the brain and not only in the areas of the frontal lobe. In any case, more brain activation is not always a suggestion of better brain functioning. Networks theory suggests that some brain areas, like the default mode network, are impaired in schizophrenia because they are overactive. For these reasons, not only activation of networks but also connectivity and synchronisation between them may become the focus of future studies.

6.4.3 Structural Findings

Eack et al. [63] published the first study trying to identify the effects of cognitive remediation in brain morphology. Using a voxel-based, morphometry methodology, they performed volumetric analysis in different regions of interest involving frontal and temporal regions. Participants were treated with the CET for 2 years, and the control group followed a supportive therapy for the same period. Noticeably, the analyses revealed some remarkable changes. Patients following cognitive intervention showed significant preservation of the grey matter in relation with the control group. The group with cognitive intervention showed less grey matter loss in fusiform gyrus and in the left parahippocampal region. Moreover, significant increases of grey matter were found in the left amygdala. All of these findings were statistically related with cognitive improvement after the cognitive intervention. The authors defended the possibility that cognitive remediation might have a protective effect on the brain in persons affected by schizophrenia.

Morimoto et al. [64] recently performed a randomised controlled study using a computer-assisted, cognitive remediation programme. They used a whole-brain, voxel-based, morphometric analysis to detect significant volume changes in cortical grey matter. Interestingly, patients who received the cognitive intervention showed statistically significant increases in right hippocampal volume. Changes in hippocampal volumes were positively correlated with verbal fluency improvements. The results suggest that cognitive remediation might be able to facilitate, hippocampal plasticity to some degree.

Finally, despite the incipient evidence for structural changes in the context of cognitive remediation, little is known about whether neuroimaging data can be used as a predictor of treatment response. This question was tested by Penadés et al. [65], who used baseline cortical thickness as a potential predictor of outcomes. The results suggested that baseline measures of cortical thickness in the frontal and temporal lobes are associated with responsiveness to cognitive remediation. In particular, improvement in non-verbal and verbal memory was associated with greater thickness in some areas of the frontal and temporal lobes at baseline. Significant differences were found in the left superior frontal, left caudal middle frontal, left precuneus and paracentral, superior frontal, right caudal middle frontal gyrus and pars opercularis. The results of this study need to be replicated, but they suggest that responsiveness to cognitive remediation may be dependent on the integrity of some brain structures located mainly in frontal and temporal lobes.

On the whole, structural studies suggest some detectable changes in the context of cognitive remediation. Some putative protective effects in the grey matter of the frontal and temporal lobes have been suggested. In addition, some sort of plasticity has been detected in the form of an increase of volume in the hippocampal region and in augmentation of white matter fibres at the corpus callosum. Finally, responsiveness to cognitive remediation may be related to the integrity of some brain areas in frontal and temporal lobes (Fig. 6.1).

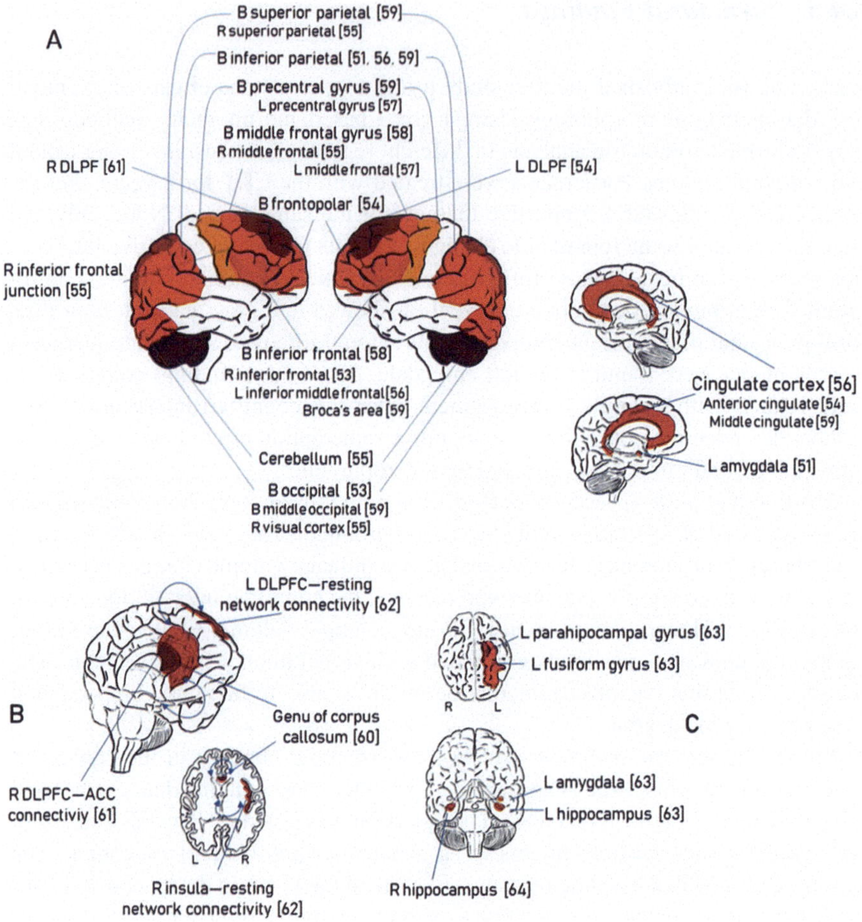

Fig. 6.1 Areas of the brain associated with cognitive remediation improvements in schizophrenia. For more detailed regions, direction of change and related studies, see Table 6.1. (**a**) Functional findings: areas that have shown activation changes. Bilateral and sagittal views of the brain. (**b**) Connectivity findings. Sagittal and superior coronal view of the brain. Arrows symbolise connectivity. (**c**) Structural findings: areas that have shown changes in terms of volume (grey and white matter). Inferior transverse views of the brain. All the figures are qualitative representations. *ACC* anterior cingulate cortex, *DLPFC* dorsolateral prefrontal cortex, *B* bilateral, *L* left, *R* right

6.5 Conclusion

Neuroimaging studies in recent years have shed some light on the topic of the neurobiological basis of cognitive function in schizophrenia. Thus, prefrontal cortex engagement appears to be a core characteristic in schizophrenia, although the results of studies investigating the neural correlates of cognition are far from consensus. Some authors suggest that some neurocognitive dysfunction could be seen as a consequence of prefrontal cortex dysfunction [40, 42]. Interestingly, cues for compensatory mechanisms are found in all the cognitive domains reviewed, suggesting that schizophrenic brains adapt their functioning to improve performance, although it is not fully efficient.

Also, neuroimaging studies of cognitive remediation in patients with schizophrenia have led to a better understanding of cognitive processes in schizophrenia. Cognitive remediation therapies have proven their proficiency to induce significant improvements on brain functioning. In particular, changes involving the prefrontal and thalamic regions are the most commonly reported results. These changes are being interpreted in terms of functional reorganisation of the neural networks. Moreover, structural changes in grey and white matter have also been described. These could be understood as the effects of neuroplasticity induced by the cognitive intervention. At the same time, these findings suggest a neuroprotective effect in response to cognitive remediation as they could be preventing, to some degree, grey matter loss. Obviously, more studies are required to confirm and clarify these results. As already suggested [5, 6], future studies should incorporate multimodal neuroimaging procedures, whole-brain analyses, brain networks theories, studies of connectivity, tractography and more complex theoretical models like graph theory.

References

1. Green MF, Nuechterlein KH, Gold JM, Barch DM, Cohen J, Essock S et al (2004) Approaching a consensus cognitive battery for clinical trials in schizophrenia: the NIMH-MATRICS conference to select cognitive domains and test criteria. Biol Psychiatry 56(5):301–307
2. Heinrichs RW (2005) The primacy of cognition in schizophrenia. Am Psychol 60(3):229–242
3. Reichenberg A (2010) The assessment of neuropsychological functioning in schizophrenia. Dialogues Clin Neurosci 12:383–392
4. Moritz S, Klein JP, Desler T, Lill H, Gallinat J, Schneider BC (2017) Neurocognitive deficits in schizophrenia. Are we making mountains out of molehills? Psychol Med 47(15):2602–2612
5. Bon L, Franck N (2018) The impact of cognitive remediation on cerebral activity in schizophrenia: systematic review of the literature. Brain Behav 8(3):e00908. https://doi.org/10.1002/brb3.908
6. Penadés R, González-Rodríguez A, Catalán R, Segura B, Bernardo M, Junqué C (2017) Neuroimaging studies of cognitive remediation in schizophrenia: a systematic and critical review. World J Psychiatry 7(1):34–43
7. Bora E, Murray RM (2014) Meta-analysis of cognitive deficits in ultra-high risk to psychosis and first-episode psychosis: do the cognitive deficits progress over, or after, the onset of psychosis? Schizophr Bull 40(4):744–755
8. Mesholam-Gately RI, Giuliano AJ, Goff KP, Faraone SV, Seidman LJ (2009) Neurocognition in first-episode schizophrenia: a meta-analytic review. Neuropsychology 23(3):315–336
9. Bilder RM, Reiter G, Bates J, Lencz T, Szeszko P, Goldman RS et al (2006) Cognitive development in schizophrenia: follow-back from the first episode. J Clin Exp Neuropsychol 28(2):270–282
10. Goghari VM (2011) Executive functioning-related brain abnormalities associated with the genetic liability for schizophrenia: an activation likelihood estimation meta-analysis. Psychol Med 41(6):1239–1252
11. Stone WS, Giuliano AJ, Tsuang MT, Braff DL, Cadenhead KS, Calkins ME et al (2011) Group and site differences on the California Verbal Learning Test in persons with schizophrenia and

their first-degree relatives: findings from the Consortium on the Genetics of Schizophrenia (COGS). Schizophr Res 128(1–3):102–110
12. Ekerholm M, Firus Waltersson S, Fagerberg T, Söderman E, Terenius L, Agartz I et al (2012) Neurocognitive function in long-term treated schizophrenia: a five-year follow-up study. Psychiatry Res 200(2–3):144–152
13. Fett A-KJ, Viechtbauer W, Dominguez M-G, Penn DL, van Os J, Krabbendam L (2011) The relationship between neurocognition and social cognition with functional outcomes in schizophrenia: a meta-analysis. Neurosci Biobehav Rev 35(3):573–588
14. Howes OD, Kapur S (2009) The dopamine hypothesis of schizophrenia: Version III--The final common pathway. Schizophr Bull 35(3):549–562
15. Sugranyes G, Kyriakopoulos M, Dima D, O'Muircheartaigh J, Corrigall R, Pendelbury G et al (2012) Multimodal analyses identify linked functional and white matter abnormalities within the working memory network in schizophrenia. Schizophr Res 138(2–3):136–142
16. Unschuld PG, Buchholz AS, Varvaris M, van Zijl PCM, Ross CA, Pekar JJ et al (2014) Prefrontal brain network connectivity indicates degree of both schizophrenia risk and cognitive dysfunctions. Schizophr Bull 40(3):653–664
17. Canu E, Agosta F, Filippi M (2015) A selective review of structural connectivity abnormalities of schizophrenic patients at different stages of the disease. Schizophr Res 161(1):19–28
18. Miller EK, Cohen JD (2001) An integrative theory of prefrontal cortex functions. Annu Rev Neurosci 24:167–202
19. Diamond A (2013) Executive functions. Annu Rev Psychol 64:135–168
20. Miyake A, Friedman NP, Emerson MJ, Witzki AH, Howerter A, Wager TD (2000) The unity and diversity of executive functions and their contributions to complex "frontal lobe" tasks: a latent variable analysis. Cogn Psychol 41(1):49–100
21. Kraguljac N, Srivastava A, Lahti A (2013) Memory deficits in schizophrenia: a selective review of functional magnetic resonance imaging (fMRI) studies. Behav Sci (Basel) 3(3):330–347
22. Yuan P, Raz N (2014) Prefrontal cortex and executive functions in healthy adults: a meta-analysis of structural neuroimaging studies. Neurosci Biobehav Rev 42:180–192
23. Bonilha L, Molnar C, Horner MD, Anderson B, Forster L, George MS et al (2008) Neurocognitive deficits and prefrontal cortical atrophy in patients with schizophrenia. Schizophr Res 101(1–3):142–151
24. Frascarelli M, Tognin S, Mirigliani A, Parente F, Buzzanca A, Torti MC et al (2015) Medial frontal gyrus alterations in schizophrenia: relationship with duration of illness and executive dysfunctions. Psychiatry Res Neuroimaging 231(2):103–110
25. Guo X, Li J, Wang J, Fan X, Hu M, Shen Y et al (2014) Hippocampal and orbital inferior frontal gray matter volume abnormalities and cognitive deficit in treatment-naive, first-episode patients with schizophrenia. Schizophr Res 152(2–3):339–343
26. Glahn DC, Ragland JD, Abramoff A, Barrett J, Laird AR, Bearden CE et al (2005) Beyond hypofrontality: a quantitative meta-analysis of functional neuroimaging studies of working memory in schizophrenia. Hum Brain Mapp 25(1):60–69
27. Callicott JH, Mattay VS, Verchinski BA, Marenco S, Egan MF, Weinberger DR (2003) Complexity of prefrontal cortical dysfunction in schizophrenia: more than up or down. Am J Psychiatry 160(12):2209–2215
28. Barbalat G, Chambon V, Domenech PJD, Ody C, Koechlin E, Franck N et al (2011) Impaired hierarchical control within the lateral prefrontal cortex in schizophrenia. Biol Psychiatry 70(1):73–80
29. Monchi O, Petrides M, Petre V, Worsley K, Dagher A (2001) Wisconsin card sorting revisited: distinct neural circuits participating in different stages of the task identified by event-related functional magnetic resonance imaging. J Neurosci 21(19):7733–7741
30. Wilmsmeier A, Ohrmann P, Suslow T, Siegmund A, Koelkebeck K, Rothermundt M et al (2010) Neural correlates of set-shifting: decomposing executive functions in schizophrenia. J Psychiatry Neurosci 35(5):321–329
31. Pedersen A, Wilmsmeier A, Wiedl KH, Bauer J, Kueppers K, Koelkebeck K et al (2012) Anterior cingulate cortex activation is related to learning potential on the WCST in schizophrenia patients. Brain Cogn 79(3):245–251

32. Carter CS, MacDonald AW, Ross LL, Stenger VA (2001) Anterior cingulate cortex activity and impaired self-monitoring of performance in patients with schizophrenia: an event-related fMRI study. Am J Psychiatry 158(9):1423–1428
33. MacDonald AW, Carter CS, Kerns JG, Ursu S, Barch DM, Holmes AJ et al (2005) Specificity of prefrontal dysfunction and context processing deficits to schizophrenia in never-medicated patients with first-episode psychosis. Am J Psychiatry 162(3):475–484
34. Minzenberg MJ, Laird AR, Thelen S, Carter CS, Glahn DC (2009) Meta-analysis of 41 functional neuroimaging studies of executive function in schizophrenia. Arch Gen Psychiatry 66(8):811–822
35. Van Petten C (2004) Relationship between hippocampal volume and memory ability in healthy individuals across the lifespan: review and meta-analysis. Neuropsychologia 42(10):1394–1413
36. Adriano F, Caltagirone C, Spalletta G (2012) Hippocampal volume reduction in first-episode and chronic schizophrenia: a review and meta-analysis. Neuroscientist 18(2):180–200
37. van Erp TG, Hibar DP, Rasmussen JM, Glahn DC, Pearlson GD, Andreassen OA et al (2016) Subcortical brain volume abnormalities in 2028 individuals with schizophrenia and 2540 healthy controls via the ENIGMA consortium. Mol Psychiatry 21(4):547–553
38. Guimond S, Chakravarty MM, Bergeron-Gagnon L, Patel R, Lepage M (2016) Verbal memory impairments in schizophrenia associated with cortical thinning. NeuroImage Clin 11:20–29
39. Antoniades M, Schoeler T, Radua J, Valli I, Allen P, Kempton MJ et al (2018) Verbal learning and hippocampal dysfunction in schizophrenia: a meta-analysis. Neurosci Biobehav Rev 86:166–175
40. Ragland JD, Laird AR, Ranganath C, Blumenfeld RS, Gonzales SM, Glahn DC (2009) Prefrontal activation deficits during episodic memory in schizophrenia. Am J Psychiatry 166(8):863–874
41. Murray LJ, Ranganath C (2007) The dorsolateral prefrontal cortex contributes to successful relational memory encoding. J Neurosci 27(20):5515–5522
42. Ragland JD, Ranganath C, Phillips J, Boudewyn MA, Kring AM, Lesh TA et al (2015) Cognitive control of episodic memory in schizophrenia: differential role of dorsolateral and ventrolateral prefrontal cortex. Front Hum Neurosci 9:604. https://doi.org/10.3389/fnhum.2015.00604
43. Achim AM, Lepage M (2005) Episodic memory-related activation in schizophrenia: meta-analysis. Br J Psychiatry 187:500–509
44. Pirnia T, Woods RP, Hamilton LS, Lyden H, Joshi SH, Asarnow RF et al (2015) Hippocampal dysfunction during declarative memory encoding in schizophrenia and effects of genetic liability. Schizophr Res 161(2–3):357–366
45. McGurk SR, Twamley EW, Sitzer DI, McHugo GJ, Mueser KT (2007) A meta-analysis of cognitive remediation in schizophrenia. Am J Psychiatry 164(12):1791–1802
46. Wykes T, Huddy V, Cellard C, McGurk SR, Czobor P (2011) A metaanalysis of cognitive remediation for schizophrenia: methodology and effect sizes. Am J Psychiatry 168(5):472–485
47. Wykes T (1998) What are we changing with neurocognitive rehabilitation? Illustrations from two single cases of changes in neuropsychological performance and brain systems as measured by SPECT. Schizophr Res 34(1–2):77–86
48. Wexler BE, Anderson M, Fulbright RK, Gore JC (2000) Preliminary evidence of improved verbal working memory performance and normalization of task-related frontal lobe activation in schizophrenia following cognitive exercises. Am J Psychiatry 157(10):1694–1697
49. Penadés R, Boget T, Lomeña F, Mateos JJ, Catalán R, Gastó C et al (2002) Could the hypofrontality pattern in schizophrenia be modified through neuropsychological rehabilitation? Acta Psychiatr Scand 105(3):202–208
50. Penadés R, Boget T, Lomeña F, Bernardo M, Mateos JJ, Laterza C et al (2000) Brain perfusion and neuropsychological changes in schizophrenic patients after cognitive rehabilitation. Psychiatry Res 98(2):127–132
51. Rowland LM, Griego JA, Spieker EA, Cortes CR, Holcomb HH (2010) Neural changes associated with relational learning in schizophrenia. Schizophr Bull 36(3):496–503
52. Subramaniam K, Luks TL, Fisher M, Simpson GV, Nagarajan S, Vinogradov S (2012) Computerized cognitive training restores neural activity within the reality monitoring network in schizophrenia. Neuron 73(4):842–583

53. Wykes T, Brammer M, Mellers J, Bray P, Reeder C, Williams C et al (2002) Effects on the brain of a psychological treatment: cognitive remediation therapy: functional magnetic resonance imaging in schizophrenia. Br J Psychiatry 181:144–152
54. Haut KM, Lim KO, MacDonald A (2010) Prefrontal cortical changes following cognitive training in patients with chronic schizophrenia: effects of practice, generalization, and specificity. Neuropsychopharmacology 35(9):1850–1859
55. Edwards BG, Barch DM, Braver TS (2010) Improving prefrontal cortex function in schizophrenia through focused training of cognitive control. Front Hum Neurosci 4:32. https://doi.org/10.3389/fnhum.2010.00032
56. Bor J, Brunelin J, d'Amato T, Costes N, Suaud-Chagny MF, Saoud M et al (2011) How can cognitive remediation therapy modulate brain activations in schizophrenia? An fMRI study. Psychiatry Res 192(3):160–166
57. Ramsay IS, Nienow TM, Marggraf MP, MacDonald AW (2017) Neuroplastic changes in patients with schizophrenia undergoing cognitive remediation: triple-blind trial. Br J Psychiatry 210(3):216–222
58. Subramaniam K, Luks TL, Garrett C, Chung C, Fisher M, Nagarajan S et al (2014) Intensive cognitive training in schizophrenia enhances working memory and associated prefrontal cortical efficiency in a manner that drives long-term functional gains. NeuroImage 99:281–292
59. Vianin P, Urben S, Magistretti P, Marquet P, Fornari E, Jaugey L (2014) Increased activation in Broca's area after cognitive remediation in schizophrenia. Psychiatry Res 221(3):204–209
60. Penadés R, Pujol N, Catalán R, Massana G, Rametti G, García-Rizo C et al (2013) Brain effects of cognitive remediation therapy in schizophrenia: a structural and functional neuroimaging study. Biol Psychiatry 73(10):1015–1023
61. Keshavan MS, Eack SM, Prasad KM, Haller CS, Cho RY (2016) Longitudinal functional brain imaging study in early course schizophrenia before and after cognitive enhancement therapy. NeuroImage 151:55–64
62. Eack SM, Newhill CE, Keshavan MS (2016) Cognitive enhancement therapy improves resting-state functional connectivity in early course schizophrenia. J Soc Social Work Res 7(2):211–230
63. Eack SM, Hogarty GE, Cho RY, Prasad KM, Greenwald DP, Hogarty SS et al (2010) Neuroprotective effects of cognitive enhancement therapy against gray matter loss in early schizophrenia: results from a 2-year randomized controlled trial. Arch Gen Psychiatry 67(7):674–682
64. Morimoto T, Matsuda Y, Matsuoka K, Yasuno F, Ikebuchi E, Kameda H et al (2018) Computer-assisted cognitive remediation therapy increases hippocampal volume in patients with schizophrenia: a randomized controlled trial. BMC Psychiatry 18(1):83. doi: 10.1186/s12888-018-1667-1
65. Penadés R, Pujol N, Catalan R, Masana G, Garcia-Rizo C, Bargallo N et al (2016) Cortical thickness in regions of frontal and temporal lobes is associated with responsiveness to cognitive remediation therapy in schizophrenia. Schizophr Res 171(1–3):110–116

Chapter 7
The Role of Biomarkers in Psychiatry

Madia Lozupone, Maddalena La Montagna, Francesca D'Urso, Antonio Daniele, Antonio Greco, Davide Seripa, Giancarlo Logroscino, Antonello Bellomo, and Francesco Panza

Abstract Psychiatric illnesses are cognitive and behavioral disorders of the brain. At present, psychiatric diagnosis is based on DSM-5 criteria. Even if endophenotype specificity for psychiatric disorders is discussed, it is difficult to study and identify psychiatric biomarkers to support diagnosis, prognosis, or clinical response to treatment. This chapter investigates the innovative biomarkers of psychiatric diseases for diagnosis and personalized treatment, in particular post-genomic data and proteomic analyses.

Madia Lozupone and Maddalena La Montagna have contributed equally to this chapter.

M. Lozupone
Neurodegenerative Disease Unit, Department of Basic Medical Sciences, Neuroscience and Sense Organs, University of Bari Aldo Moro, Bari, Italy

M. La Montagna · F. D'Urso · A. Bellomo
Psychiatric Unit, Department of Clinical and Experimental Medicine, University of Foggia, Foggia, Italy

A. Daniele
Institute of Neurology, Catholic University of Sacred Heart, Rome, Italy

Fondazione Policlinico Universitario A. Gemelli IRCCS, Rome, Italy

A. Greco · D. Seripa
Geriatric Unit, Fondazione IRCCS Casa Sollievo della Sofferenza, Foggia, Italy

G. Logroscino
Neurodegenerative Disease Unit, Department of Basic Medical Sciences, Neuroscience and Sense Organs, University of Bari Aldo Moro, Bari, Italy

Department of Clinical Research in Neurology, University of Bari Aldo Moro, Lecce, Italy

F. Panza (✉)
Neurodegenerative Disease Unit, Department of Basic Medical Sciences, Neuroscience and Sense Organs, University of Bari Aldo Moro, Bari, Italy

Geriatric Unit, Fondazione IRCCS Casa Sollievo della Sofferenza, Foggia, Italy

Department of Clinical Research in Neurology, University of Bari Aldo Moro, Lecce, Italy
e-mail: geriat.dot@geriatria.uniba.it

P. C. Guest (ed.), *Reviews on Biomarker Studies in Psychiatric and Neurodegenerative Disorders*, Advances in Experimental Medicine and Biology 1118,
https://doi.org/10.1007/978-3-030-05542-4_7

Keywords Proteomics · Biomarker · Post-genomic diagnostics · Psychiatric disease · Epigenetics · Microbiota · Lifestyle · Depressive disorder · Schizophrenia · Bipolar disorder

7.1 Introduction

7.1.1 The Significance of "Biomarker"

One of the most important goals of psychiatry research is to find appropriate biomarkers for mental illnesses [1]. According to the National Institute of Health Biomarkers Definitions Working Group, a "biomarker" is defined as "a characteristic that is objectively measured and evaluated as an indicator of normal biological processes, pathogenic processes or pharmacological responses to a therapeutic intervention" [2]. Another definition of biomarker made by the International Program on Chemical Safety is "any substance, structure, or process that can be measured in the body (or its products) and their influence or prediction about the incidence of outcome or disease" [3]. Based on these ideas, a biomarker can be used to identify the presence or progression of a disease or the effectiveness of a given treatment from a clinical viewpoint [4].

The use of biomarkers in medicine is a common and valuable approach in several clinical fields [5], and biomarker analyses are growing in number and providing potential targets for several medical conditions, such as diabetes and cancer. However, clinical applications of biomarkers for neuropsychiatric illnesses and possible use for clinical diagnosis and prognosis have not consistently led to better quantifiable outcomes [6]. In the present chapter, we discuss the innovative biomarkers of psychiatric diseases for diagnosis and personalized treatment, with a focus on post-genomic data and proteomics analyses.

7.1.2 Biomarker Potential Role in Psychiatric Setting

Certainly, finding consistent biomarkers for early discovery of psychiatric illnesses has been an attractive topic for researchers, in particular with the study of the brain (postmortem, neuroimaging, and neurophysiological studies), of cerebrospinal fluid (CSF), and of serum and plasma biomarkers (cytokines, neurotrophins, neurotransmitters, and genes).

The goals of biomarker applications in psychiatry are diagnosis, prognosis (risk), prediction and assessment of responses to treatment (therapeutic failures), prevention of adverse drug reactions, classification within diagnostic categories, and prediction of intervention effects [7, 8]. Biomarkers could also define the staging of psychiatric illnesses, risk vulnerability across stages, syndrome progression, and epiphenomena [9]. Network neuroscience pursues new ways to model, analyze, map, and record the elements and interactions of neurobiological systems, considering the multi-scale nature of brain networks [10]. In this regard, a primary focus is the neuropsychological construct and the analysis of cognitive functions (attention, working memory, processing speed, learning and memory, executive functions, and global intellectual functions, including social cognition) as endophenotypes for psychiatric illnesses [11, 12]. From recent evidences, social dysfunction and its most evident clinical expression (i.e., social withdrawal) may represent an innovative transdiagnostic domain, with the potential of being an independent entity in terms of biological roots, with the prospect of targeted interventions not only in psychiatric but also in neurodegenerative disorders [13, 14].

Actually, the need to categorize and validate biomarkers has grown to enable clinicians to match specific individual patient treatments to increase the probability of an optimal, personalized outcome. Thanks to genotyping, it could be possible to assess factors that predict antidepressant or antipsychotic drug response [15, 16]. There is a need for characterizing patient variability, for example, to guide pharmacological dosing according to specific phenotypes [17, 18]. The goal of personalized medicine is important in the case of psychiatric diseases to reduce side effects of inappropriate medication or to enable detection of an efficacy signal quickly without potential toxicity [19].

The success of disease-specific biomarkers or endophenotypes is still fragmentary, based on neuroimaging, neuropsychological, biological, biochemical, and genetic aspects. The interest for the psychiatric setting is to go beyond this, to ensure a consistent value of their actual contribution in disease, also through application of post-genomics techniques [20]. One of the most important advances in psychiatry has been the sequencing of human genome in the 1990s [21], but genomic methods cannot differentiate splice variants or proteins with posttranslational modifications (PTMs). Moreover, gene expression is regulated at the post-transcriptional level by microRNAs (miRNAs), small noncoding RNAs. The most important targets of epigenetic regulation in psychiatric processes are synapse development, plasticity, neurogenesis, dendritic extension, and dendritic spine formation [22]. Furthermore, brain imaging, neurotrophic and electrophysiological factors, neurotransmitters, epigenetics, epigenomics, pharmacogenomics, and proteomics are complementary to yield a more complete understanding of the biological basis and appropriate treatments of psychiatric disorders (Figs. 7.1, 7.2, 7.3 and 7.4) [23–27].

7.2 Brain Imaging Biomarkers

Neuroimaging techniques have the power to capture the structure and function of the brain in health and disease. This has revolutionized the study of the organization of the human brain and how its structure and function are changed in psychiatric illnesses. Advances in neuroimaging techniques have made it possible to more clearly elucidate the neural basis of psychiatric disorders. In the past few decades, neuroimaging analyses have served as the main tools for exploring the

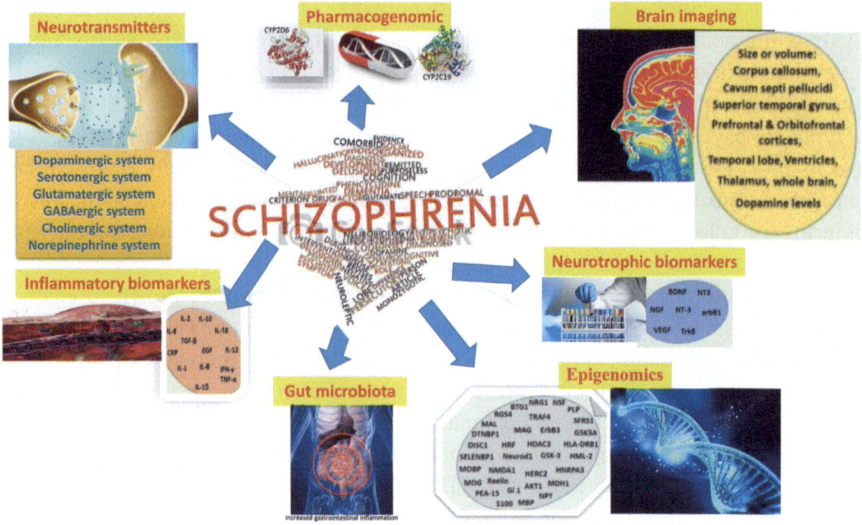

Fig. 7.1 Innovative biomarkers in schizophrenia

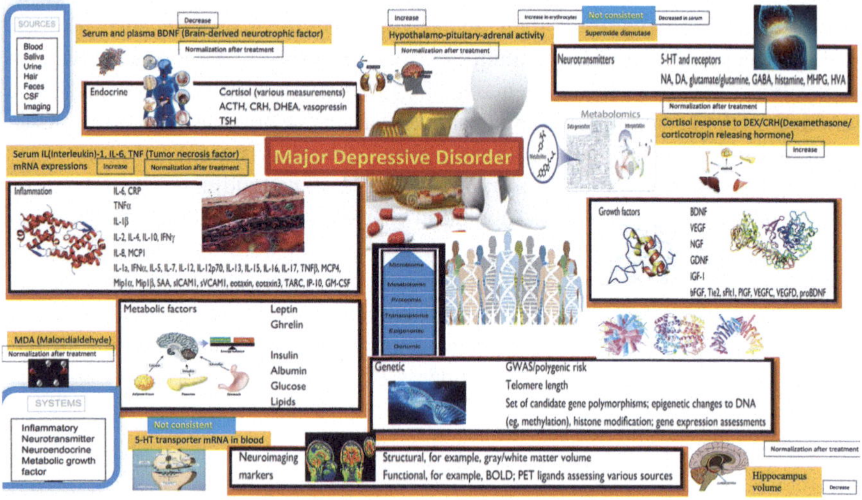

Fig. 7.2 Innovative biomarkers in major depressive disorders

neurobiological etiology of psychiatric disorders [28]. The most important brain imaging techniques are magnetic resonance imaging (MRI), computed tomography (CT), diffusion tensor imaging (DTI), functional magnetic resonance imaging (fMRI), positron-emission tomography (PET), and single photon emission computed tomography (SPECT) [29–33].

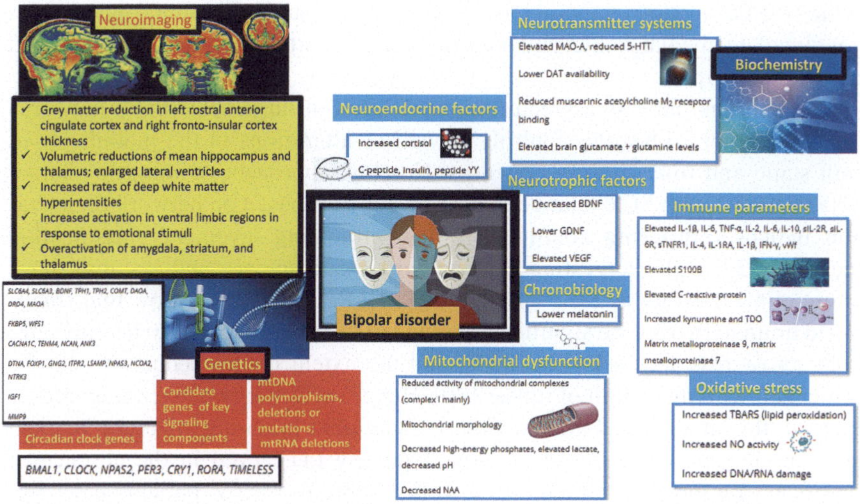

Fig. 7.3 Innovative biomarkers in bipolar disorders

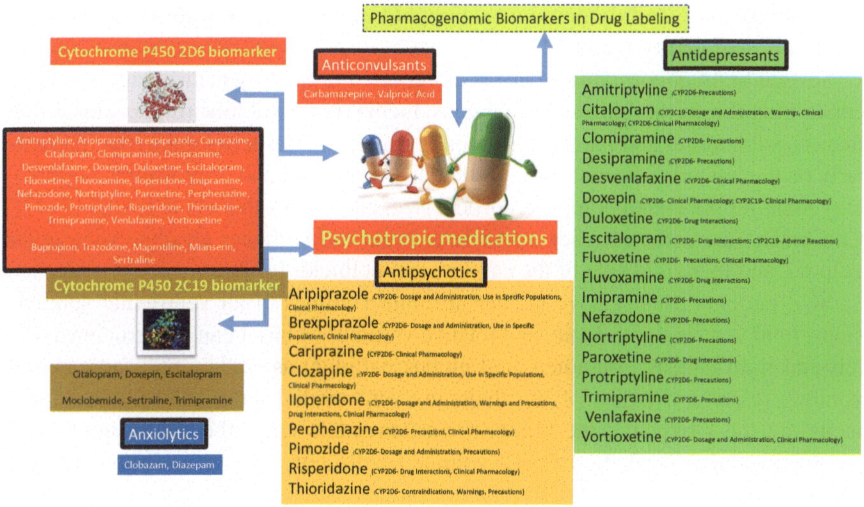

Fig. 7.4 Psychotropic medications: pharmacogenomic biomarkers in drug labeling

7.2.1 Schizophrenia

Patients with schizophrenia (SCZ) have differences in brain structure, brain volume, glucose metabolism, and blood flow at rest and during the performance of cognitive tasks [34] (Fig. 7.1). In SCZ, there are reduced activation in the dorsolateral prefrontal cortex (DLPFC) and the right temporal and ventral prefrontal cortices during the performance of working memory tasks [35] and abnormalities in the DLPFC, medial temporal lobe, hippocampus, parahippocampal gyrus, anterior cingulate, medial frontal and posterior parietal cortex, striatum, thalamus, and cerebellum [36].

The findings indicate a greater randomization of large-scale brain networks in SCZ relative to healthy controls as well as alterations in the modularity of both static and time-varying networks. Notably, approaches aiming to characterize patients with SCZ relative to healthy controls based on network organization indices (clustering coefficient) show promising levels of classification accuracy [37, 38], suggesting that network neuroscience indices may have future clinical utility as SCZ biomarkers [10]. Recent studies show that small-world brain networks are significantly reduced in SCZ compared to healthy controls across rest and task states [39], and the extent of this reduction may be associated with the length of illness [40]. Hence, SCZ is characterized by differences in the small-world architecture of functional brain organization, marked by a subtle randomization of network topology [41], even in the absence of significant findings for structural networks. Moreover, SCZ patients have significant reductions in connectivity [37, 41]. Large-scale organization features (small-world organization) seem to be less impacted in major depressive disorder (MDD) compared to SCZ. Instead MDD is characterized by disconnectivity across both static and dynamic measures of connectivity, and this is a potential future clinical utility factor [42].

Several brain abnormalities have been reported in SCZ by neuroimaging studies concerning the corpus callosum, thalamus, medial temporal lobe (hippocampal formation, subiculum, and parahippocampal gyrus), superior temporal gyrus (particularly on the left side), frontal lobe (particularly prefrontal and orbitofrontal regions), amygdala-hippocampal complex, cortical size, and size of the whole brain [43–47]. Conversely, there are a higher ventricle-to-brain ratio, greater absolute ventricular volume, and increased size of the cavum septi pellucidi [46].

SCZ patients usually have greater absolute volumes of all ventricular subdivisions, total ventricular volume, and relative volumes of basal ganglia structures (the left and right caudate, putamen, and globus pallidus) as well as reduced cerebral volume, relative volumes of the thalamus, and medial temporal lobe structures including the amygdala, the hippocampus/amygdala ratio, the hippocampus and parahippocampus, and the relative volume of the left anterior superior temporal gyrus [47].

Duration of untreated psychosis (DUP) has been associated with poor outcome in SCZ [48]. Recently, a naturalistic longitudinal study with matched healthy controls highlighted the function of hippocampal volume loss as a biomarker of DUP [49]. This leads to the idea that early hippocampal volume loss may play a role in mediating the association between DUP and poor outcomes in SCZ. Therefore, accelerated hippocampal volume loss could be associated to DUP and poor response in SCZ. Finally, in SCZ, there is also white matter disorganization in prefrontal and temporal white matter, corpus callosum, and uncinate fasciculus [50].

7.2.2 Major Depressive Disorders

Structural imaging works show anatomical and neuropathological abnormalities concerning the disruptions to cortico-striatal-limbic circuits in patients with MDD [51]. MDD patients have reduced metabolism or hypoactivity with "hypofrontality" of the DLPFC, in the left central executive network, along with increased activity in the subcallosal cingulate cortex and limbic regions, such as the amygdala and the insula. The increased inter-functional connectivity between the salience network and right executive network, and the decreased inter-functional connectivity between the anterior default mode network and right central executive network, could be considered as biomarkers of MDD (Fig. 7.2) [52, 53]. Yang et al. suggested a paradigm using a multiple classifier evaluation with external validation by diffusion MRI, to evaluate orientation and diffusion characteristics of white matter and, by inference, white matter microstructure. Although four features (mean fractional anisotropy in the right cuneus and left insula, asymmetry in the volume of the pars triangularis and cerebellum) were implicated across all analyses, low classification and prediction accuracy using these features indicated that they cannot represent the entire pathophysiology of MDD. However, they may be relevant for future investigations of MDD neurobiology in conjunction with other methods [54]. MDD patients show volumetric reductions in the hippocampus, basal ganglia, subcallosal cingulate cortex, and orbitofrontal cortex in patients with more severe or chronic forms of disease [55]. A neuroimaging meta-analysis highlighted reduced volumes of the right hippocampus and reduced gray matter volumes in the left DLPFC as structural imaging predictors of nonresponse to treatment [56].

7.2.3 Bipolar Disorders

The International Society for Bipolar Disorders Biomarkers Network Task Force has described the results of neuroimaging biomarker studies in bipolar disorder (BD) patients as loss of gray matter in cortical-cognitive brain network, as well as increased activation in ventral limbic regions in response to emotional stimuli [24]. Specifically as shown by morphometric measures, BD patients have amplification of the lateral and third ventricles after several manic episodes (Fig. 7.3) [57]; progressive decline in hippocampal, fusiform, and cerebellar gray matter density after frequent episodes; subregion-specific gray matter volume reductions in the prefrontal cortex; and increased rates of deep white matter hyperintensities [58]. BD patients have gray matter reductions in the left rostral anterior cingulate cortex and right fronto-insular cortex thickness, above all in anterior limbic regions (executive control and emotional processing abnormalities) [59], volumetric reductions in hippocampus and thalamus, and enlarged lateral ventricles [60]. Although gray/white matter changes appear early in BD development, the brain volume may be altered by environmental factors such as drugs [61].

Studies of fMRI point out excessive activation in numerous corticolimbic pathways, including overactivation of the amygdala, striatum, and thalamus [24, 62, 63]. Decreased activity in prefrontal cortical areas shown by imaging data underscores an insufficient modulation of limbic/subcortical regions, related to depressed mood and poor cognitive coping in BD [62, 63]. Recently, Li et al. [64] studied cortical thickness and subcortical volume alterations in euthymic BD type I patients treated with lithium and valproate. In particular, patients treated with lithium had increased cortical thickness of the left rostral middle frontal cortex and right superior frontal cortex compared with valproate, while cortical thickness was not different between BD patients on lithium treatment compared to healthy controls in the bilateral rostral middle frontal cortex. Moreover, there were no differences observed in subcortical volume. These data indicate that lithium and valproate have different effects on cortical thinning of the prefrontal cortex in BD but an analogous effect on subcortical volumes [64]. However, neuroimaging could be used as a potential biomarker for lithium response prediction in BD [65, 66]. In MRI studies, patients exposed to lithium treatment showed a bigger volume of gray matter mainly in the hippocampus as a direct consequence of the drug (neurotrophic and neuroprotective influence) or secondary to better symptomatic outcome [67, 68].

7.3 Inflammatory Biomarkers

There is increasing evidence on the involvement of inflammatory pathways in the pathophysiology of major psychiatric disorders including MDD, SCZ, and BD. Elevated levels of cytokines and C-reactive protein and alterations in serum molecules involved in pro-inflammatory and oxidative stress response and immune molecules, including hyperactivation of the hypothalamic-pituitary-adrenal (HPA) axis, have been demonstrated in these major psychiatric illnesses (Figs. 7.1, 7.2 and 7.3) [69, 70]. According to the recent scientific literature, anomalies in the immune system (blood or CSF levels of certain cytokines) are involved in the pathogenesis of SCZ, MDD, and BD and may be useful as biomarkers for diagnosis and treatment monitoring. Studies have also shown increased levels of peripheral pro-inflammatory markers related to the genes involved in regulation of the immune system in both SCZ and MDD [71–75].

In particular, increased levels of C-reactive protein (CRP) [23, 76] and increased levels of IL-1β, IL-6, IL-8, IL-10, IL-12, IL-15, IL18, endogenous IL-1 receptor antagonist (IL-1RA), and soluble IL-2 receptor (sIL-2R) in the blood, CSF, and serum have been found in SCZ patients [69, 77, 78] (Fig. 7.1). It should be noted that the potential of IL-2 has been a matter of controversy as it was found to be elevated in some studies and diminished in others [79, 80]. Other cytokines [tumor necrosis factor-α (TNF-α), transforming growth factor-β (TGF-β), and interferon γ] have been shown to be altered in SCZ, while epidermal growth factor (EGF) has been associated with an increased risk of developing SCZ [81].

More generally, according to recent meta-analysis, all patients with severe mental disorders have increased CSF levels of interleukin 1β (IL-1β), IL-6, and IL-8 [82]. Moreover, autoimmune dysregulation has been found to occur in BD II and MDD as underlined by proteomic analysis based on two-dimensional electrophoresis coupled with matrix-assisted laser desorption/ionization time-of-flight/time-of-flight tandem mass spectrometry analysis of plasma samples [83]. An area of particular attention in mental disorders is immunology linked to infections and autoimmune diseases with a larger risk identified for SCZ and affective disorders [84, 85].

The CSF/serum albumin ratio was known to be increased in SCZ, and affective disorders and total CSF protein levels were elevated, indicating increased blood-brain barrier (BBB) permeability [86]. Furthermore, the IgG ratio, IL-6 levels, and IL-8 levels are increased in the CSF of SCZ but not in the case of affective disorders [87–90]. A correlation of the levels of inflammation markers and symptoms has been found and also between albumin and IgG levels and the Scale for the Assessment of Negative Symptoms [91] and between IL-8 levels and the Montgomery-Asberg Depression Rating Scale [92]. Furthermore, altered chemokine levels were found in the CSF and plasma of suicide attempters [93].

A number of studies have found that HPA axis activation in MDD may be linked to the severity of illness. Moreover, MDD patients are at specific risk for cardiovascular syndromes, because of higher levels of inflammatory biomarkers such as the high sensitivity C-reactive protein and pro-inflammatory acute phase cytokines interleukin-1β and interleukin-6. For this reason, cytokines could be considered as biomarkers of depression severity [94]. The heat shock proteins CPN10, CPN60, and CPN70 might have potential as biomarkers for BD, and CPN60 blood level might distinguish patients with abnormal and normal HPA axis activities [95]. Among other biomarkers in BD, increased pro-inflammatory cytokines could be considered markers of mitochondrial dysfunction and oxidative stress (Fig. 7.3) [96].

7.4 Neurotrophic Biomarkers

The etiology of major psychiatric disorders has often been linked to altered intracellular signaling, synaptogenesis, and neuroplasticity. Over the last years, the role of brain-derived neurotrophic factor (BDNF) in cognitive impairments in psychiatric patients has become a focus of interest. BDNF is the most common neurotrophin in the human brain and is involved in the synthesis, differentiation, maintenance, and survival of neurons, both in the central and in the peripheral nervous systems [97]. According to some genomic studies, there is a correlation between the BDNF gene polymorphism (Val66Met) and SCZ as found by whole-blood polymerase chain reaction (PCR) studies [98], and this association was correlated with cognition [99, 100]. Additionally, BDNF Met alleles are associated with age of onset and with phenotype of aggressive behavior in SCZ [25, 101] (Fig. 7.1).

Postmortem studies have shown that the mRNA levels of BDNF and TrkB and BDNF protein levels are decreased in the hippocampus and prefrontal cortex of SCZ and major psychiatric disorders [102, 103]. Also, the levels of other neurotrophins such as nerve growth factor (NGF) and NGF receptor, vascular endothelial growth factor (VEGF), and neurotrophin-3 (NT-3) have been found to be reduced [25, 104–110]. Moreover, serum levels of BDNF can be influenced by pharmacotherapy. Generally, BDNF levels were found to be decreased in treated SCZ [111–115].

Recently, the differential levels of neuregulin-1 (NRG1), its receptor ErbB4, BDNF, DNA methyltransferases 1 (DNMT1), and ten-eleven translocation 1 (TET1) proteins in peripheral blood have exhibited promising efficiency for diagnosis of first episode psychosis [116].

7.5 Neurotransmitters Biomarkers

Considering the classical monoamine hypothesis of MDD, several studies conducted on CSF biomarkers for affective disorders have focused on the levels of 5-hydroxytryptamine (serotonin), dopamine, and noradrenaline and on the respective enzymes monoamine oxidases and catechol-*O*-methyltransferase involved in their degradation to 5-hydroxyindoleacetic acid (5-HIAA), homovanillic acid (HVA), and 3-methoxy-4-hydroxyphenylglycol (MHPG) [117–122]. In addition, peripheral metabolic disturbances have been found in MDD, suggesting that characteristic metabolic alterations associated with the pathogenesis of MDD may generate a detectable molecular phenotype in the blood using metabonomic methods [123]. Previous studies have also shown that perturbations in central and peripheral neurotransmitters are a hallmark of MDD. In particular, MDD patients showed disturbances in several neurotransmitters in the periphery and brain, including dopamine, glutamate, γ-aminobutyric acid (GABA), and serotonin which were thought to be involved in the pathogenesis of the disorder [124]. In this regard, plasma metabolite biomarkers (GABA, dopamine, tyramine, kynurenine) could be used to distinguish MDD subjects from healthy controls and BD patients with high accuracy [123, 124].

7.5.1 Dopaminergic System

The levels of dopamine uptake have been investigated as a potential biomarker in SCZ [125]. In addition, tyrosine hydroxylase (TH), dopamine transporter (DAT) mRNA [126, 127], HVA (a major metabolite of dopamine), and the dopamine D3 receptor (DRD3) mRNAs were found to be increased and DRD4 mRNA levels decreased in SCZ [25, 128]. Also brain functional imaging conducted with SPECT in SCZ patients showed elevated synaptic dopamine levels [129], increased numbers of postsynaptic dopamine receptors and signal transduction, and striatal amphetamine-induced dopamine release [130]. Regarding dopaminergic metabolites in MDD, a recent meta-analysis concluded that only CSF levels of HVA, and not those of 5-HIAA or MHPG, are reduced in MDD. Therefore, the potential utility of CSF HVA concentrations as a potential biomarker in MDD should be investigated further (Fig. 7.2) [131].

Another area of specific interest for neurotransmitter biomarkers is the association between neuroreceptor density and self-reported personality dimensions, to examine the neurobiology of the underlying behavioral phenotypes. As shown from recent molecular imaging studies, there are significant correlations between dopaminergic markers and specific behavioral traits. In particular, correlations were found between striatal D2R density and detachment, a measure of social avoidance and withdrawal [132–134]. Conversely, psychosis-related traits do not appear to be linked to D2R, but striatal amphetamine-induced dopamine release was found to be related to schizotypal personality traits [135]. Similarly an increase in dopamine release was reported in SCZ patients using the presynaptic marker [18F]DOPA [136].

7.5.2 Serotonergic System

Alterations in the cortical serotonergic system have been reported in SCZ patients [137], such as the findings of decreased levels of the 5-HTT receptors in the frontal cortex [138–140]. Altered levels of 5-HT1A and 5-HT1B and reduced 5HT2A receptors have been reported in prefrontal cortex and hippocampus of BD and MDD patients [141]. In addition, plasma serotonin levels have been found to be decreased, while platelet serotonin levels were found to be higher in SCZ [142].

The study of the biological underpinnings of personality traits with the use of molecular imaging techniques has several advantages for the early stages, evolution, and treatment of psychiatric diseases. In particular, these methods can be used to examine the relationship between serotonin receptor availability, social trust, and status as potential novel biomarkers in psychiatry. Molecular imaging studies of associations between serotonin receptors and transporters with personality traits, such as neuroticism, have not been clear. Although the association between the 5-HT1A receptor and neuroticism was found to have a strong negative correlation, there were no associations with the serotonin transporter [143].

7.5.3 Glutamate and Other Amino Acid Systems

SCZ patients show decreased levels of glutamate, glycine, and D-serine in the CSF and plasma, but increased homocysteine [144–146]. An important focus in SCZ is glutamatergic dysfunction, in particular *N*-methyl-D-aspartate (NMDA) receptor hypofunction, as this can be informative about several SCZ symptoms linked to excitatory-to-inhibitory imbalance. In this way, administration of the NMDA receptor antagonist ketamine leads to SCZ-like positive, negative, and cognitive symptoms [147].

7.5.4 GABAergic System and Neurosteroids

SCZ patients have been found to display decreased plasma levels of GABA, with downregulation of the GABA-A receptor alpha 5 subunit in prefrontal regions and polymorphisms and haplotypes in the GABA-A receptor β2 subunit gene [148–150]. According to the specific role in modulating the GABA receptor, the deficiency of the biosynthesis of allopregnanolone, a positive allosteric modulator of GABA action at GABA-A receptors, was found in several neuropsychiatric disorders such as MDD, post-traumatic stress disorder (PTSD), epilepsy, postpartum depression, and anorexia nervosa, as well as in premenstrual syndrome and obesity [151–156]. The special focus on neurosteroids, inhibitors of NMDA-mediated tonic neurotransmission [157], was confirmed in women with post-traumatic stress disorder (PTSD) through an association with a block in conversion of progesterone to the GABAergic neurosteroids allopregnanolone and pregnanolone [158]. This is important for potential therapeutics in PTSD considering the role of the endocannabinoid system and associated neurosteroids in this condition [159].

7.5.5 Cholinergic System

A number of studies have demonstrated involvement of the cholinergic system in psychiatric disorders. For example, studies have shown that the nicotinic and muscarinic receptors are reduced in thalamus and frontal regions of SCZ [160–162]. Thus, studies of these systems may also lead to identification of novel biomarkers and drug targets in these diseases.

7.6 Epigenetics

Epigenetics or epigenomics is a modification of the genome expression without changes in the DNA sequence and can result in alterations of gene expression, allowing for differential expression of common genetic information [163]. New techniques such as genomics, epigenomics, transcriptomics, and proteomics guarantee a more global examination of stress-related dysregulation, allowing the discovery of novel biomarkers and targets for new therapies, compared to standard biochemical analyses. Many psychiatric patients have alterations in stress response and stress reactivity levels, influenced by biological moderating factors such as the HPA axis and early life trauma [164]. Stress hormones (glucocorticoids) and immune mediators (cytokines) provide a connection between the peripheral and central pathways and have exemplified functional biomarkers of stress response, as found in PTSD [165]. The link is further demonstrated by the finding that affective and psychotic patients have elevated cortisol secretion and an enlarged pituitary gland volume, with hyperactivity of the HPA axis [166–168].

Recently, molecular examinations have discovered aberrant microRNA expression in different biological samples from psychiatric patients, including brain tissue, plasma, serum, and peripheral blood mononuclear cells. Such microRNA alterations may be useful biomarkers in studies of MDD, SCZ, or BD [169–171] as certain gene expression patterns are present in subgroups of patients [172–183]. A recent meta-analysis found that the utilization of blood-derived microRNAs, especially those from peripheral blood mononuclear cells isolated from patients, may lead to a useful set of biomarkers for SCZ diagnosis [184]. Also, the candidate gene targets of these microRNAs have been linked to increased risk for developing BD, including pathways such as circadian rhythm, neuronal development, and calcium metabolism [25].

MicroRNAs are ~22-nucleotide-long, noncoding RNA molecules, which are important regulators of posttranscriptional gene expression. They may lead to increased or decreased regulation of the translational stage of mRNA processing or render it stable or unstable [185].

MicroRNA-16 is a posttranscriptional repressor of the serotonin transporter (SERT) and acts as a central regulator of SERT expression. It provides a mechanism for adaptive changes in SERT expression in monoaminergic neurons, which can differentiate into either serotonergic (1C115-HT) or noradrenergic (1C11NE neuroectodermal cell line) neuronal cells [186]. MicroRNA-134 represses the translation of the Limk1 mRNA, a protein kinase that influences dendritic spine development. The miRNA-mediated repression of translation occurs via exogenous stimuli like BDNF, which has emerged as a key mediator for synaptic efficacy, neuronal connectivity, and neuroplasticity [187]. Interestingly, one study showed that microRNA-134 levels in BD were inversely correlated with severity of manic symptoms [187].

Chromosome 8p, which contains at least seven transcribed microRNAs, has been linked to neurodevelopmental disorders such as autism and SCZ. Patients with DiGeorge syndrome and 22q11.2 deletion have a deficiency in DGCR8 microprocessor complex subunit expression, resulting in decreased microRNA biosynthesis and leading to a 30-fold increased risk of SCZ [187, 188]. The functional targets of these microRNAs include a number of genes that have been implicated in SCZ, such as BDNF, the dopamine D1 receptor, the synaptic protein neuregulin-1 (NRG1), and the early growth response gene 3 (EGR3) [188]. Furthermore microRNA-219 has been found to negatively regulate the function of NMDA receptors, serving as an integral component of the NMDA receptor signaling cascade. MicroRNA-219 may directly modulate NMDA receptor signaling by regulating the expression of components in this cascade [188].

7.7 Pharmacogenomic Biomarkers

An important type of pharmacogenomic biomarker individuation in psychotropic drug classification relates to the cytochrome P450 enzyme family [12, 13, 189]. These enzymes play a critical role in drug metabolism and therefore may be important in efficacy- and toxicity-related issues. Interestingly, a majority of the commercially available pharmacogenomic testing resources assay for CYP2D6 and CYP2C19, considering that these enzymes are involved in metabolism of many commercial drugs and variants, exist which could affect their activities with respect to specific drugs. Pharmacogenomics could be useful in determining dosage and administration, warnings, precautions, or other areas listed on the package insert of commercially available drugs. This will be helpful in providing information at the personalized level to minimize adverse events, to provide genotype-specific dosing, and to identify polymorphic drug targets and genes [15, 27, 190] (Fig. 7.2).

7.8 Electrophysiological Biomarkers

The autonomous nervous system (ANS) and its imbalance is important in physiological and pathological disorders [191, 192], including stress. Accordingly, resting heart rate (RHR), heart rate variability (HRV), respiration rate (RR), skin temperature (ST), and skin conductance (SC) are common clinical methods to measure ANS activity, and HRV is the most established parameter to evaluate the sympatho-vagal balance [193–197]. Recent studies show that useful stress indexes may also be obtained from electroencephalogram (EEG)-based features [198].

Psychiatric patients have an ANS imbalance, especially in psychosis [199–201] and depression [202–204]. There are also sympatho-vagal alterations in patients affected by anxiety and phobic anxiety, social anxiety and somatoform disorders [205–207], alcohol dependence [208, 209], and cognitive impairment [210]. Considering intraindividual variability, electrophysiological parameters could be possible biomarkers in psychiatry, even if some parameters (RR, RHR, LF, and HF parameters of HRV) are more robust and stable over time than others (SC, ST, time domain parameters of HRV), and RHR and RR are easy to obtain in everyday clinical practice and can be used as measures of ANS dysregulation [211]. Certainly, two or more different parameters should be evaluated to moderate intraindividual variability [211].

Electrophysiological changes, including the components of sensory gating, mismatch negativity (MMN), and P300 of the evoked potentials are consistently reported to be abnormal in SCZ [212].

7.9 Gut Microbiota

The microbiota is composed of over 100 trillion of commensal bacteria in symbiosis with human body, in the distal gut and fecal metabolites, and can be examined with metabolomic analysis [nuclear magnetic resonance (NMR) spectroscopy] of fecal water [213].

The gut and the brain are strictly connected through bidirectional signaling pathways [214]. Bacteria can produce GABA, tryptophan, 5-HT, and several neurotransmitters and monoamines. Therefore, the gut microbiota could regulate many activities within the brain including hippocampal neurogenesis, myelin-related gene expression in the PFC (an important brain region involved in anxiety and social behavior), CNS serotonergic neurotransmission, and stress and antidepressant treatment response [215–217]. The gut microbiota could also control brain functional pathways through inflammasome signaling and could therefore be useful as both biomarkers and potential drug targets in psychiatry [218]. Moreover, in epigenetic studies of SCZ, the impact of microbiota should also be taken into consideration [219, 220] (Fig. 7.1).

MDD patients have an increase in gut microbiota alpha diversity, in the genera *Eggerthella*, *Holdemania*, *Gelria*, *Turicibacter*, *Paraprevotella*, and *Anaerofilm*, with overrepresentation of *Bacteroidales*, *Oscillibacter*, and *Alistipes*, reductions in *Prevotella* and *Dialister*, and lower numbers of *Bifidobacterium*, *Lachnospiraceae*, and *Lactobacillus* [221–223], with high levels of serum IgM and IgA against lipopolysaccharide of gram-negative gut commensals. This is coherent with the pathophysiology of psychiatric illnesses linked to bacterial translocation, through increased gut permeability [224]. Interestingly, diet and depression are strongly linked through the gut microbiota. Dietary fiber can modify the composition of the intestinal flora and affect brain and behavior [225]. Indeed, higher intake of dietary fiber (fruits and vegetables) leads to a lower prevalence of MDD [226]. Specifically, the Mediterranean diet could be protective, while the Western diet could increase risk of MDD through effects on the microbiota [227]. A probiotic combination of *Lactobacillus helveticus* R0052 and *Bifidobacterium longum* R0175 has been shown to have beneficial effects in human resilience to stress [228]. A recent systematic review on the fecal microbiota concluded that Archaeon *Methanobrevibacter smithii* is increased in anorexia nervosa patients [229]. *Methanobrevibacter smithii* may be a benchmark biomarker for future studies.

7.10 Conclusions

Psychiatry is in need of an objective, valid diagnostic classification that transcends the Diagnostic and Statistical Manual (DSM) model of symptom clusters. The US National Institute of Mental Health (NIMH) Research Domain Criteria (RDoC) [230] has called for the inclusion of biological markers for either diagnosis or treatment outcomes. However, there have been many criticisms, and, at present, there are no accepted specific biomarkers in psychiatry [230, 231].

A focal point of medicine is the search of biomarkers to aid correct diagnosis, risk prognosis, and prediction of response to treatment. In the case of psychiatric disorders, it is important to have clear criteria for distinguishing pathological behaviors and appropriate methods to categorize these diseases and facilitate earlier intervention for better outcomes. And one of the most important aims in psychiatric medicine is that of personalized treatment for prediction of response and therapeutic or adverse effects at the level of the individual [232]. In summary, we should view with optimism our capabilities to develop biomarkers that will ultimately lead to new interventions and personalized medicines and transform our ability to prevent illness onset and treat complex psychiatric disorders more effectively [232].

Considering the complex interactions among genotype, lifestyle, diet, pharmacological therapy, environmental exposure, and gut microflora, the most ambitious goals could be the discovery of novel pharmacological targets and to rationalize the utilization of known drugs. Finally, this chapter underlines important advices for future studies, to create a link between several types of biomarkers considering that psychiatric disorders are complex diseases. Thus, the use of a single biomarker is not advised, but rather a combination of diverse biomarker types. This could lead to improved treatment of psychiatric patients on a personalized level for the best possible outcomes.

References

1. Katahira K, Yamashita Y (2017) A theoretical framework for evaluating psychiatric research strategies. Comput Psychiatr 1:184–207
2. Biomarkers Definitions Working Group (2001) Biomarkers and surrogate endpoints: preferred definitions and conceptual framework. Clin Pharmacol Ther 69(3):89–95
3. Biomarkers and risk assessment: concepts and principles/published under the joint sponsorship of the United Nations environment Programme, the International Labour Organisation, and the World Health Organization (1993) http://www.who.int/iris/handle/10665/39037
4. Boksa P (2013) A way forward for research on biomarkers for psychiatric disorders. J Psychiatry Neurosci 38(2):75–77
5. Quevedo J, Yatham LN (2018) Biomarkers in mood disorders: are we there yet? J Affect Disord 233:1–2
6. Scarr E, Millan MJ, Bahn S, Bertolino A, Turck CW, Kapur S et al (2015) Biomarkers for psychiatry: the journey from fantasy to fact, a report of the 2013 CINP Think Tank. Int J Neuropsychopharmacol 18(10):pyv042. https://doi.org/10.1093/ijnp/pyv042
7. Singh I, Rose N (2009) Biomarkers in psychiatry. Nature 460(7252):202–207
8. Dean B (2011) Dissecting the syndrome of schizophrenia: progress toward clinically useful biomarkers. Schizophr Res Treat 2011:614730. https://doi.org/10.1155/2011/614730
9. McGorry P, Keshavan M, Goldstone S, Amminger P, Allott K, Berk M et al (2014) Biomarkers and clinical staging in psychiatry. World Psychiatry 13(3):211–223
10. Lydon-Staley DM, Bassett DS (2018) Network neuroscience: a framework for developing biomarkers in psychiatry. Curr Top Behav Neurosci 40:79-109
11. Bora E, Yucel M, Pantelis C (2009) Cognitive endophenotypes of bipolar disorder: a meta-analysis of neuropsychological deficits in euthymic patients and their first-degree relatives. J Affect Disord 113(1–2):1–20

12. Snyder HR (2013) Major depressive disorder is associated with broad impairments on neuropsychological measures of executive function: a meta-analysis and review. Psychol Bull 139(1):81–132
13. Porcelli S, Van Der Wee N, van der Werff S, Aghajani M, Glennon JC, van Heukelum S et al (2018) Social brain, social dysfunction and social withdrawal. Neurosci Biobehav Rev Sep 20. doi: 10.1016/j.neubiorev.2018.09.012. [Epub ahead of print]
14. Lozupone M, Panza F, Piccininni M, Copetti M, Sardone R, Imbimbo BP et al (2018) Social dysfunction in older age and relationships with cognition, depression, and apathy: the GreatAGE Study. J Alzheimers Dis 65(3):989–1000
15. Tansey KE, Guipponi M, Perroud N, Bondolfi G, Domenici E, Evans D et al (2012) Genetic predictors of response to serotonergic and noradrenergic antidepressants in major depressive disorder: a genome-wide analysis of individual-level data and a meta-analysis. PLoS Med 9(10):e1001326. https://doi.org/10.1371/journal.pmed.1001326
16. Rabinowitz J, Werbeloff N, Caers I, Mandel FS, Stauffer V, Ménard F et al (2014) Determinants of antipsychotic response in schizophrenia: implications for practice and future clinical trials. J Clin Psychiatry 75(4):e308–e316
17. Panza F, Lozupone M, Stella E, Lofano L, Gravina C, Urbano M (2016) Psychiatry meets pharmacogenetics for the treatment of revolving door patients with psychiatric disorders. Expert Rev Neurother 16(12):1357–1369
18. Panza F, Lozupone M, Stella E, Miscio G, La Montagna M, Daniele A (2016) The pharmacogenetic road to avoid adverse drug reactions and therapeutic failures in revolving door patients with psychiatric illnesses: focus on the CYP2D6 isoenzymes. Expert Rev Precis Med Drug Dev 1(5):431–442
19. Hurko O (2009) The uses of biomarkers in drug development. Ann N Y Acad Sci 1180:1–10
20. Lozupone M, Seripa D, Stella E, La Montagna M, Solfrizzi V, Quaranta N et al (2017) Innovative biomarkers in psychiatric disorders: a major clinical challenge in psychiatry. Expert Rev Proteomics 14(9):809–824
21. Petterson RF (1990) Mapping and sequencing the human genome. Ann Med 22(6):371–373
22. Higa GS, De Sousa E, Walter LT, Kinjo ER, Resende RR, Kihara AH (2014) MicroRNAs in neuronal communication. Mol Neurobiol 49(3):1309–1326
23. Mohammadi A, Rashidi E, Amooeian VG (2018) Brain, blood, cerebrospinal fluid, and serum biomarkers in schizophrenia. Psychiatry Res 265:25–38
24. Sigitova E, Fišar Z, Hroudová CT, Raboch J (2017) Biological hypotheses and biomarkers of bipolar disorder. Psychiatry Clin Neurosci 71(2):77–103
25. Hacimusalar Y, Eşel E (2018) Suggested biomarkers for major depressive disorder. Noro Psikiyatr Ars 55(3):280–290
26. Strawbridge R, Young AH, Cleare AJ (2017) Biomarkers for depression: recent insights, current challenges and future prospects. Neuropsychiatr Dis Treat 13:1245–1262
27. Alawieh A, Zaraket FA, Li JL, Mondello S, Nokkari A, Razafsha M et al (2012) Systems biology, bioinformatics, and biomarkers in neuropsychiatry. Front Neurosci 6:187. https://doi.org/10.3389/fnins.2012.00187
28. Han KM, Kim YK (2018) Promising neural diagnostic biomarkers and predictors of treatment outcomes for psychiatric disorders: novel neuroimaging approaches. Prog Neuropsychopharmacol Biol Psychiatry. pii: S0278-5846(18)30773-5. https://doi.org/10.1016/j.pnpbp.2018.10.001
29. Pasternak O, Kelly S, Sydnor VJ (2018) Advances in microstructural diffusion neuroimaging for psychiatric disorders. NeuroImage 182:259–282. https://doi.org/10.1016/j.neuroimage.2018.04.051
30. Shenton ME, Javadpour A, Mohammadi A (2016) Improving brain magnetic resonance image (MRI) segmentation via a novel algorithm based on genetic and regional growth. J Biomed Phys Eng 6(2):95–108
31. Zipursky RB (2007) Imaging mental disorders in the 21st century. Can J Psychiatr 52(3):133–134
32. Bailey DL, Pichler BJ, Gückel B, Barthel H, Beer AJ, Botnar R et al (2016) Combined PET/MRI: from Status Quo to Status Go. Summary report of the fifth international work-

shop on PET/MR imaging, February 15–19, 2016, Tübingen, Germany. Mol Imaging Biol 18(5):637–650
33. Ngounou Wetie AG, Sokolowska I, Wormwood K, Beglinger K, Michel TM, Thome J (2013) Mass spectrometry for the detection of potential psychiatric biomarkers. J Mol Psychiatry 1(1):8. https://doi.org/10.1186/2049-9256-1-8
34. Ross CA, Margolis RL, Reading SA, Pletnikov M, Coyle JT (2006) Neurobiology of schizophrenia. Neuron 52(1):139–153
35. Niznikiewicz MK, Kubicki M, Shenton ME (2003) Recent structural and functional imaging findings in schizophrenia. Curr Opin Psychiatry 16:123–147
36. Cannon TD, Glahn DC, Kim J, Van Erp TG, Karlsgodt K, Cohen MS et al (2005) Dorsolateral prefrontal cortex activity during maintenance and manipulation of information in working memory in patients with schizophrenia. Arch Gen Psychiatry 62:1071–1080
37. Wang L, Metzak PD, Honer WG, Woodward TS (2010) Impaired efficiency of functional networks underlying episodic memory-for-context in schizophrenia. J Neurosci 30(39):13171–13179
38. Anderson A, Cohen MS (2013) Decreased small-world functional network connectivity and clustering across resting state networks in schizophrenia: an fMRI classification tutorial. Front Hum Neurosci 7:520. https://doi.org/10.3389/fnhum.2013.00520
39. Liu Y, Liang M, Zhou Y, He Y, Hao Y, Song M et al (2008) Disrupted small-world networks in schizophrenia. Brain 131(4):945–961
40. Fornito A, Yoon J, Zalesky A, Bullmore ET, Carter CS (2011) General and specific functional connectivity disturbances in first-episode schizophrenia during cognitive control performance. Biol Psychiatry 70(1):64–72
41. He H, Sui J, Yu Q, Turner JA, Ho BC, Sponheim SR et al (2012) Altered small-world brain networks in schizophrenia patients during working memory performance. PLoS One 7(6):e38195. https://doi.org/10.1371/journal.pone.0038195
42. Demirtaş M, Tornador C, Falcon C, López-Solà M, Hernández-Ribas R, Pujol J et al (2016) Dynamic functional connectivity reveals altered variability in functional connectivity among patients with major depressive disorder. Hum Brain Mapp 37(8):2918–2930
43. Lawrie SM, Abukmeil SS (1998) Brain abnormality in schizophrenia. A systematic and quantitative review of volumetric magnetic resonance imaging studies. Br J Psychiatry 172:110–120
44. Nelson MD, Saykin AJ, Flashman LA, Riordan HJ (1998) Hippocampal volume reduction in schizophrenia as assessed by magnetic resonance imaging: a meta-analytic study. Arch Gen Psychiatry 55:433–440
45. Honea R, Crow TJ, Passingham D, Mackay CE (2005) Regional deficits in brain volume in schizophrenia: a meta-analysis of voxel-based morphometry studies. Am J Psychiatry 162:2233–2245
46. Van Horn JD, Mcmanus IC (1992) Ventricular enlargement in schizophrenia. A metaanalysis of studies of the ventricle: brain ratio (VBR). Br J Psychiatry 160:687–697
47. Wright IC, Rabe-Hesketh S, Woodruff PW, David AS, Murray RM, Bullmore ET (2000) Meta-analysis of regional brain volumes in schizophrenia. Am J Psychiatry 157:16–25
48. Penttila M, Jaaskelainen E, Hirvonen N, Isohanni M, Miettunen J (2014) Duration of untreated psychosis as predictor of long-term outcome in schizophrenia: systematic review and meta-analysis. Br J Psychiatry 205(2):88–94
49. Goff DC, Zeng B, Ardekani BA, Diminich ED, Tang Y, Fan X et al (2018) Association of hippocampal atrophy with duration of untreated psychosis and molecular biomarkers during initial antipsychotic treatment of first-episode psychosis. JAMA Psychiat 75(4):370–378
50. Kubicki M, Mccarley R, Westin CF, Park HJ, Maier S, Kikinis R et al (2007) A review of diffusion tensor imaging studies in schizophrenia. J Psychiatr Res 41:15–30
51. Mayberg HS (2003) Modulating dysfunctional limbic-cortical circuits in depression: towards development of brain-based algorithms for diagnosis and optimized treatment. Br Med Bull 65:193–207
52. Dong D, Ming Q, Zhong X, Pu W, Zhang X, Jiang Y et al (2018) State-independent alterations of intrinsic brain network in current and remitted depression. Prog Neuropsychopharmacol Biol Psychiatry 89:475-480

53. Ressler KJ, Mayberg HS (2007) Targeting abnormal neural circuits in mood and anxiety disorders: from the laboratory to the clinic. Nat Neurosci 10:1116–1124
54. Yang J, Zhang M, Ahn H, Zhang Q, Jin TB, Li IA (2018) Development and evaluation of a multimodal marker of major depressive disorder. Hum Brain Mapp 39(11):4420–4439
55. Lorenzetti V, Allen NB, Fornito A, Yucel M (2009) Structural brain abnormalities in major depressive disorder: a selective review of recent MRI studies. J Affect Disord 117:1–17
56. Fu CH, Steiner H, Costafreda SG (2012) Predictive neural biomarkers of clinical response in depression: a meta-analysis of functional and structural neuroimaging studies of pharmacological and psychological therapies. Neurobiol Dis 52:75–83
57. Strakowski SM, DelBello MP, Zimmerman ME, Getz GE, Mills NP, Ret J et al (2002) Ventricular and periventricular structural volumes in first- versus multiple-episode bipolar disorder. Am J Psychiatry 159(11):1841–1847
58. Moorhead TW, McKirdy J, Sussmann JE, Hall J, Lawrie SM, Johnstone EC et al (2007) Progressive gray matter loss in patients with bipolar disorder. Biol Psychiatry 62(8):894–900
59. Bora E, Fornito A, Yücel M, Pantelis C (2010) Voxelwise meta-analysis of gray matter abnormalities in bipolar disorder. Biol Psychiatry 67:1097–1105
60. Hibar DP, Westlye LT, van Erp TG, Rasmussen J, Leonardo CD, Faskowitz J et al (2016) Subcortical volumetric abnormalities in bipolar disorder. Mol Psychiatry 21(12):1710–1171
61. Maletic V, Raison C (2014) Integrated neurobiology of bipolar disorder. Front Psych 5:98. https://doi.org/10.3389/fpsyt.2014.00098
62. Cerullo MA, Adler CM, Delbello MP, Strakowski SM (2009) The functional neuroanatomy of bipolar disorder. Int Rev Psychiatry 21:314–322
63. Vargas C, López-Jaramillo C, Vieta E (2013) A systematic literature review of resting state network—functional MRI in bipolar disorder. J Affect Disord 150:727–735
64. Li L, Ji E, Han X, Tang F, Bai Y, Peng D et al (2018) Cortical thickness and subcortical volumes alterations in euthymic bipolar I patients treated with different mood stabilizers. Brain Imaging Behav Aug 25. https://doi.org/10.1007/s11682-018-9950-9. [Epub ahead of print]
65. Alda M (2015) Lithium in the treatment of bipolar disorder: pharmacology and pharmacogenetics. Mol Psychiatry 20(6):661–670
66. Fleck DE, Ernest N, Adler CM, Cohen K, Eliassen JC, Norris M et al (2017) Prediction of lithium response in first-episode mania using the LITHium Intelligent Agent (LITHIA): pilot data and proof-of-concept. Bipolar Disord 19(4):259–272
67. Houenou J, Boisgontier J, Henrion A, d'Albis MA, Dumaine A, Linke J et al (2017) A multilevel functional study of a SNAP25 at-risk variant for bipolar disorder and schizophrenia. J Neurosci 37(43):10389–10397
68. Keener MT, Phillips ML (2007) Neuroimaging in bipolar disorder: a critical review of current findings. Curr Psychiatry Rep 9(6):512–520
69. De Witte L, Tomasik J, Schwarz E, Guest PC, Rahmoune H, Kahn RS et al (2014) Cytokine alterations in first episode schizophrenia patients before and after antipsychotic treatment. Schizophr Res 154(1–4):23–29
70. Bauer IE, Pascoe MC, Wollenhaupt-Aguiar B, Kapczinski F, Soares JC (2014) Inflammatory mediators of cognitive impairment in bipolar disorder. J Psychiatr Res 56:18–27
71. Nikkila H, Müller K, Ahokas A, Miettinen K, Andersson LC, Rimón R (1995) Abnormal distributions of T-lymphocyte subsets in the cerebrospinal fluid of patients with acute schizophrenia. Schizophr Res 14:215–221
72. Müller N, Ackenheil M (1995) Immunoglobulin and albumin content of cerebrospinal fluid in schizophrenic patients: relationship to negative symptomatology. Schizophr Res 14:223–228
73. Howren MB, Lamkin DM, Suls J (2009) Associations of depression with C-reactive protein, IL-1, and IL-6: a meta-analysis. Psychosom Med 71:171–186
74. Haapakoski R, Mathieu J, Ebmeier KP, Alenius H, Kivimäki M (2015) Cumulative meta-analysis of interleukins 6 and 1β, tumour necrosis factorα and C-reactive protein in patients with major depressive disorder. Brain Behav Immun 49:206–215
75. Ripke S, Neale BM, Corvin A, Walters JTR, Farh K-H, Holmans PA et al (2014) Biological insights from 108 schizophrenia-associated genetic loci. Nature 511:421–427

76. Miller BJ, Culpepper N, Rapaport MH (2014) C-reactive protein levels in schizophrenia: a review and meta-analysis. Clin Schizophr Relat Psychoses 7(4):223–230
77. Wan C, La Y, Zhu H, Yang Y, Jiang L, Chen Y et al (2007) Abnormal changes of plasma acute phase proteins in schizophrenia and the relation between schizophrenia and haptoglobin (Hp) gene. Amino Acids 32:101–108
78. Van Kammen DP, Mcallister-Sistilli CG, Kelley ME, Gurklis JA, Yao JK (1999) Elevated interleukin-6 in schizophrenia. Psychiatry Res 87:129–136
79. Zhang XY, Zhou DF, Cao LY, Zhang PY, Wu GY (2002) Decreased production of interleukin-2 (IL-2), IL-2 secreting cells and CD4+ cells in medication-free patients with schizophrenia. J Psychiatr Res 36:331–336
80. Asevedo E, Rizzo LB, Gadelha A, Mansur RB, Ota VK, Berberian AA et al (2014) Peripheral interleukin-2 level is associated with negative symptoms and cognitive performance in schizophrenia. Physiol Behav 129:194–198
81. Nawa H, Takahashi M, Patterson PH (2000) Cytokine and growth factor involvement in schizophrenia–support for the developmental model. Mol Psychiatry 5:594–603
82. Wang AK, Miller BJ (2018) Meta-analysis of cerebrospinal fluid cytokine and tryptophan catabolite alterations in psychiatric patients: comparisons between schizophrenia, bipolar disorder, and depression. Schizophr Bull 44(1):75–83
83. Chen J, Huang C, Song Y, Shi H, Wu D, Yang Y et al (2015) Comparative proteomic analysis of plasma from bipolar depression and depressive disorder: identification of proteins associated with immune regulatory. Protein Cell 6(12):908–911
84. Orlovska-Waast S, Köhler-Forsberg O, Brix SW, Nordentoft M, Kondziella D, Krogh J et al (2018) Cerebrospinal fluid markers of inflammation and infections in schizophrenia and affective disorders: a systematic review and meta-analysis. Mol Psychiatry. https://doi.org/10.1038/s41380-018-0220-4. [Epub ahead of print]
85. Benros ME, Mortensen PB, Eaton WW (2012) Autoimmune diseases and infections as risk factors for schizophrenia. Ann N Y Acad Sci 1262:56–66
86. Kirch DG, Alexander RC, Suddath RL, Papadopoulos NM, Kaufmann CA, Daniel DG et al (1992) Blood-CSF barrier permeability and central nervous system immunoglobulin G in schizophrenia. J Neural Transm 89:219–232
87. Stich O, Andres TA, Gross CM, Gerber SI, Rauer S, Langosch JM (2015) An observational study of inflammation in the central nervous system in patients with bipolar disorder. Bipolar Disord 17:291–302
88. Endres D, Perlov E, Baumgartner A, Hottenrott T, Dersch R, Stich O et al (2015) Immunological findings in psychotic syndromes: a tertiary care hospital's CSF sample of 180 patients. Front Hum Neurosci 9:476. https://doi.org/10.3389/fnhum.2015.00476
89. Kirch DG, Kaufmann CA, Papadopoulos NM, Martin B, Weinberger DR (1985) Abnormal cerebrospinal fluid protein indices in schizophrenia. Biol Psychiatry 20:1039–1046
90. Müller N, Dobmeier P, Empl M, Riedel M, Schwarz M, Ackenheil M (1997) Soluble IL-6 receptors in the serum and cerebrospinal fluid of paranoid schizophrenic patients. Eur Psychiatry 12:294–299
91. Pazzaglia PJ, Post RM, Rubinow D, Kling MA, Huggins TS, Sunderland T (1995) Cerebrospinal fluid total protein in patients with affective disorders. Psychiatry Res 57:259–266
92. Kern S, Skoog I, Börjesson-Hanson A, Blennow K, Zetterberg H, Östling S et al (2014) Higher CSF interleukin-6 and CSF interleukin-8 in current depression in older women. Results from a population-based sample. Brain Behav Immun 41:55–58
93. Janelidze S, Ventorp F, Erhardt S, Hansson O, Minthon L, Flax J et al (2013) Altered chemokine levels in the cerebrospinal fluid and plasma of suicide attempters. Psychoneuroendocrinology 38:853–862
94. Teixeira AL, Barbosa IG, Machado-Vieira R, Rizzo LB, Wieck A, Bauer ME (2013) Novel biomarkers for bipolar disorder. Expert Opin Med Diagn 7(2):147–159
95. Baghai TC, Varallo-Bedarida G, Born C, Häfner S, Schüle C, Eser D et al (2018) Classical risk factors and inflammatory biomarkers: one of the missing biological links between car-

diovascular disease and major depressive disorder. Int J Mol Sci 19(6):E1740. https://doi.org/10.3390/ijms19061740
96. Cheng Y, Li Z, He S, Tian Y, He F, Li W (2018) Elevated heat shock proteins in bipolar disorder patients with hypothalamic pituitary adrenal axis dysfunction. Medicine (Baltimore) 97(27):e11089. https://doi.org/10.1097/MD.0000000000011089
97. Kuipers SD, Bramham CR (2006) Brain-derived neurotrophic factor mechanisms and function in adult synaptic plasticity: new insights and implications for therapy. Curr Opin Drug Discov Devel 9:580–586
98. Neves-Pereira M, Cheung JK, Pasdar A, Zhang F, Breen G, Yates P et al (2005) BDNF gene is a risk factor for schizophrenia in a Scottish population. Mol Psychiatry 10:208–212
99. Numata S, Ueno S, Iga J, Yamauchi K, Hongwei S, Ohta K et al (2006) Brain derived neurotrophic factor (BDNF) Val66Met polymorphism in schizophrenia is associated with age at onset and symptoms. Neurosci Lett 401:1–5
100. Rybakowski JK, Borkowska A, Skibinska M, Szczepankiewicz A, Kapelski P, Leszczynska-Rodziewicz A et al (2006) Prefrontal cognition in schizophrenia and bipolar illness in relation to Val66Met polymorphism of the brain-derived neurotrophic factor gene. Psychiatry Clin Neurosci 60:70–76
101. Spalletta G, Morris DW, Angelucci F, Rubino IA, Spoletini I, Bria P et al (2010) BDNF Val66Met polymorphism is associated with aggressive behavior in schizophrenia. Eur Psychiatry 25:311–313
102. Issa G, Wilson C, Terry AV Jr, Pillai A (2010) An inverse relationship between cortisol and BDNF levels in schizophrenia: data from human postmortem and animal studies. Neurobiol Dis 39:327–333
103. Thompson Ray M, Weickert CS, Wyatt E, Webster MJ (2011) Decreased BDNF, trkBTK+ and GAD67 mRNA expression in the hippocampus of individuals with schizophrenia and mood disorders. J Psychiatry Neurosci 36:195–203
104. Durany N, Michel T, Zochling R, Boissl KW, Cruz-Sanchez FF, Riederer P et al (2001) Brain-derived neurotrophic factor and neurotrophin 3 in schizophrenic psychoses. Schizophr Res 52:79–86
105. Fulzele S, Pillai A (2009) Decreased VEGF mRNA expression in the dorsolateral prefrontal cortex of schizophrenia subjects. Schizophr Res 115:372–373
106. Kale A, Joshi S, Pillai A, Naphade N, Raju M, Nasrallah H et al (2009) Reduced cerebrospinal fluid and plasma nerve growth factor in drug-naive psychotic patients. Schizophr Res 115:209–214
107. Zakharyan R, Atshemyan S, Gevorgyan A, Boyajyan A (2014) Nerve growth factor and its receptor in schizophrenia. BBA Clin 1:24–29
108. Lee BH, Hong JP, Hwang JA, Ham BJ, Na KS, Kim WJ et al (2015) Alterations in plasma vascular endothelial growth factor levels in patients with schizophrenia before and after treatment. Psychiatry Res 228:95–99
109. Rao S, Martinez-Cengotitabengoa M, Yao Y, Guo Z, Xu Q, Li S et al (2017) Peripheral blood nerve growth factor levels in major psychiatric disorders. J Psychiatr Res 86:39–45
110. Toyooka K, Asama K, Watanabe Y, Muratake T, Takahashi M, Someya T et al (2002) Decreased levels of brain-derived neurotrophic factor in serum of chronic schizophrenic patients. Psychiatry Res 110:249–257
111. Pirildar S, Gonul AS, Taneli F, Akdeniz F (2004) Low serum levels of brain-derived neurotrophic factor in patients with schizophrenia do not elevate after antipsychotic treatment. Prog Neuropsychopharmacol Biol Psychiatry 28:709–713
112. Tan YL, Zhou DF, Cao LY, Zou YZ, Zhang XY (2005) Decreased BDNF in serum of patients with chronic schizophrenia on long-term treatment with antipsychotics. Neurosci Lett 382:27–32
113. Grillo RW, Ottoni GL, Leke R, Souza DO, Portela LV, Lara DR (2007) Reduced serum BDNF levels in schizophrenic patients on clozapine or typical antipsychotics. J Psychiatr Res 41:31–35

114. Xiu MH, Hui L, Dang YF, Hou TD, Zhang CX, Zheng YL et al (2009) Decreased serum BDNF levels in chronic institutionalized schizophrenia on long-term treatment with typical and atypical antipsychotics. Prog Neuropsychopharmacol Biol Psychiatry 33:1508–1512
115. Rizos EN, Papadopoulou A, Laskos E, Michalopoulou PG, Kastania A, Vasilopoulos D et al (2010) Reduced serum BDNF levels in patients with chronic schizophrenic disorder in relapse, who were treated with typical or atypical antipsychotics. World J Biol Psychiatry 11:251–255
116. Li C, Tao H, Yang X, Zhang X, Liu Y, Tang Y et al (2018) Assessment of a combination of Serum Proteins as potential biomarkers to clinically predict Schizophrenia. Int J Med Sci 15(9):900–906
117. Schildkraut JJ (1965) The catecholamine hypothesis of affective disorders: a review of supporting evidence. Am J Psychiatr 122(5):509–522
118. Hashimoto K, Bruno D, Nierenberg J, Marmar CR, Zetterberg H, Blennow K et al (2016) Abnormality in glutamine-glutamate cycle in the cerebrospinal fluid of cognitively intact elderly individuals with major depressive disorder: a 3-year follow-up study. Transl Psychiatry 6:e744. https://doi.org/10.1038/tp.2016.8
119. Hattori K, Ota M, Sasayama D, Yoshida S, Matsumura R, Miyakawa T et al (2015) Increased cerebrospinal fluid fibrinogen in major depressive disorder. Sci Rep 5:11412. https://doi.org/10.1038/srep11412
120. Ogawa S, Hattori K, Sasayama D, Yokota Y, Matsumura R, Matsuo J et al (2015) Reduced cerebrospinal fluid ethanolamine concentration in major depressive disorder. Sci Rep 5:7796. https://doi.org/10.1038/srep07796
121. Hyland K (2007) Inherited disorders affecting dopamine and serotonin: critical neurotransmitters derived from aromatic amino acids. J Nutr 137(6 Suppl 1):1568S–1572S; discussion 1573S–1575S
122. Bowers MB (1974) Lumbar CSF 5-hydroxyindoleacetic acid and homovanillic acid in affective syndromes. J Nerv Ment Dis 158(5):325–330
123. Ren J, Zhao G, Sun X, Liu H, Jiang P, Chen J et al (2017) Identification of plasma biomarkers for distinguishing bipolar depression from major depressive disorder by iTRAQ-coupled LC-MS/MS and bioinformatics analysis. Psychoneuroendocrinology 86:17–24
124. Pan JX, Xia JJ, Deng FL, Liang WW, Wu J, Yin BM et al (2018) Diagnosis of major depressive disorder based on changes in multiple plasma neurotransmitters: a targeted metabolomics study. Transl Psychiatry 8(1):130. https://doi.org/10.1038/s41398-018-0183-x
125. Dean B, Kulkarni J, Copolov DL, Shrikanthan P, Malone V, Hill C (1992) Dopamine uptake by platelets from subjects with schizophrenia: a correlation with the delusional state of the patient. Psychiatry Res 41:17–24
126. Liu L, Jia F, Yuan G, Chen Z, Yao J, Li H et al (2010) Tyrosine hydroxylase, interleukin-1beta and tumor necrosis factor-alpha are overexpressed in peripheral blood mononuclear cells from schizophrenia patients as determined by semi-quantitative analysis. Psychiatry Res 176:1–7
127. Liu L, Yuan G, Cheng Z, Zhang G, Liu X, Zhang H (2013) Identification of the mRNA expression status of the dopamine D2 receptor and dopamine transporter in peripheral blood lymphocytes of schizophrenia patients. PLoS One 8:e75259. https://doi.org/10.1371/journal.pone.0075259
128. Boneberg EM, Von Seydlitz E, Propster K, Watzl H, Rockstroh B, Illges H (2006) D3 dopamine receptor mRNA is elevated in T cells of schizophrenic patients whereas D4 dopamine receptor mRNA is reduced in CD4+-T cells. J Neuroimmunol 173:180–187
129. Abi-Dargham A, Rodenhiser J, Printz D, Zea-Ponce Y, Gil R, Kegeles LS et al (2000) Increased baseline occupancy of D2 receptors by dopamine in schizophrenia. Proc Natl Acad Sci U S A 97:8104–8109
130. Laruelle M, Abi-Dargham A, Van Dyck CH, Gil R, D'souza CD, Erdos J et al (1996) Single photon emission computerized tomography imaging of amphetamine induced dopamine release in drug-free schizophrenic subjects. Proc Natl Acad Sci U S A 93:9235–9240

131. Ogawa S, Tsuchimine S, Kunugi H (2018) Cerebrospinal fluid monoamine metabolite concentrations in depressive disorder: a meta-analysis of historic evidence. J Psychiatr Res 105:137–146
132. Farde L, Gustavsson JP, Jönsson E (1997) D2 dopamine receptors and personality traits. Nature 385:590. https://doi.org/10.1038/385590a0
133. Farde L, Plavén-Sigray P, Borg J, Cervenka S (2018) Brain neuroreceptor density and personality traits: towards dimensional biomarkers for psychiatric disorders. Philos Trans R Soc Lond Ser B Biol Sci 373(1744). https://doi.org/10.1098/rstb.2017.0156
134. Woodward ND, Cowan RL, Park S, Ansari MS, Baldwin RM, Li R et al (2011) Correlation of individual differences in schizotypal personality traits with amphetamine induced dopamine release in striatal and extrastriatal brain regions. Am J Psychiatry 168:418–426
135. Breier A, Su TP, Saunders R, Carson RE, Kolachana BS, de Bartolomeis A et al (1997) Schizophrenia is associated with elevated amphetamine-induced synaptic dopamine concentrations: evidence from a novel positron emission tomography method. Proc Natl Acad Sci U S A 94(6):2569–2574
136. Howes OD, Montgomery AJ, Asselin MC, Murray RM, Valli I, Tabraham P et al (2009) Elevated striatal dopamine function linked to prodromal signs of schizophrenia. Arch Gen Psychiatry 66(1):13–20
137. Joyce JN, Shane A, Lexow N, Winokur A, Casanova MF, Kleinman JE (1993) Serotonin uptake sites and serotonin receptors are altered in the limbic system of schizophrenics. Neuropsychopharmacology 8:315–336
138. Gurevich EV, Joyce JN (1997) Alterations in the cortical serotonergic system in schizophrenia: a postmortem study. Biol Psychiatry 42:529–545
139. Laruelle M, Abi-Dargham A, Casanova MF, Toti R, Weinberger DR, Kleinman JE (1993) Selective abnormalities of prefrontal serotonergic receptors in schizophrenia. A postmortem study. Arch Gen Psychiatry 50:810–818
140. Sumiyoshi T, Stockmeier CA, Overholser JC, Dilley GE, Meltzer HY (1996) Serotonin1A receptors are increased in postmortem prefrontal cortex in schizophrenia. Brain Res 708:209–214
141. Lopez-Figueroa AL, Norton CS, Lopez-Figueroa MO, Armellini-Dodel D, Burke S, Akil H et al (2004) Serotonin 5-HT1A, 5-HT1B, and 5-HT2A receptor mRNA expression in subjects with major depression, bipolar disorder, and schizophrenia. Biol Psychiatry 55:225–233
142. Muck-Seler D, Pivac N, Jakovljevic M (1999) Sex differences, season of birth and platelet 5-HT levels in schizophrenic patients. J Neural Transm 106(3–4):337–347
143. Tuominen L, Salo J, Hirvonen J, Någren K, Laine P, Melartin T et al (2013) Temperament, character and serotonin activity in the human brain: a positron emission tomography study based on a general population cohort. Psychol Med 43(4):881–894
144. Kim JS, Kornhuber HH, Schmid-Burgk W, Holzmüller B (1980) Low cerebrospinal fluid glutamate in schizophrenic patients and a new hypothesis on schizophrenia. Neurosci Lett 20:379–382
145. Neeman G, Blanaru M, Bloch B, Kremer I, Ermilov M, Javitt DC et al (2005) Relation of plasma glycine, serine, and homocysteine levels to schizophrenia symptoms and medication type. Am J Psychiatry 162:1738–1740
146. Hashimoto K, Engberg G, Shimizu E, Nordin C, Lindstrom LH, Iyo M (2005) Reduced D-serine to total serine ratio in the cerebrospinal fluid of drug naive schizophrenic patients. Prog Neuropsychopharmacol Biol Psychiatry 29:767–769
147. Haaf M, Leicht G, Curic S, Mulert C (2018) Glutamatergic deficits in schizophrenia - biomarkers and pharmacological interventions within the ketamine model. Curr Pharm Biotechnol 19(4):293–307
148. Duncan CE, Webster MJ, Rothmond DA, Bahn S, Elashoff M, Shannon Weickert C (2010) Prefrontal GABA(A) receptor alpha-subunit expression in normal postnatal human development and schizophrenia. J Psychiatr Res 44:673–681
149. Arrue A, Davila R, Zumarraga M, Basterreche N, Gonzalez-Torres MA, Goienetxea B et al (2010) GABA and homovanillic acid in the plasma of Schizophrenic and bipolar I patients. Neurochem Res 35:247–253

150. Lo WS, Harano M, Gawlik M, Yu Z, Chen J, Pun FW et al (2007) GABRB2 association with schizophrenia: commonalities and differences between ethnic groups and clinical subtypes. Biol Psychiatry 61:653–660
151. Pinna G (2018) Biomarkers for PTSD at the interface of the endocannabinoid and neurosteroid axis. Front Neurosci 12:482. https://doi.org/10.3389/fnins.2018.00482
152. Rasmusson AM, King M, Gregor K, Scioli-Salter E, Pineles S, Valovski I et al (2018) GABAergic neurosteroids in cerebrospinal fluid are negatively associated with PTSD severity in men. Biol Psychiatry 83:S15–S16
153. Nemeroff CB (2008) Understanding the pathophysiology of postpartum depression: implications for the development of novel treatments. Neuron 59:185–186
154. Lovick T (2013) SSRIs and the female brain–potential for utilizing steroid-stimulating properties to treat menstrual cycle-linked dysphorias. J Psychopharmacol 27:1180–1185
155. Trivisano M, Lucchi C, Rustichelli C, Terracciano A, Cusmai R, Ubertini GM et al (2017) Reduced steroidogenesis in patients with PCDH19-female limited epilepsy. Epilepsia 58:e91–e95
156. Dichtel LE, Lawson EA, Schorr M, Meenaghan E, Paskal ML, Eddy KT et al (2018) Neuroactive steroids and affective symptoms in women across the weight spectrum. Neuropsychopharmacology 43(6):1436–1444
157. Belelli D, Lambert JJ (2005) Neurosteroids: endogenous regulators of the GABA(A) receptor. Nat Rev Neurosci 6:565–575
158. Pineles SL, Nillni YI, Pinna G, Irvine J, Webb A, Hall A et al (2018) PTSD in women is associated with a block in conversion of progesterone to the GABAergic neurosteroids allopregnanolone and pregnanolone: confirmed in plasma. Psychoneuroendocrinology 93:133–141
159. Wilker S, Pfeiffer A, Elbert T, Ovuga E, Karabatsiakis A, Krumbholz A et al (2016) Endocannabinoid concentrations in hair are associated with PTSD symptom severity. Psychoneuroendocrinology 67:198–206
160. Court J, Spurden D, Lloyd S, Mckeith I, Ballard C, Cairns N et al (1999) Neuronal nicotinic receptors in dementia with Lewy bodies and schizophrenia: alpha-bungarotoxin and nicotine binding in the thalamus. J Neurochem 73:1590–1597
161. Guan ZZ, Zhang X, Blennow K, Nordberg A (1999) Decreased protein level of nicotinic receptor alpha7 subunit in the frontal cortex from schizophrenic brain. Neuroreport 10:1779–1782
162. Mancama D, Arranz MJ, Landau S, Kerwin R (2003) Reduced expression of the muscarinic 1 receptor cortical subtype in schizophrenia. Am J Med Genet B Neuropsychiatr Genet 119b:2–6
163. Rozek LS, Dolinoy DC, Sartor MA, Omenn GS (2014) Epigenetics: relevance and implications for public health. Annu Rev Public Health 35:105–122
164. Zhang TY, Labonte B, Wen XL, Turecki G, Meaney MJ (2013) Epigenetic mechanisms for the early environmental regulation of hippocampal glucocorticoid receptor gene expression in rodents and humans. Neuropsychopharmacology 38(1):111–123
165. Daskalakis NP, Cohen H, Nievergelt CM, Baker DG, Buxbaum JD, Russo SJ et al (2016) New translational perspectives for blood-based biomarkers of PTSD: from glucocorticoid to immune mediators of stress susceptibility. Exp Neurol 284(Pt B):133–140
166. Watson S, Gallagher P, Ritchie JC, Ferrier IN, Young AH (2004) Hypothalamic-pituitary adrenal axis function in patients with bipolar disorder. Br J Psychiatry 184:496–502
167. Pariante CM, Dazzan P, Danese A, Morgan KD, Brudaglio F, Morgan C et al (2005) Increased pituitary volume in antipsychotic-free and antipsychotic-treated patients of the Aesop first-onset psychosis study. Neuropsychopharmacology 30(10):1923–1931
168. Ryan MC, Sharifi N, Condren R, Thakore JH (2004) Evidence of basal pituitary adrenal overactivity in first episode, drug naive patients with schizophrenia. Psychoneuroendocrinology 29(8):1065–1070
169. Perkins DO, Jeffries CD, Jarskog LF, Thomson JM, Woods K, Newman MA et al (2007) microRNA expression in the prefrontal cortex of individuals with schizophrenia and schizoaffective disorder. Genome Biol 8(2):R27. https://doi.org/10.1186/gb-2007-8-2-r27

170. Kim AH, Reimers M, Maher B, Williamson V, McMichael O, McClay JL et al (2010) MicroRNA expression profiling in the prefrontal cortex of individuals affected with schizophrenia and bipolar disorders. Schizophr Res 124(1–3):183–191
171. Moreau MP, Bruse SE, David-Rus R, Buyske S, Brzustowicz LM (2011) Altered microRNA expression profiles in postmortem brain samples from individuals with schizophrenia and bipolar disorder. Biol Psychiatry 69(2):188–193
172. Beveridge NJ, Tooney PA, Carroll AP, Gardiner E, Bowden N, Scott RJ et al (2008) Dysregulation of miRNA 181b in the temporal cortex in schizophrenia. Hum Mol Genet 17(8):1156–1168
173. Xu Y, Yue W, Yao Shugart Y, Li S, Cai L, Li Q et al (2016) Exploring transcription factors-microRNAs co-regulation networks in schizophrenia. Schizophr Bull 42:1037–1045
174. Burmistrova OA, Goltsov AY, Abramova LI, Kaleda VG, Orlova VA, Rogaev EI (2007) MicroRNA in schizophrenia: genetic and expression analysis of miR-130b (22q11). Biochemistry (Mosc) 72:578–582
175. Beveridge NJ, Gardiner E, Carroll AP, Tooney PA, Cairns MJ (2010) Schizophrenia is associated with an increase in cortical microRNA biogenesis. Mol Psychiatry 15(12):1176–1189
176. Santarelli DM, Beveridge NJ, Tooney PA, Cairns MJ (2011) Upregulation of dicer and microRNA expression in the dorsolateral prefrontal cortex Brodmann area 46 in schizophrenia. Biol Psychiatry 69(2):180–187
177. Miller BH, Zeier Z, Xi L, Lanz TA, Deng S, Strathmann J et al (2012) MicroRNA-132 dysregulation in schizophrenia has implications for both neurodevelopment and adult brain function. Proc Natl Acad Sci U S A 109(8):3125–3130
178. Shi W, Du J, Qi Y, Liang G, Wang T, Li S et al (2012) Aberrant expression of serum miRNAs in schizophrenia. J Psychiatr Res 46(2):198–204
179. Banigan MG, Kao PF, Kozubek JA, Winslow AR, Medina J, Costa J et al (2013) Differential expression of exosomal microRNAs in prefrontal cortices of schizophrenia and bipolar disorder patients. PLoS One 8(1):e48814. https://doi.org/10.1371/journal.pone.0048814
180. Fan HM, Sun XY, Niu W, Zhao L, Zhang QL, Li WS et al (2015) Altered microRNA expression in peripheral blood mononuclear cells from young patients with schizophrenia. J Mol Neurosci 56(3):562–571
181. Sun XY, Lu J, Zhang L, Song HT, Zhao L, Fan HM et al (2015) Aberrant microRNA expression in peripheral plasma and mononuclear cells as specific blood-based biomarkers in schizophrenia patients. J Clin Neurosci 22(3):570–574
182. Lai CY, Lee SY, Scarr E, Yu YH, Lin YT, Liu CM (2016) Aberrant expression of microRNAs as biomarker for schizophrenia: from acute state to partial remission, and from peripheral blood to cortical tissue. Transl Psychiatry 6:e717. https://doi.org/10.1038/tp.2015.213
183. Yao Y, Schröder J, Karlsson H (2008) Verification of proposed peripheral biomarkers in mononuclear cells of individuals with schizophrenia. J Psychiatr Res 42:639–643
184. Liu S, Zhang F, Wang X, Shugart YY, Zhao Y, Li X et al (2017) Diagnostic value of blood-derived microRNAs for schizophrenia: results of a meta-analysis and validation. Sci Rep 7(1):15328. https://doi.org/10.1038/s41598-017-15751-5
185. Wang X, Wang X (2006) Systematic identification of microRNA functions by combining target prediction and expression profiling. Nucleic Acids Res 34(5):1646–1652
186. Schratt GM, Tuebing F, Nigh EA, Kane CG, Sabatini ME, Kiebler M et al (2006) A brain-specific microRNA regulates dendritic spine development. Nature 439(7074):283–289
187. Padula MC, Scariati E, Schaer M, Eliez S (2018) A mini review on the contribution of the anterior cingulate cortex in the risk of psychosis in 22q11.2 deletion syndrome. Front Psych 9:372. https://doi.org/10.3389/fpsyt.2018.00372
188. Narahari A, Hussain M, Sreeram V (2017) MicroRNAs as biomarkers for psychiatric conditions: a review of current research. Innov Clin Neurosci 14(1–2):53–55
189. Seripa D, Lozupone M, Miscio G, Stella E, La Montagna M, Gravina C et al (2018) CYP2D6 genotypes in revolving door patients with bipolar disorders: a case series. Medicine (Baltimore) 97(37):e11998. https://doi.org/10.1097/MD.0000000000011998

190. Seripa D, Lozupone M, Stella E, Paroni G, Bisceglia P, La Montagna M et al (2017) Psychotropic drugs and CYP2D6 in late-life psychiatric and neurological disorders. What do we know? Expert Opin Drug Saf 16(12):1373–1385
191. Moodithaya S, Avadhany ST (2012) Gender differences in age-related changes in cardiac autonomic nervous function. J Aging Res 2012:679345. https://doi.org/10.1155/2012/679345
192. Sztajzel J (2004) Heart rate variability: a noninvasive electrocardiographic method to measure the autonomic nervous system. Swiss Med Wkly 134(35–36):514–522
193. Lin HP, Lin HY, Lin WL, Huang AC (2011) Effects of stress, depression, and their interaction on heart rate, skin conductance, finger temperature, and respiratory rate: sympathetic-parasympathetic hypothesis of stress and depression. J Clin Psychol 67(10):1080–1091
194. Shah LB, Torres S, Kannusamy P, Chng CM, He HG, Klainin-Yobas P (2015) Efficacy of the virtual reality-based stress management program on stress-related variables in people with mood disorders: the feasibility study. Arch Psychiatr Nurs 29(1):6–13
195. Barua S, Begum S, Ahmed MU (2015) Supervised machine learning algorithms to diagnose stress for vehicle drivers based on physiological sensor signals. Stud Health Technol Inform 211:241–248
196. Teisala T, Mutikainen S, Tolvanen A, Rottensteiner M, Leskinen T, Kaprio J (2014) Associations of physical activity, fitness, and body composition with heart rate variability-based indicators of stress and recovery on workdays: a cross-sectional study. J Occup Med Toxicol 9:16. https://doi.org/10.1186/1745-6673-9-16
197. Bootsma M, Swenne CA, Van Bolhuis HH, Chang PC, Cats VM, Bruschke AV (1994) Heart rate and heart rate variability as indexes of sympathovagal balance. Am J Phys 266(4 Pt 2):H1565–H1571
198. Alonso JF, Romero S, Ballester MR, Antonijoan RM, Mañanas MA et al (2015) Stress assessment based on EEG univariate features and functional connectivity measures. Physiol Meas 36(7):1351–1365
199. Valkonen-Korhonen M, Tarvainen MP, Ranta-Aho P, Karjalainen PA, Partanen J, Karhu J et al (2003) Heart rate variability in acute psychosis. Psychophysiology 40(5):716–726
200. Rachow T, Berger S, Boettger MK, Schulz S, Guinjoan S, Yeragani VK et al (2011) Nonlinear relationship between electrodermal activity and heart rate variability in patients with acute schizophrenia. Psychophysiology 48(10):1323–1332
201. Olbrich R, Kirsch P, Pfeiffer H, Mussgay L (2001) Patterns of recovery of autonomic dysfunctions and neurocognitive deficits in schizophrenics after acute psychotic episodes. J Abnorm Psychol 110(1):142–150
202. Yeragani VK, Rao KA, Smitha MR, Pohl RB, Balon R, Srinivasan K (2002) Diminished chaos of heart rate time series in patients with major depression. Biol Psychiatry 51(9):733–744
203. Yeragani VK (2000) Major depression and long-term heart period variability. Depress Anxiety 12(1):51–52
204. Won E, Kim YK (2016) Stress, the autonomic nervous system, and the immune-kynurenine pathway in the etiology of depression. Curr Neuropharmacol 14(7):665–673
205. Alvares GA, Quintana DS, Kemp AH, Van Zwieten A, Balleine BW, Hickie IB et al (2013) Reduced heart rate variability in social anxiety disorder: associations with gender and symptom severity. PLoS One 8(7):e70468. https://doi.org/10.1371/journal.pone.0070468
206. Pollatos O, Herbert BM, Wankner S, Dietel A, Wachsmuth C, Henningsen P et al (2011) Autonomic imbalance is associated with reduced facial recognition in somatoform disorders. J Psychosom Res 71(4):232–239
207. Kawachi I, Sparrow D, Vokonas PS, Weiss ST (1995) Decreased heart rate variability in men with phobic anxiety (data from the Normative Aging Study). Am J Cardiol 75(14):882–885
208. Karpyak VM, Romanowicz M, Schmidt JE, Lewis KA, Bostwick JM (2014) Characteristics of heart rate variability in alcohol-dependent subjects and nondependent chronic alcohol users. Alcohol Clin Exp Res 38(1):9–26
209. Garland EL, Franken IH, Sheetz JJ, Howard MO (2012) Alcohol attentional bias is associated with autonomic indices of stress-primed alcohol cue-reactivity in alcohol-dependent patients. Exp Clin Psychopharmacol 20(3):225–235

210. Frewen J, Finucane C, Savva GM, Boyle G, Coen RF, Kenny RA (2013) Cognitive function is associated with impaired heart rate variability in ageing adults: the Irish longitudinal study on ageing wave one results. Clin Auton Res 23(6):313–323
211. Sarlon J, Plaszczyk S, Engel S, Oertel-Knöchel V (2018) Electrophysiological parameters as biomarkers for psychiatry: intra-individual variability and influencing factors. Int J Psychophysiol 123:42–47
212. Boutros NN, Mucci A, Vignapiano A, Galderisi S (2014) Electrophysiological aberrations associated with negative symptoms in schizophrenia. Curr Top Behav Neurosci 21:129–156
213. Claesson MJ, Jeffery IB, Conde S, Power SE, O'Connor EM, Cusack S et al (2012) Gut microbiota composition correlates with diet and health in the elderly. Nature 488(7410):178–184
214. Mayer EA (2011) Gut feelings: the emerging biology of gut–brain communication. Nat Rev Neurosci 12(8):453–466
215. Hoban AE, Stilling RM, Ryan FJ, Shanahan F, Dinan TG, Claesson MJ et al (2016) Regulation of prefrontal cortex myelination by the microbiota. Transl Psychiatry 6:e774. https://doi.org/10.1038/tp.2016.42
216. Ogbonnaya ES, Clarke G, Shanahan F, Dinan TG, Cryan JF, O'Leary OF (2015) Adult hippocampal neurogenesis is regulated by the microbiome. Biol Psychiatry 78(4):e7–e9. https://doi.org/10.1016/j.biopsych.2014.12.023
217. Clarke G, Grenham S, Scully P, Fitzgerald P, Moloney RD, Shanahan F et al (2013) The microbiome-gut-brain axis during early life regulates the hippocampal serotonergic system in a sex-dependent manner. Mol Psychiatry 18(6):666–673
218. Cenit MC, Sanz Y, Codoñer-Franch P (2017) Influence of gut microbiota on neuropsychiatric disorders. World J Gastroenterol 23(30):5486–5498
219. Fabi E, Fusco A, Valiante M, Celli R (2013) Genetics and epigenetics of schizophrenia. Clin Ter 164:e319–e324
220. Dinan TG, Borre YE, Cryan JF (2014) Genomics of schizophrenia: time to consider the gut microbiome? Mol Psychiatry 19:1252–1257
221. Dinan TG, Cryan JF (2013) Melancholic microbes: a link between gut microbiota and depression? Neurogastroenterol Motil 25:713–719
222. Aizawa E, Tsuji H, Asahara T, Takahashi T, Teraishi T, Yoshida S et al (2016) Possible association of Bifidobacterium and Lactobacillus in the gut microbiota of patients with major depressive disorder. J Affect Disord 202:254–257
223. Hughes C, Davoodi-Semiromi Y, Colee JC, Culpepper T, Dahl WJ, Mai V et al (2011) Galactooligosaccharide supplementation reduces stress-induced gastrointestinal dysfunction and days of cold or flu: a randomized, double-blind, controlled trial in healthy university students. Am J Clin Nutr 93:1305–1311
224. Kelly JR, Borre Y, O'Brien C, Patterson E, El Aidy S, Deane J et al (2016) Transferring the blues: depression-associated gut microbiota induces neurobehavioural changes in the rat. J Psychiatr Res 82:109–118
225. Solfrizzi V, Custodero C, Lozupone M, Imbimbo BP, Valiani V, Agosti P (2017) Relationships of dietary patterns, foods, and micro- and macronutrients with Alzheimer's disease and late-life cognitive disorders: a systematic review. J Alzheimers Dis 59(3):815–849
226. Jørgensen BP, Hansen JT, Krych L, Larsen C, Klein AB, Nielsen DS et al (2014) A possible link between food and mood: dietary impact on gut microbiota and behavior in BALB/c mice. PLoS One 9(8):e103398. https://doi.org/10.1371/journal.pone.0103398
227. Zhernakova A, Kurilshikov A, Bonder MJ, Tigchelaar EF, Schirmer M, Vatanen T et al (2016) Population-based metagenomics analysis reveals markers for gut microbiome composition and diversity. Science 352:565–569
228. Messaoudi M, Violle N, Bisson JF, Desor D, Javelot H, Rougeot C (2011) Beneficial psychological effects of a probiotic formulation (Lactobacillus helveticus R0052 and Bifidobacterium longum R0175) in healthy human volunteers. Gut Microbes 2:256–261

229. Schwensen HF, Kan C, Treasure J, Høiby N, Sjögren M (2018) A systematic review of studies on the faecal microbiota in anorexia nervosa: future research may need to include microbiota from the small intestine. Eat Weight Disord 23(4):399–418
230. Insel T, Cuthbert B, Garvey M, Heinssen R, Pine DS, Quinn K et al (2010) Research domain criteria (RDoC): toward a new classification framework for research on mental disorders. Am J Psychiatry 167(7):748–751
231. Wakefield JC (2014) Wittgenstein's nightmare: why the RDoC grid needs a conceptual dimension. World Psychiatry 13(1):38–40
232. Pratt J, Hall J (2018) Biomarkers in neuropsychiatry: a prospect for the twenty-first century? Curr Top Behav Neurosci 40:3-10

Chapter 8
Interactome Studies of Psychiatric Disorders

Dong Ik Park and Christoph W. Turck

Abstract High comorbidity and complexity have precluded reliable diagnostic assessment and treatment of psychiatric disorders. Impaired molecular interactions may be relevant for underlying mechanisms of psychiatric disorders but by and large remain unknown. With the help of a number of publicly available databases and various technological tools, recent research has filled the paucity of information by generating a novel dataset of psychiatric interactomes. Different technological platforms including yeast two-hybrid screen, co-immunoprecipitation-coupled with mass spectrometry-based proteomics, and transcriptomics have been widely used in combination with cellular and molecular techniques to interrogate the psychiatric interactome. Novel molecular interactions have been identified in association with different psychiatric disorders including autism spectrum disorders, schizophrenia, bipolar disorder, and major depressive disorder. However, more extensive and sophisticated interactome research needs to be conducted to overcome the current limitations such as incomplete interactome databases and a lack of functional information among components. Ultimately, integrated psychiatric interactome databases will contribute to the implementation of biomarkers and therapeutic intervention.

Keywords Interactome · Psychiatric disorders · Psychiatric interactome · Protein-protein interaction · Co-immunoprecipitation-coupled mass spectrometry-based proteomics · Yeast two-hybrid screen · Transcriptomics

D. I. Park (✉)
Danish Research Institute of Translational Neuroscience (DANDRITE), Department of Biomedicine, Aarhus University, Aarhus, Denmark
e-mail: park@biomed.au.dk

C. W. Turck
Proteomics and Biomarkers, Department of Translational Research in Psychiatry, Max Planck Institute of Psychiatry, Munich, Germany

P. C. Guest (ed.), *Reviews on Biomarker Studies in Psychiatric and Neurodegenerative Disorders*, Advances in Experimental Medicine and Biology 1118,
https://doi.org/10.1007/978-3-030-05542-4_8

8.1 Introduction

Increased knowledge of the interactome, a collection of biological interactions, is important for a better understanding of complex disease mechanisms. The basic units of the interactome are protein-protein interactions (PPIs), which represent physical contact of two or more proteins that can stimulate biochemical reactions. PPIs mediate many essential molecular processes including signal transduction and transcriptional regulation. PPIs have been implicated in various diseases including neurodegenerative disorders, leukemia, cervical cancer, and bacterial infection [1]. While it is estimated that more than 650,000 PPIs exist in the human interactome [2], our current knowledge on disease-relevant interactomes is incomplete.

The study of psychiatric disorders has been challenging due to highly complex and heterogeneous disease mechanisms. This complexity has hampered the development of reliable clinical diagnostics and therapeutics. The difficulties in biomarker and therapeutic development may also be the result of lacking knowledge of the psychiatric disease interactome. With the help of -omics methods and computational analyses that take into account protein interaction information, we can now acquire novel and useful information on the interactome of several psychiatric disorders. Here, we present and discuss the progress made in interactome studies and state-of-the-art approaches that will advance our knowledge in psychiatry.

8.2 Complexity and Comorbidity in Psychiatric Disorders

Genome-wide association studies (GWAS) of psychiatric disorders have found many heritable genetic variants which increase the risk of schizophrenia (SCZ), autism spectrum disorders (ASD), major depressive disorder (MDD), and bipolar disorder (BD) [3]. These studies have revealed that psychiatric disorders are polygenic, i.e., influenced by many genetic variants with small effect sizes. GWAS identified 108 and 44 genetic risk loci associated with SCZ and MDD, respectively [4, 5]. While an individual genetic variant has a small effect, the combined effects of multiple common risk variants may profoundly contribute to the aetiology of psychiatric disorders. Therefore, polygenic risk score, a weighted sum of trait-associated risk variants, has been applied to predict cumulative influence on phenotypic traits in psychiatric disorders [6, 7].

Interestingly, a substantial number of genetic risk loci are shared among different psychiatric disorders. In addition, patients with different psychiatric diagnoses often display common symptoms and phenotypes. For instance, phenotypes in anxiety, MDD, panic disorder, and posttraumatic stress disorder (PTSD) are common in SCZ patients [8]. Furthermore, ASD was found to have high comorbidity with MDD and anxiety [9–11].

Complexity and comorbidity for psychiatric disorders may result from global molecular pathway dysfunction in the central nervous system. Consistent with the polygenic nature of the illness, ASD, SCZ, and MDD patients show alterations in various biological pathways including synaptic transmission, calcium homeostasis, energy metabolism, oxidative stress, cytoskeleton, and immune system [12–14]. Interactome analysis can address the high complexity and comorbidity among psychiatric disorders by identifying disease-specific molecular interaction networks. Distinct molecular interaction signatures may provide novel insights which will allow more precise diagnosis and treatment of comorbid psychiatric disorders.

8.3 Psychiatric Interactome Studies

8.3.1 Protein-Protein Interaction Database

Large-scale experiments and advances in computational bioinformatics have generated large amounts of molecular interaction data. As a consequence, several PPI databases have been established based on manual curation of the scientific literature with experimentally verified data, computational prediction, and automated text mining. These include the Biomolecular Interaction Network Database (BIND), the Biological General Repository for Interaction Datasets (BioGRID, https://thebiogrid.org), the Database of Interacting Proteins (DIP, http://dip.doe-mbi.ucla.edu/dip/), the Human Protein Reference Database (HPRD, http://www.hprd.org), the Molecular INTeraction database (MINT, https://mint.bio.uniroma2.it), the IntAct molecular interaction database (IntAct, https://www.ebi.ac.uk/intact/), the Human Protein Interaction Database (HPID, http://wilab.inha.ac.kr/hpid/), and the Search Tool for the Retrieval of Interacting Genes/Proteins (STRING, https://string-db.org). PPI databases can be used to map gene and protein interaction networks and the relevant pathways and are an important resource to investigate altered molecular interactions in psychiatric disorders.

8.3.2 Yeast Two-Hybrid Screen

The yeast two-hybrid (Y2H) system has been used extensively to identify in vivo binary PPIs.

Camargo et al. have investigated protein complexes of Disrupted in Schizophrenia 1 (DISC1) [15] and reported on multiple interactions with proteins involved in cytoskeletal stability and organization, intracellular transport, and cell cycle/division. They also found interactions with several synaptic proteins, which may help to explain the synaptic pathology and cognitive deficits seen in SCZ. DISC1 gene single-nucleotide polymorphisms are associated with multiple psychiatric disorders including BD, SCZ, MDD, ASD, and Asperger syndrome [16–18].

Zhou et al. identified novel interacting proteins of another SCZ risk gene, ZNF804A, using a Y2H screen [19]. Proteins involved in translation and mitochondrial regulation were found to interact with the mouse homolog of ZNF804A. Among the ZNF804A-associated proteins was neurogranin, and the gene encoding this protein has been associated with SCZ. Based on their results, the authors concluded that ZFP804A regulates neuronal migration, progenitor cell proliferation, and protein translation efficiency.

Sakai et al. performed Y2H screen to identify interaction partners of ASD proteins or ASD-associated proteins [20]. The authors identified hundreds of PPIs with ASD (-associated) proteins and validated the interaction candidates using glutathione-sepharose affinity co-purification and co-immunoprecipitation (Co-IP). While the ASD interactome data confirmed previously identified interactions, the authors also found new partners including tuberous sclerosis 1 (TSC1) and tuberous sclerosis 2 (TSC2) proteins. TSC1 and TSC2 genes are involved in tuberous sclerosis complex (TSC), a rare disease associated with ASD [21]. Furthermore, interactions of SH3 and multiple ankyrin repeat domains (Shank) proteins and TSC1 were identified. The authors used previously obtained microarray data and incorporated literature-curated interaction data to analyze the ASD interactome. De novo lesions of three network genes, deletions of pyruvate kinase M2 (PKM2) and N-terminal EF-hand calcium-binding protein 2 (NECAB2), and duplication of filamin A (FLNA) were identified in patients with idiopathic ASD.

8.3.3 Interactome Studies by Co-IP-Coupled Mass Spectrometry-Based Proteomics

Co-IP-coupled with mass spectrometry-based proteomics is a powerful method used to identify constituents of protein complexes. Martins-de-Souza et al. interrogated the collapsin response mediator protein-2 (CRMP2) interactome by combining Co-IP and shotgun proteomics [22]. CRMP2, also known as dihydropyrimidinase-like 2 (DPYSL2), is a protein enriched in the central nervous system. It plays diverse roles in cytoskeleton dynamics, vesicle trafficking, synaptic transmission, neurite outgrowth, neurotransmitter release, and Ca^{2+} homeostasis. The findings of many studies have indicated that CRMP2 abundance or function may be altered in SCZ. CRMP2 functional variants were found to increase SCZ risk [23]. CRMP2 protein activity was altered in SCZ and other psychiatric disorders. Specific CRMP2 polymorphisms are associated with reduced susceptibility to paranoid-type SCZ [24]. In addition, significant CRMP2 protein-level differences were found in postmortem brains from SCZ patients [25, 26]. The authors identified 78 novel partner proteins involved in 7 biological pathways and 32 molecular functions in mouse frontal brain cortices. In silico pathway analysis further identified the most overrepresented functions of these proteins including semaphoring interaction, axon guidance, and WNT5A signaling, suggesting a critical role of the CRMP2 interactome in the regulation of neuronal and synaptic functions.

Alfieri et al. performed Co-IP-coupled mass spectrometry-based proteomics to investigate the synaptic interactome associated with the p140Cap protein [27]. p140Cap has been shown to be critical for synaptogenesis, synaptic transmission, and plasticity. The p140cap knockout mouse displayed defects in memory consolidation and cognitive functions along with impaired synaptic maturation and plasticity [28]. The authors analyzed p140Cap protein-associated synaptosomal interactome in the mouse telencephalon. They identified 351 interacting partners that are involved in key synaptic processes. Interestingly, the p140Cap protein interactome network showed gene enrichment associated with SCZ, ASD, and BD, indicating that the p140Cap interactome is relevant for multiple psychiatric disorders.

K^+-Cl^- cotransporter 2 (KCC2) is a chloride potassium symporter present exclusively in neurons of the central nervous system, and KCC2 dysfunction has been associated with psychiatric disorders. One study showed that the postmortem hippocampus of SCZ patients had significantly decreased full-length KCC2 protein expression levels [29]. EXON6B transcript, one of the splice variants of KCC2, was found to be differentially expressed in dorsolateral prefrontal cortex of SCZ and MDD patients [30]. Mahadevan et al. analyzed KCC2 interactome in whole-brain membrane fractions using Co-IP followed by mass spectrometry-based proteomics [31]. The authors identified 150 KCC2 interacting proteins. They revealed that a novel interacting protein, protein kinase C and casein kinase substrate in neurons protein 1 (PACSIN1), plays a negative role in KCC2 expression and function in neurons. PACSIN1 has been shown to regulate endocytosis and recycling of presynaptic vesicles [32, 33] and postsynaptic α-amino-3-hydroxy-5-methyl-4-isoxazolepropionic acid (AMPA), *N*-methyl-D-aspartate (NMDA), and glycine receptors [34–36].

The MET receptor tyrosine kinase interactome was examined for ASD. MET receptor tyrosine kinase regulates various synaptic structures including dendritic complexity, spine morphogenesis, and glutamatergic synapse maturation in the hippocampus, thus controlling neuronal growth and functional maturation [37]. Human genetic studies have indicated MET as an ASD risk gene [38, 39]. Xie et al. performed Co-IP followed by mass spectrometry-based proteomics using developing mouse neocortical synaptosomes. The authors found that the neurodevelopmental disorder-associated candidates including Shank3, synaptic Ras GTPase-activating protein 1 (SynGAP1), and glutamate ionotropic receptor NMDA type subunit 2B (GRIN2B) are highly enriched in MET interactome networks, supporting a role for MET in ASD pathobiology.

8.3.4 Interactome Studies with Transcriptomics

Lee et al. interrogated the Shank3-mammalian target of rapamycin (mTOR) interactome associated with BD and SCZ [40]. The authors previously reported that Shank3-overexpressing transgenic mice showed behavioral abnormalities such as locomotor hyperactivity, amphetamine hypersensitivity, acoustic startle increase, reduced prepulse inhibition, and abnormal circadian rhythms, which can be observed in BD, ASD, and SCZ [41]. Shank3 gene duplications, deletions, and various point mutations have been detected in patients suffering from mania-like hyperkinetic disorders, ASD, mental retardation, and SCZ [41–44]. As a scaffolding protein found in neuronal excitatory synapses, Shank3 has been shown to organize the post-synaptic density by forming complexes with postsynaptic receptors, signaling molecules, and cytoskeletal proteins [45, 46]. mTOR has also been connected with various psychiatric conditions including ASD, drug addiction, intellectual disability, MDD, and SCZ [47, 48]. In addition, ketamine's rapid-acting antidepressant-like effect was found to be associated with mTOR-dependent synapse formation [49]. The authors performed transcriptomic analysis in Shank3-overexpressed striatum. They built interaction networks using interactome data of Shank3, TSC1/TSC2, and Ras homolog enriched in striatum (Rhes) which were obtained from previous studies. The data reanalyses revealed a strong connection between Shank3 and mTORC1 upstream regulators including TSC1/TSC2 and Rhes. The authors found 94 proteins in Shank3-mTORC1 interactome associated with BD and SCZ. They demonstrated that Shank3-mTORC1 interactome may contribute to the abnormal mTORC1 activity and manic-like behaviors of Shank3-overexpressing transgenic mice.

Fryland et al. investigated Bromodomain containing 1 (BRD1) protein interactome [50]. BRD1 genetic variants have been associated with SCZ and BD [51, 52]. BRD1 protein has been shown to be responsible for histone H3K14 acetylation during embryonic development in mice [53]. To identify BRD1 interaction network, the authors used HEK293T human cells that stably expressed epitope tagged BRD1 long and short splice variants. The BRD1 splice variants have been shown to be differentially expressed in the hippocampus after chronic restrained stress in rats [54]. The study reported on 20 and 13 newly identified PPIs for the BRD1 short and long isoforms, respectively. Several PPIs were shared between isoforms. Newly identified PPIs, including 14-3-3 tyrosine monooxygenase proteins YWHAE, YWHAH, YWHAZ, and PBRM1, have been associated with SCZ and BD [55–58].

8.4 Conclusions and Future Directions in the Interactome Study of Psychiatric Disorders

While advanced analytical and computational tools are accelerating interaction data generation, we presently have only a partial and limited map of the psychiatric interactome. A more complete map will ultimately assist efforts aimed at more precise diagnosis and treatment.

Future studies need to consider diverse factors including age, family history, lifestyle, diet, and ethnicity. General and disease-specific effects of environmental and lifestyle factors on interactome components may further help molecular network characterizations relevant for psychiatric disorders.

A peripheral interactome study is also required to fill the gap of knowledge in the systemic interactions between the periphery and the brain. Close relationships between psychiatric and peripheral conditions have been repeatedly reported. PTSD patients have accelerated physiological aging indicated by altered N-glycosylation in blood plasma [59]. A meta-analysis showed that individuals with severe mental illnesses including SCZ, BD, and MDD had a significantly increased risk of cardiovascular disease and related mortality [60]. Diabetes has also been associated with psychiatric disorders. BD patients were shown to have an up to three times higher risk of type 2 diabetes mellitus (T2DM) compared to the general population [61, 62]. Furthermore, BD patients with T2DM were found to have greater chronicity and disability [63] and less favorable treatment response [64]. T2DM was highly prevalent in SCZ [65]. MDD patients have a 60% increased risk for developing T2DM [66]. A prospective interactome study in the periphery is required to investigate underlying molecular mechanisms to bridge the knowledge gap between psychiatric and physiological illnesses.

Identifying protein-small molecule interactions in cells is also important to further expand our knowledge of psychiatric disorders. Small molecules such as metabolites interact with proteins including enzymes and allosteric regulators in many biological processes. Integrated protein-protein and protein-small molecule interactomes will generate novel information about systemically interconnected molecular interaction networks.

Several technology platforms including mass spectrometry-coupled limited proteolysis (LiP) and thermal proteome profiling (TPP) allow systemic and high-throughput analyses of protein-small molecule interactomes. The LiP method employs proteases such as proteinase K [67]. Metabolite binding to a protein significantly alters its structure and protease accessibility resulting in different peptide profiles during limited proteolysis. The method generates information on protein-metabolite interactions and metabolite binding sites. The TPP method is based on protein thermal stability changes upon binding of small molecules [68]. The proteome thermal stability profile changes induced by binding of a small molecule provide unbiased metabolite-protein interaction networks. These methods will further benefit the interactome studies of psychiatric disorders by providing information on interactions and functions of proteins and small molecules.

The current interactome studies have focused to dissect complex mechanisms of psychiatric diseases. However, pharmacological effects of antipsychotics/antidepressants have not yet been investigated. In this regard it has been shown that antipsychotic and antidepressant drugs result in changes of molecular pathways such as purine/pyrimidine metabolism, glutamate, ubiquitin-proteasome, and energy metabolism [69–73]. Interactome characterization of drug-treated cells or animals may provide important insights on mechanisms involved in treatment response and resistance.

Extensive functional studies are vital to explore relationships between interactome components to obtain mechanistic insights. Ultimately, there is a great need to integrate acquired interactome data and establish the "psychiatric interactome platform". The integrated psychiatric interactome database will advance clinical diagnosis and treatment of psychiatric disorders.

References

1. Ryan DP, Matthews JM (2005) Protein-protein interactions in human disease. Curr Opin Struct Biol 15(4):441–446
2. Stumpf MP, Thorne T, de Silva E, Stewart R, An HJ, Lappe M et al (2008) Estimating the size of the human interactome. Proc Natl Acad Sci U S A 105(19):6959–6964
3. Geschwind DH, Flint J (2015) Genetics and genomics of psychiatric disease. Science 349(6255):1489–1494
4. Schizophrenia Working Group of the Psychiatric Genomics Consortium (2014) Biological insights from 108 schizophrenia-associated genetic loci. Nature 511(7510):421–427
5. Wray NR, Ripke S, Mattheisen M, Trzaskowski M, Byrne EM, Abdellaoui A et al (2018) Genome-wide association analyses identify 44 risk variants and refine the genetic architecture of major depression. Nat Genet 50(5):668–681
6. Mullins N, Ingason A, Porter H, Euesden J, Gillett A, Ólafsson S et al (2017) Reproductive fitness and genetic risk of psychiatric disorders in the general population. Nat Commun 8:15833. https://doi.org/10.1038/ncomms15833
7. Wang T, Zhang X, Li A, Zhu M, Liu S, Qin W et al (2017) Polygenic risk for five psychiatric disorders and cross-disorder and disorder-specific neural connectivity in two independent populations. Neuroimage Clin 14:441–449
8. Buckley PF, Miller BJ, Lehrer DS, Castle DJ (2009) Psychiatric comorbidities and schizophrenia. Schizophr Bull 35(2):383–402
9. Maddox BB, White SW (2015) Comorbid social anxiety disorder in adults with autism spectrum disorder. J Autism Dev Disord 45(12):3949–3960
10. Magnuson KM, Constantino JN (2011) Characterization of depression in children with autism spectrum disorders. J Dev Behav Pediatr 32(4):332–340
11. Zaboski BA, Storch EA (2018) Comorbid autism spectrum disorder and anxiety disorders: a brief review. Future Neurol 13(1):31–37
12. Careaga M, Van de Water J, Ashwood P (2010) Immune dysfunction in autism: a pathway to treatment. Neurotherapeutics 7(3):283–292
13. Malki K, Pain O, Tosto MG, Du Rietz E, Carboni L, Schalkwyk LC (2015) Identification of genes and gene pathways associated with major depressive disorder by integrative brain analysis of rat and human prefrontal cortex transcriptomes. Transl Psychiatry 5:e519. https://doi.org/10.1038/tp.2015.15
14. Nascimento JM, Martins-de-Souza D (2015) The proteome of schizophrenia. NPJ Schizophr 1:14003. https://doi.org/10.1038/npjschz.2014.3
15. Camargo LM, Collura V, Rain JC, Mizuguchi K, Hermjakob H, Kerrien S et al (2007) Disrupted in Schizophrenia 1 Interactome: evidence for the close connectivity of risk genes and a potential synaptic basis for schizophrenia. Mol Psychiatry 12(1):74–86
16. Hashimoto R, Numakawa T, Ohnishi T, Kumamaru E, Yagasaki Y, Ishimoto T et al (2006) Impact of the DISC1 Ser704Cys polymorphism on risk for major depression, brain morphology and ERK signaling. Hum Mol Genet 15(20):3024–3033
17. Kilpinen H, Ylisaukko-Oja T, Hennah W, Palo OM, Varilo T, Vanhala R et al (2008) Association of DISC1 with autism and Asperger syndrome. Mol Psychiatry 13(2):187–196
18. Thomson PA, Wray NR, Millar JK, Evans KL, Hellard SL, Condie A et al (2005) Association between the TRAX/DISC locus and both bipolar disorder and schizophrenia in the Scottish population. Mol Psychiatry 10(7):657–668, 616
19. Zhou Y, Dong F, Lanz TA, Reinhart V, Li M, Liu L et al (2018) Interactome analysis reveals ZNF804A, a schizophrenia risk gene, as a novel component of protein translational machinery critical for embryonic neurodevelopment. Mol Psychiatry 23(4):952–962
20. Sakai Y, Shaw CA, Dawson BC, Dugas DV, Al-Mohtaseb Z, Hill DE et al (2011) Protein interactome reveals converging molecular pathways among autism disorders. Sci Transl Med 3(86):86ra49. https://doi.org/10.1126/scitranslmed.3002166

21. Vignoli A, La Briola F, Peron A, Turner K, Vannicola C, Saccani M et al (2015) Autism spectrum disorder in tuberous sclerosis complex: searching for risk markers. Orphanet J Rare Dis 10:154. https://doi.org/10.1186/s13023-015-0371-1
22. Martins-de-Souza D, Cassoli JS, Nascimento JM, Hensley K, Guest PC, Pinzon-Velasco AM et al (2015) The protein interactome of collapsin response mediator protein-2 (CRMP2/DPYSL2) reveals novel partner proteins in brain tissue. Proteomics Clin Appl 9(9–10):817–831
23. Liu Y, Pham X, Zhang L, Chen PL, Burzynski G, McGaughey DM et al (2014) Functional variants in DPYSL2 sequence increase risk of schizophrenia and suggest a link to mTOR signaling. G3 (Bethesda) 5(1):61–72
24. Nakata K, Ujike H, Sakai A, Takaki M, Imamura T, Tanaka Y et al (2003) The human dihydropyrimidinase-related protein 2 gene on chromosome 8p21 is associated with paranoid-type schizophrenia. Biol Psychiatry 53(7):571–576
25. Martins-de-Souza D, Gattaz WF, Schmitt A, Maccarrone G, Hunyadi-Gulyás E, Eberlin MN et al (2009) Proteomic analysis of dorsolateral prefrontal cortex indicates the involvement of cytoskeleton, oligodendrocyte, energy metabolism and new potential markers in schizophrenia. J Psychiatr Res 43(11):978–986
26. Martins-de-Souza D, Gattaz WF, Schmitt A, Novello JC, Marangoni S, Turck CW et al (2009) Proteome analysis of schizophrenia patients Wernicke's area reveals an energy metabolism dysregulation. BMC Psychiatry 9:17. https://doi.org/10.1186/1471-244X-9-17
27. Alfieri A, Sorokina O, Adrait A, Angelini C, Russo I, Morellato A et al (2017) Synaptic interactome mining reveals p140Cap as a new hub for PSD proteins involved in psychiatric and neurological disorders. Front Mol Neurosci 10:212. https://doi.org/10.3389/fnmol.2017.00212
28. Repetto D, Camera P, Melani R, Morello N, Russo I, Calcagno E et al (2014) p140Cap regulates memory and synaptic plasticity through Src-mediated and citron-N-mediated actin reorganization. J Neurosci 34(4):1542–1553
29. Hyde TM, Lipska BK, Ali T, Mathew SV, Law AJ, Metitiri OE et al (2011) Expression of GABA signaling molecules KCC2, NKCC1, and GAD1 in cortical development and schizophrenia. J Neurosci 31(30):11088–11095
30. Tao R, Li C, Newburn EN, Ye T, Lipska BK, Herman MM et al (2012) Transcript-specific associations of SLC12A5 (KCC2) in human prefrontal cortex with development, schizophrenia, and affective disorders. J Neurosci 32(15):5216–5222
31. Mahadevan V, Khademullah CS, Dargaei Z, Chevrier J, Uvarov P, Kwan J et al (2017) Native KCC2 interactome reveals PACSIN1 as a critical regulator of synaptic inhibition. Elife 6:e28270. https://doi.org/10.7554/eLife.28270
32. Andersson F, Jakobsson J, Löw P, Shupliakov O, Brodin L (2008) Perturbation of syndapin/PACSIN impairs synaptic vesicle recycling evoked by intense stimulation. J Neurosci 28(15):3925–3933
33. Anggono V, Smillie KJ, Graham ME, Valova VA, Cousin MA, Robinson PJ (2006) Syndapin I is the phosphorylation-regulated dynamin I partner in synaptic vesicle endocytosis. Nat Neurosci 9(6):752–760
34. Anggono V, Koç-Schmitz Y, Widagdo J, Kormann J, Quan A, Chen CM et al (2013) PICK1 interacts with PACSIN to regulate AMPA receptor internalization and cerebellar long-term depression. Proc Natl Acad Sci U S A 110(34):13976–13981
35. Del Pino I, Koch D, Schemm R, Qualmann B, Betz H, Paarmann I (2014) Proteomic analysis of glycine receptor beta subunit (GlyRbeta)-interacting proteins: evidence for syndapin I regulating synaptic glycine receptors. J Biol Chem 289(16):11396–11409
36. Perez-Otano I, Luján R, Tavalin SJ, Plomann M, Modregger J, Liu XB et al (2006) Endocytosis and synaptic removal of NR3A-containing NMDA receptors by PACSIN1/syndapin1. Nat Neurosci 9(5):611–621
37. Qiu S, Lu Z, Levitt P (2014) MET receptor tyrosine kinase controls dendritic complexity, spine morphogenesis, and glutamatergic synapse maturation in the hippocampus. J Neurosci 34(49):16166–16179

38. Campbell DB, Li C, Sutcliffe JS, Persico AM, Levitt P (2008) Genetic evidence implicating multiple genes in the MET receptor tyrosine kinase pathway in autism spectrum disorder. Autism Res 1(3):159–168
39. Campbell DB, Sutcliffe JS, Ebert PJ, Militerni R, Bravaccio C, Trillo S et al (2006) A genetic variant that disrupts MET transcription is associated with autism. Proc Natl Acad Sci U S A 103(45):16834–16839
40. Lee Y, Kim SG, Lee B, Zhang Y, Kim Y, Kim S et al (2017) Striatal transcriptome and interactome analysis of Shank3-overexpressing mice reveals the connectivity between Shank3 and mTORC1 signaling. Front Mol Neurosci 10:201. https://doi.org/10.3389/fnmol.2017.00201
41. Han K, Holder JL Jr, Schaaf CP, Lu H, Chen H, Kang H et al (2013) SHANK3 overexpression causes manic-like behaviour with unique pharmacogenetic properties. Nature 503(7474):72–77
42. Bonaglia MC, Giorda R, Borgatti R, Felisari G, Gagliardi C, Selicorni A et al (2001) Disruption of the ProSAP2 gene in a t(12;22)(q24.1;q13.3) is associated with the 22q13.3 deletion syndrome. Am J Hum Genet 69(2):261–268
43. Durand CM, Betancur C, Boeckers TM, Bockmann J, Chaste P, Fauchereau F et al (2007) Mutations in the gene encoding the synaptic scaffolding protein SHANK3 are associated with autism spectrum disorders. Nat Genet 39(1):25–27
44. Failla P, Romano C, Alberti A, Vasta A, Buono S, Castiglia L et al (2007) Schizophrenia in a patient with subtelomeric duplication of chromosome 22q. Clin Genet 71(6):599–601
45. Boeckers TM, Bockmann J, Kreutz MR, Gundelfinger ED (2002) ProSAP/Shank proteins - a family of higher order organizing molecules of the postsynaptic density with an emerging role in human neurological disease. J Neurochem 81(5):903–910
46. Naisbitt S, Kim E, Tu JC, Xiao B, Sala C, Valtschanoff J et al (1999) Shank, a novel family of postsynaptic density proteins that binds to the NMDA receptor/PSD-95/GKAP complex and cortactin. Neuron 23(3):569–582
47. Costa-Mattioli M, Monteggia LM (2013) mTOR complexes in neurodevelopmental and neuropsychiatric disorders. Nat Neurosci 16(11):1537–1543
48. Jernigan CS, Goswami DB, Austin MC, Iyo AH, Chandran A, Stockmeier CA et al (2011) The mTOR signaling pathway in the prefrontal cortex is compromised in major depressive disorder. Prog Neuropsychopharmacol Biol Psychiatry 35(7):1774–1779
49. Li N, Lee B, Liu RJ, Banasr M, Dwyer JM, Iwata M et al (2010) mTOR-dependent synapse formation underlies the rapid antidepressant effects of NMDA antagonists. Science 329(5994):959–964
50. Fryland T, Christensen JH, Pallesen J, Mattheisen M, Palmfeldt J, Bak M et al (2016) Identification of the BRD1 interaction network and its impact on mental disorder risk. Genome Med 8(1):53. https://doi.org/10.1186/s13073-016-0308-x
51. Nyegaard M, Severinsen JE, Als TD, Hedemand A, Straarup S, Nordentoft M et al (2010) Support of association between BRD1 and both schizophrenia and bipolar affective disorder. Am J Med Genet B Neuropsychiatr Genet 153b(2):582–591
52. Severinsen JE, Bjarkam CR, Kiaer-Larsen S, Olsen IM, Nielsen MM, Blechingberg J et al (2006) Evidence implicating BRD1 with brain development and susceptibility to both schizophrenia and bipolar affective disorder. Mol Psychiatry 11(12):1126–1138
53. Mishima Y, Miyagi S, Saraya A, Negishi M, Endoh M, Endo TA et al (2011) The Hbo1-Brd1/Brpf2 complex is responsible for global acetylation of H3K14 and required for fetal liver erythropoiesis. Blood 118(9):2443–2453
54. Christensen JH, Elfving B, Müller HK, Fryland T, Nyegaard M, Corydon TJ et al (2012) The Schizophrenia and Bipolar Disorder associated BRD1 gene is regulated upon chronic restraint stress. Eur Neuropsychopharmacol 22(9):651–656
55. Grover D, Verma R, Goes FS, Mahon PL, Gershon ES, McMahon FJ et al (2009) Family-based association of YWHAH in psychotic bipolar disorder. Am J Med Genet B Neuropsychiatr Genet 150b(7):977–983
56. Williams HJ, Craddock N, Russo G, Hamshere ML, Moskvina V, Dwyer S et al (2011) Most genome-wide significant susceptibility loci for schizophrenia and bipolar disorder reported to date cross-traditional diagnostic boundaries. Hum Mol Genet 20(2):387–291

57. Wong AH, Likhodi O, Trakalo J, Yusuf M, Sinha A, Pato CN et al (2005) Genetic and post-mortem mRNA analysis of the 14-3-3 genes that encode phosphoserine/threonine-binding regulatory proteins in schizophrenia and bipolar disorder. Schizophr Res 78(2–3):137–146
58. Wong AH, Macciardi F, Klempan T, Kawczynski W, Barr CL, Lakatoo S et al (2003) Identification of candidate genes for psychosis in rat models, and possible association between schizophrenia and the 14-3-3eta gene. Mol Psychiatry 8(2):156–166
59. Moreno-Villanueva M, Morath J, Vanhooren V, Elbert T, Kolassa S, Libert C et al (2013) N-glycosylation profiling of plasma provides evidence for accelerated physiological aging in post-traumatic stress disorder. Transl Psychiatry 3:e320. https://doi.org/10.1038/tp.2013.93
60. Correll CU, Solmi M, Veronese N, Bortolato B, Rosson S, Santonastaso P et al (2017) Prevalence, incidence and mortality from cardiovascular disease in patients with pooled and specific severe mental illness: a large-scale meta-analysis of 3,211,768 patients and 113,383,368 controls. World Psychiatry 16(2):163–180
61. Cassidy F, Ahearn E, Carroll BJ (1999) Elevated frequency of diabetes mellitus in hospitalized manic-depressive patients. Am J Psychiatry 156(9):1417–1420
62. McIntyre RS, Konarski JZ, Misener VL, Kennedy SH (2005) Bipolar disorder and diabetes mellitus: epidemiology, etiology, and treatment implications. Ann Clin Psychiatry 17(2):83–93
63. Hajek T, Slaney C, Garnham J, Ruzickova M, Passmore M, Alda M (2005) Clinical correlates of current level of functioning in primary care-treated bipolar patients. Bipolar Disord 7(3):286–291
64. McIntyre RS, Danilewitz M, Liauw SS, Kemp DE, Nguyen HT, Kahn LS et al (2010) Bipolar disorder and metabolic syndrome: an international perspective. J Affect Disord 126(3):366–387
65. Annamalai A, Kosir U, Tek C (2017) Prevalence of obesity and diabetes in patients with schizophrenia. World J Diabetes 8(8):390–396
66. Mezuk B, Eaton WW, Albrecht S, Golden SH (2008) Depression and type 2 diabetes over the lifespan: a meta-analysis. Diabetes Care 31(12):2383–2390
67. Piazza I, Kochanowski K, Cappelletti V, Fuhrer T, Noor E, Sauer U et al (2018) A map of protein-metabolite interactions reveals principles of chemical communication. Cell 172(1–2):358–372.e23
68. Huber KV, Olek KM, Müller AC, Tan CS, Bennett KL, Colinge J et al (2015) Proteome-wide drug and metabolite interaction mapping by thermal-stability profiling. Nat Methods 12(11):1055–1057
69. Cassoli JS, Iwata K, Steiner J, Guest PC, Turck CW, Nascimento JM et al (2016) Effect of MK-801 and clozapine on the proteome of cultured human oligodendrocytes. Front Cell Neurosci 10:52. https://doi.org/10.3389/fncel.2016.00052
70. Park DI, Dournes C, Sillaber I, Ising M, Asara JM, Webhofer C et al (2017) Delineation of molecular pathway activities of the chronic antidepressant treatment response suggests important roles for glutamatergic and ubiquitin-proteasome systems. Transl Psychiatry 7(4):e1078. https://doi.org/10.1038/tp.2017.39
71. Park DI, Dournes C, Sillaber I, Uhr M, Asara JM, Gassen NC et al (2016) Purine and pyrimidine metabolism: convergent evidence on chronic antidepressant treatment response in mice and humans. Sci Rep 6:35317. https://doi.org/10.1038/srep35317
72. Steiner J, Martins-de-Souza D, Schiltz K, Sarnyai Z, Westphal S, Isermann B et al (2014) Clozapine promotes glycolysis and myelin lipid synthesis in cultured oligodendrocytes. Front Cell Neurosci 8:384. https://doi.org/10.3389/fncel.2014.00384
73. Weckmann K, Deery MJ, Howard JA, Feret R, Asara JM, Dethloff F (2017) Ketamine's antidepressant effect is mediated by energy metabolism and antioxidant defense system. Sci Rep 7(1):15788. https://doi.org/10.1038/s41598-017-16183-x

Chapter 9
MicroRNAs in Major Depressive Disorder

Gabriel R. Fries, Wei Zhang, Deborah Benevenuto, and Joao Quevedo

Abstract Major depressive disorder (MDD) is a severe and chronic psychiatric disorder with a high prevalence in the population. Although our understanding of its pathophysiological mechanisms has significantly increased over the years, available treatments still present several limitations and are not effective to all MDD patients. Epigenetic mechanisms have recently been suggested to play key roles in MDD pathogenesis and treatment, including the effects of small noncoding RNAs known as microRNAs (miRNAs). miRNAs can modulate gene expression posttranscriptionally by interfering with the stability and translation of messenger RNA molecules and are also known to cross-talk with other epigenetic mechanisms. In this review, we will summarize and discuss recent findings of alterations in miRNAs in tissues of patients with MDD and evidence of treatment-induced effects in these molecules.

Keywords Major depressive disorder · Patients · Pathological mechanisms · Treatment-induced effects · Biomarkers · MicroRNAs · miRNAs

G. R. Fries · W. Zhang · D. Benevenuto
Translational Psychiatry Program, Department of Psychiatry and Behavioral Sciences, McGovern Medical School, The University of Texas Health Science Center at Houston (UTHealth), Houston, TX, USA

J. Quevedo (✉)
Translational Psychiatry Program, Department of Psychiatry and Behavioral Sciences, McGovern Medical School, The University of Texas Health Science Center at Houston (UTHealth), Houston, TX, USA

Center of Excellence on Mood Disorders, Department of Psychiatry and Behavioral Sciences, McGovern Medical School, The University of Texas Health Science Center at Houston (UTHealth), Houston, TX, USA

Neuroscience Graduate Program, The University of Texas MD Anderson Cancer Center UTHealth Graduate School of Biomedical Sciences, Houston, TX, USA

Translational Psychiatry Laboratory, Graduate Program in Health Sciences, University of Southern Santa Catarina (UNESC), Criciúma, SC, Brazil
e-mail: Joao.L.DeQuevedo@uth.tmc.edu

P. C. Guest (ed.), *Reviews on Biomarker Studies in Psychiatric and Neurodegenerative Disorders*, Advances in Experimental Medicine and Biology 1118,
https://doi.org/10.1007/978-3-030-05542-4_9

9.1 Introduction

Major depressive disorder (MDD) is a chronic, debilitating, and life-threatening psychiatric illness. It is characterized by the presence of at least 2 weeks of depressed mood associated with changes in appetite and/or weight, sleep pattern, interest, concentration, and energy, besides psychomotor agitation or retardation, feelings of guilt, and suicidal behavior [1]. According to the World Health Organization, the global prevalence of MDD was 4.4% in 2015, with a rise of 18.4% in the number of cases since 2005. MDD is a major health concern as it affects more than 300 million people worldwide, with females having almost double the prevalence of males [2]. It is also known that MDD is associated with an increased probability of developing other medical conditions, such as diabetes mellitus, cardiovascular diseases, cancer, and cognitive impairment [3]. In addition, MDD is a major contributor to suicidal ideation and attempts, and the risk of suicide in this population has been reported to be approximately 20-fold higher than in the general population [4].

Recent evidence has shown significant advances in the understanding of the pathophysiological mechanisms of MDD. The etiology of MDD is multifactorial, involving a complex combination of biological, psychological, and socio-cultural determinants. Evidence suggests that MDD has a polygenic heritage with an estimated hereditability of approximately 35% [5], with its genetic background thought to interact with many environmental triggers throughout life to increase the risk for MDD. These include events involving financial issues and unemployment, bereavement, exposure to violence, lack of social support, childhood trauma, and major illnesses [6].

Among pathophysiological mechanisms, alterations in the hypothalamic-pituitary-adrenal (HPA) axis, which is known as the main neurobiological system related to stress response, have been closely related to MDD, with depressed patients showing increased levels of cortisol compared to controls [7]. Dysregulation of the immune system and microglial cells may also play an important role, which is suggested by evidence of increased levels of cytokines in patients with MDD and the reported higher risk of depression after infections and autoimmune diseases [8]. Additionally, downregulation of brain-derived neurotrophic factor (BDNF) and abnormalities in neurogenesis, synaptic plasticity, and dendritic morphology have also been associated with MDD [9, 10]. As a consequence, neuroimaging studies of patients with MDD have shown reduced volumes of different areas in the central nervous system (CNS), including amygdala, prefrontal cortex, and raphe nuclei, with the most consistent change being a reduced hippocampal volume compared to that in healthy controls [9].

Over the past decades, the monoamine hypothesis dominated the understanding of the MDD etiology and contributed to the development of most antidepressants currently in use by patients, such as monoamine oxidase (MAO) inhibitors, tricyclic antidepressants (TCA), and selective serotonin reuptake inhibitors (SSRI) [11]. However, besides all of the advances in knowledge related to the MDD pathophysiology, its current pharmacological treatment still relies vastly on the aforementioned monoaminergic antidepressants, presenting limited efficacy, delayed clinical effects, and considerable side effects. Therefore, there is an urging need for new therapeutic modalities that will effectively approach the complex pathophysiological underpinnings of MDD [12].

Based on this and the potential role of gene vs. environment interactions in MDD, epigenetic mechanisms have been proposed as novel targets to be explored in its treatment [13]. Small noncoding molecules of RNA known as microRNAs (miRNA) have gained special attention in recent studies. miRNAs play an important role in regulating neuronal physiology by disrupting messenger RNA (mRNA) expression, and miRNA dysregulation has been implicated as an important contributor to disorders of the CNS, including MDD [14]. In this chapter, we aim to review studies of miRNAs in MDD and provide an overview of the future perspectives and limitations in the field.

9.2 Biology of miRNAs

miRNAs comprise a large family of small noncoding RNAs that are typically ~22 nucleotides in length and act as key posttranscriptional regulators of gene expression [15]. In mammals, miRNAs are predicted to control the transcription of approximately 50% of all protein-coding genes, thus being involved in nearly all developmental and pathological processes.

9.2.1 MicroRNA Biogenesis

miRNAs are processed from primary miRNA (pri-miRNA) transcripts, most of which are transcribed by RNA polymerase II from independent miRNA genes, while others are located in the introns of protein-coding genes [16]. The pri-miRNAs are initially 5′-capped, spliced, and polyadenylated, after which they fold into one or more hairpin structures with a stem and a terminal loop. miRNAs in the genome may encode a single miRNA hairpin precursor or clusters of multiple precursors.

In the canonical pathway, the microprocessor complex processes the pri-miRNA into a hairpin-shaped molecule (pre-miRNA) of approximately 70 nucleotides in the nucleus, which is exported to the cytoplasm by exportin-5 via a ran-GTP-dependent mechanism [16]. The core components of the microprocessor complex are the Drosha, an RNase III enzyme, and the DGCR8/Pasha, a double-stranded RNA-binding domain protein. There are a variety of cofactors acting as components of the microprocessor, such as heterogeneous nuclear ribonucleoproteins (hnRNPs), the DEAD box RNA helicases p68 (DDX5) and p72 (DDX17). Some pre-miRNAs are produced from very short introns (mirtrons) as a result of splicing and debranching, in which case they can bypass the Drosha-DGCR8 processing step [16]. As a next step, cleavage by an enzyme called DICER leads to formation of a miRNA/miRNA* duplex in the cytoplasm. After the two strands of the miRNA/

miRNA* duplex are separated, one strand (the "guide" strand) associates with an argonaute (AGO) protein within the RNA-induced silencing complex (RISC or miRISC) and acts as a guide to repress target messages. The other strand (passenger or miRNA*) is released and degraded. AGO proteins are core components of the miRISC, as they directly associate with the miRNA molecule.

9.2.2 *MicroRNA Functions*

The guide strand is able to bind to the target mRNA by base pairing, after which the miRNA-RISC can inhibit the targeted coding gene by three different mechanisms: (A) site-specific cleavage, (B) enhanced mRNA degradation, or (C) translational inhibition. The decision-making step of this process depends on the degree of base-pairing complementarity between the mRNA molecule and the "seed" region at the 5′ end of the miRNA. Of note, the microRNA-mRNA binding site is short (6–8 base pairs), which indicates that one single miRNA may have the potential to bind and target multiple mRNAs. miRNAs mainly recognize and bind to the complementary sequences in the 3′-untranslated regions (UTRs) of their target mRNAs, but recent reports have suggested that they can bind to the 5′-UTR or the open reading frame (ORF), as well.

9.3 miRNAs as Biomarkers of Major Depression

Several studies have investigated alterations in miRNA systems in patients diagnosed with MDD or during major depressive episodes. As summarized in Table 9.1, the majority of these studies have focused on peripheral blood alterations, with only a few investigating other peripheral tissues, such as cerebrospinal fluid (CSF) and dermal fibroblasts, or postmortem brain tissues. Overall, possibly due to the heterogeneity of samples, the findings thus far have rarely been replicated between studies. Only a few miRNAs have been reported to be altered in more than one study, among which miR-132 [17–19], miR-451a [20–22], and miR-34a-5p [21–23] seem to be the most consistent. In addition, replication of significant findings (although not necessarily in the same direction between different studies) has also been observed for let7b [24, 25], miR-182 [18, 26], miR-124 [27, 28], miR-345 [25, 29], miR-146b-5p [29, 30], miR-146a [30, 31], miR-494 [25, 31], miR-376a [25, 31], miR-107 [19, 25], miR-33a [19, 25], and miR-221-3p [21, 22].

Among all studied miRNAs, miR-132 is one of the most consistently shown to be altered in patients with MDD. Specifically, MDD patients have been reported to show increased levels of this miRNA in whole blood [17], serum [18], and fibroblasts [19] compared to healthy controls. Moreover, miR-132 expression levels have been negatively correlated with visual memory parameters in patients [17] and with serum brain-derived neurotrophic factor (BDNF) levels in patients and controls [18], suggesting a key role for this miRNA alteration in cognitive processes. In this

Table 9.1 MicroRNA alterations in major depressive disorder (MDD) patients

Tissue	Sample	Detection method	Main findings	References
Whole blood	*Discovery sample*: adolescents (controls, risk for depression, depressed, $n = 93$) *Replication sample*: adolescents (controls, depressed, $n = 93$)	Microarray (Illumina 450k methylation chip)	Methylation at two CpG sites predicted depression (cg13227623 and cg04102384). cg04102384, which is located in the promoter region of microRNA 4646 (*MIR4646*), was shown to be hypomethylated and negatively correlated with expression levels of MIR4646-3p in controls	[67]
Whole blood	Depression/anxiety patients ($n = 169$) and healthy controls ($n = 52$). Initial screening performed with only 11 patients	Serum/Plasma Focus microRNA Panel (179 miRNA primer sets) and qPCR	Differential expression of six miRNAs between groups (miR-144-5p, miR-92b-3p, miR-885-5p, miR-30a-5p, miR-29a-5p, and miR-29b-2-5p). Negative correlation between plasma miR-144-5p expression and depression scores. Validated reduced plasma levels of miR-144-5p in patients compared to controls	[68]
Whole blood	MDD patients ($n = 50$) and healthy controls ($n = 41$)	qPCR	Downregulation of miR-320a and upregulation of miR-451a, miR-17-5p, and miR-223-3p in patients compared to controls	[20]
Whole blood	MDD patients ($n = 62$) and healthy controls ($n = 73$)	qPCR	Upregulation of miR-132 expression in MDD patients. Correlation between miR-132 and visual memory	[17]
Whole blood	Treatment-resistant depression patients ($n = 40$) and healthy controls ($n = 20$)	qPCR	Reduced expression of let-7b in patients compared to controls. No difference in let-7c expression	[24]
Whole blood	MDD patients ($n = 18$) and healthy controls ($n = 18$)	qPCR	Increased expression of miR-644, miR-450b, miR-328, and miR-182 in patients compared to controls. Lower expression of miR-335, miR-583, miR-650, miR-708, and miR-654 in patients compared to controls. miR-335 was downregulated the most	[26]
Whole blood	MDD patients ($n = 20$) and healthy controls ($n = 20$)	Microarray	Differential expression of hsa-let-7a-5p, hsa-let-7d-5p, hsa-let-7f-5p, hsa-miR-24-3p, hsa-miR-425-3p, hsa-miR-330-3p, and hsa-miR-345-5p in MDD patients compared to controls	[69]
Peripheral blood leukocytes	MDD patients ($n = 32$) and healthy controls ($n = 32$)	qPCR	Increased expression of miR-34b-5p and miR-34c-5p in patients compared to controls. Lower expression of miR-34b-5p and miR-369-3p in patients with suicide ideation. Positive correlation between N1 latency of P300 and miR-34c-5p, miR-107, and miR-381. Negative correlation between P2 latency of P300 and miR-34c-5p, miR-107, and miR-381	[70]

(continued)

Table 9.1 (continued)

Tissue	Sample	Detection method	Main findings	References
PBMCs	MDD patients ($n = 32$) and healthy controls ($n = 30$)	qPCR	Increased expression of miR-124 in patients compared to controls. ROC analysis of miR-124 showed an area under the curve of 0.762, sensitivity of 83.3%, and specificity of 66.67% to distinguish both groups	[28]
PBMCs	MDD patients ($n = 91$) and healthy controls ($n = 46$)	Microarray screening (3 patients vs. 3 controls) and qPCR (81 patients vs. 46 controls)	Changes in 26 miRNAs in patients compared to controls. Alterations were validated in five miRNAs (miR-26b, miR-1972, miR-4485, miR-4498, and miR4743)	[29]
PBMCs	Severe major depressive episode patients ($n = 16$) and healthy controls ($n = 13$)	Microarray and qPCR	Differential expression of 14 miRNAs between patients and controls both at baseline and after naturalistic treatment with different medications (9 upregulated and 5 downregulated; total of 26 unique miRNAs). Stable overexpression of miR-941 and miR-589 in patients during an 8-week follow-up	[25]
Serum	MDD patients ($n = 40$) and healthy controls ($n = 40$)	qPCR	Increased serum levels of miR-132 and miR-182 in MDD patients (in parallel to decreased BDNF serum levels). Positive correlation between miR-182 levels and depressive rating scores. Negative correlation between serum BDNF and miR-132/miR-182 levels	[18]
Serum	MDD patients ($n = 84$) and healthy controls ($n = 78$)	qPCR	Lower miR-451a and higher miR-34a-5p and miR-221-3p levels in patients compared to controls. Negative correlation between miR-451a levels and depressive symptoms. Positive correlations between miR-34a-5p and miR-221-3p and depressive symptoms	[21]
CSF and whole blood	Drug-free MDD patients ($n = 36$) and healthy controls ($n = 30$)	qPCR	Reduced miR-16 levels in the CSF of MDD patients. Negative correlation between CSF miR-16 and depressive symptoms. No alterations in blood miR-16 levels	[71]
CSF and serum	CSF: MDD patients ($n = 6$) and controls ($n = 6$) Blood: MDD patients ($n = 32$) and controls ($n = 21$)	qPCR	Higher levels of 11 miRNAs and decreased levels of 5 miRNAs in the CSF of MDD patients compared to controls. Upregulation of miR-221-3p, miR-34a-5p, and let-7d-3p and downregulation of miR-451a were validated in serum samples, as well	[22]

PFC (BA46), whole blood	Prefrontal cortex: MDD subjects ($n = 15$) and nonpsychiatric controls ($n = 15$) Blood: antidepressant-free MDD patients ($n = 18$) and healthy controls ($n = 17$)	qPCR	Increased expression of miR-124-3p in the PFC of MDD subjects compared to controls. Higher levels of miR-124-3p in the serum of MDD patient compared to controls	[27]
PFC (BA44) and whole blood	Prefrontal cortex: (1) discovery sample, MDD patients ($n = 14$) and nonpsychiatric controls ($n = 11$); (2) replication sample, MDD patients ($n = 25$), MDD patients with a history of antidepressant use ($n = 25$), controls ($n = 29$) Blood: untreated MDD patients ($n = 32$), healthy controls ($n = 18$)	Microarray and qPCR	Decreased expression of miR-1202 in depressed brains. Difference in brain miR-1202 expression between depressed subjects with and without antidepressant history. Decreased blood levels of miR-1202 in depressed patients compared to controls	[57]
PFC (BA10)	MDD patients ($n = 15$) and nonpsychiatric controls ($n = 15$)	High-throughput qPCR	Altered levels of two miRNAs in depressed subjects compared to controls (miR-508-3p and miR-152, both downregulated)	[72]
PFC (BA9)	Antidepressant-free depressed suicide subjects ($n = 18$) and nonpsychiatric controls ($n = 17$)	Multiplex qPCR	Downregulation of overall miRNA expression in depressed subjects. Lower expression of 21 miRNAs in patients	[31]
PFC	Depressed suicide completers ($n = 32$) and non-suicide controls ($n = 20$)	qPCR	Altered expression (upregulation) of miR-146a-5p, miR-146b-5p, miR-24-3p, and miR-425-3p compared to controls	[30]
Basolateral amygdala	MDD suicide subjects ($n = 16$) and nonpsychiatric controls ($n = 21$)	qPCR	Increase of miR-511 and strong tendency toward increased miR-340 levels in depressed subjects compared to controls. Significant correlation between miR-340 ad miR-511 levels	[73]
Anterior cingulate cortex	MDD patients ($n = 15$) and nonpsychiatric controls ($n = 14$)	qPCR	Differential expression (downregulation) of miR-184 and miR-34a in MDD (although not corrected for multiple correction testing)	[23]
Dermal fibroblasts	MDD patients ($n = 16$) and healthy controls ($n = 16$)	qPCR	Altered expression of 38 miRNAs in MDD patients compared to controls	[19]

BA Brodmann's area, *BDNF* brain-derived neurotrophic factor, *CSF* cerebrospinal fluid, *MDD* major depressive disorder, *PBMC* peripheral blood mononuclear cell, *PCR* polymerase chain reaction, *PFC* prefrontal cortex, *qPCR* quantitative real-time PCR, *ROC* receiver operating characteristic same manner, preclinical studies have demonstrated that brain overexpression of miR-132 induces impairment of memory mechanisms in animal models [32, 33] and reduces the expression of BDNF in cell culture [18], supporting the association between miR-132 expression and cognition in MDD patients [17]. Of note, previous studies have identified associations between miR-132 levels and other neuropsychiatric and neurodegenerative disorders, such as Alzheimer's disease [34], Huntington's disease [35], substance abuse disorders [36, 37], bipolar disorder [23, 38], and schizophrenia [39]. These findings suggest that alterations in this specific miRNA might not be specific to MDD.

Another alteration that has been replicated in different MDD studies is related to levels of miR-451a, although not always showing a consistent direction. miR-451a levels have been shown to be both upregulated in plasma [20] and downregulated in serum [21] and CSF [22] of MDD patients compared to healthy controls. Moreover, significant correlations have been reported between miR-451a levels and the duration of the depressive episode [20], the severity of depressive symptoms [21], and history of suicide attempts [21]. In addition, miR-451a serum levels were shown to be significantly increased after 8 weeks of treatment with paroxetine [21] and even demonstrated high sensitivity (84.85%) and specificity (90.48%) for the diagnosis of MDD [22]. Similar to miR-132, alterations in miR-451 have also been reported in other neuropsychiatric and neurodegenerative disorders, such as Alzheimer's disease [40] and autism spectrum disorders [41].

Finally, the third miRNA to have shown consistent and somewhat replicable alterations in MDD is the miR-34a-5p. Its levels have been shown to be upregulated in peripheral tissues of MDD patients in two studies [21, 22], while one study found a downregulation (of nominal significance) in the postmortem anterior cingulate cortex of patients [23]. The clinical significance of this miRNA is also suggested by evidence of a positive correlation between miR-34a-5p levels and the severity of depressive symptoms [21], showing significant associations with the disease course and history of suicide attempts. Similar to miR-451a, miR-34a-5p levels have been shown to present relatively high specificity (95.24%) and sensitivity (96.88%) values for MDD diagnosis [22], which is particularly exciting in terms of using this miRNA as a potential biomarker in the future. Of note, miR-34a-5p has also been linked previously to schizophrenia [42], bipolar disorder [23], Alzheimer's disease [43], and Huntington's disease [44].

9.4 miRNAs in the Pathogenesis of Major Depressive Disorder

There are many published studies using cell lines and animal models to investigate the potential role of miRNAs in MDD. Among different approaches to the modeling of MDD in the laboratory setting, the chronic unpredictable mild stress (CUMS) model is widely used to mimic major depression in rodents [45]. The CUMS is an artificial stress model consisting of food and water deprivation, light and space limitation, temperature manipulation, and predator sounds, among other stimuli.

Overall, CUMS-induced depressed-like symptoms and the associated treatment have been linked to several alterations in miRNA and related molecules. The findings include an upregulation of miR-132 in the hippocampus of CUMS-exposed rats along with downregulation of MeCP2 and BDNF, which suggests that miR-132 may play a role in neuroplasticity and neuronal survival in depression [46]. Moreover, not only has this model been associated with decreased levels of miR-101 in the ventrolateral orbital cortex of mice, but restoring miR-101 levels in the same brain region has been shown to reverse the depressive-like behavior caused by CUMS [47]. Similarly, a study found that CUMS induces a downregulation of miR-124 in the hippocampus of mice, which was reversed by antidepressant treatment [48]. The same study found that histone deacetylase (HDAC) 4, HDAC5, and glycogen synthase kinase 3β (GSK3β) are direct targets of miR-124 and that the combination of a selective HDAC4/5 inhibitor or GSK3 inhibitor can relieve the depressive-like behaviors in mice [48]. Evidence also suggests a CUMS-induced reduction in miR-326 levels in the Edinger-Westphal nucleus of rats [49] and a key role for miR-132 in the protective effects of oleanolic acid in the model [50]. Finally, elevated serum and mesocortical levels of miR-16 after 7 weeks of CUMS have been associated with resilience to the effects of the model, suggesting an interesting role for this particular miRNA in stress coping mechanisms [51].

9.5 miRNAs in the Treatment of Major Depression

One of the most studied facets of the field is the involvement of miRNAs in the mechanisms of action of antidepressants and their potential role in the response to treatment in MDD patients. This has been repeatedly reviewed by other groups [12, 52–54], offering a valuable insight into the development of novel therapeutics.

A summary of studies investigating treatment-induced alterations in the levels of miRNAs in clinical populations can be found in Table 9.2. Although several alterations have been reported, only a few of them have been replicated by independent studies. Among these, miR-1202 changes after treatment seems to be the most consistent finding [55–57], but significant results have also been replicated for let-7b [24, 58], let-7c [24, 58], let-7d [59, 60], miR-132 [58, 60], miR-151-3p [58, 59], miR-221 [21, 59, 61], miR-26a [59, 60], miR-335 [26, 60], miR-361-5p [30, 60], and miR-433 [25, 59].

9.6 Perspectives and Future Directions

Based on the concepts previously discussed in this review, there are several findings supporting a key role for miRNAs in MDD and its treatment. The field has been evolving during the years and generating more consistent and reliable results as

Table 9.2 Clinical evidence of microRNA mechanisms involved in the treatment of depression

Treatment	Tissue/model	Sample	Main findings	References
Several (8 weeks; naturalistic design)	PBMCs	MDD patients ($n = 32$) pre- and posttreatment with antidepressants for 8 weeks	Downregulation of miR-124 expression after 8 weeks of treatment. Significant reduction of miR-124 after treatment in responders compared to non-responder patients	[28]
Escitalopram (12 weeks)	Whole blood	MDD patients ($n = 10$)	Altered expression of 30 miRNAs after escitalopram treatment (28 upregulated and 2 downregulated)	[60]
Antidepressants and tranquilizers, non-specified (8 weeks)	Whole blood	Depression/anxiety patients ($n = 169$)	Increase of miR-144-5p and miR-30a-5p levels after treatment. No significant associations between the changes in miRNAs and MADRS-S score changes	[68]
Antidepressants (12 weeks)	Whole blood	MDD patients ($n = 5$)	Differential expression of 40 miRNAs after treatment (23 up- and 17 downregulated)	[55]
Desvenlafaxine (8 weeks)	Whole blood	MDD patients ($n = 20$)	Negative correlation between changes in miR-1202 levels following treatment and changes in brain activity and connectivity	[55]
Paroxetine	LCLs	Healthy controls ($n = 80$) for initial screening. Eight LCLs with high or low sensitivity to paroxetine were further analyzed for their miRNAs profile	Increased basal expression of miR-151-3p in paroxetine-sensitive LCLs. Altered expression of miR-212, miR-132, miR-30b, let-7b, and let-7c between the two groups	[58]
Cohort 1: escitalopram or desvenlafaxine (8 weeks) Cohort 2: duloxetine (up to 8 weeks)	Whole blood	MDD patients (cohort 1, $n = 55$; cohort 2, $n = 124$)	Lower baseline miR-1202 levels in responders compared to non-responders. Increase in miR-1202 levels following treatment	[56]
Paroxetine (8 weeks)	Serum	MDD patients ($n = 84$)	Increase of miR-451a levels and decrease of miR-34a-5p and miR-221-3p levels after paroxetine treatment	[21]

Discovery cohort: duloxetine (8 weeks) Replication cohort 1: escitalopram or nortriptyline (8 weeks) Replication cohort 2: escitalopram (8 weeks)	Whole blood	MDD patients from three cohorts: (1) discovery ($n = 258$), replication 1 ($n = 61$), and replication 2 ($n = 158$)	Discovery cohort: altered expression of 16 miRNAs specifically after duloxetine treatment (in responders). Downregulation of miR-146a-5p, miR-146b-5p, miR-24-3p, and miR-425-3p was validated in replication cohort 1. Downregulation of miR-146a-5p, miR-146b-5p, and miR-24-3p was further validated in replication cohort 2	[30]
Ketamine (once a week for up to three sessions) and ECT (bi-weekly)	Whole blood	Treatment-resistant depression patients ($n = 40$) and healthy controls ($n = 20$)	Reduction in the expression of let-7b and let-7c in patients who received ECT (responders and non-responders). No baseline miRNAs predicted response to ECT or ketamine.	[24]
Citalopram (4 weeks)	Human neuroprogenitor cells (in vitro) and whole blood	MDD patients ($n = 18$)	Upregulation of the expression of miR-335 after treatment of human neuroprogenitor cells with citalopram for 7 days. Increase of blood miR-335 expression after citalopram treatment	[26]
Citalopram (8 weeks) and imipramine	Human neuroprogenitor cells (in vitro) and whole blood	MDD patients ($n = 32$), divided into remitters and non-responders ($n = 16$/group)	Upregulation of miR-1202 after chronic treatment with either imipramine or citalopram in neuroprogenitor cells. Increase of blood miR-1202 levels after citalopram treatment in remitter patients. Negative correlation between changes in depression severity and changes in miR-1202 expression	[57]
ECT	Whole blood	Discovery sample: MDD patients ($n = 16$) Validation sample: MDD patients ($n = 37$) and healthy controls ($n = 34$)	Differential miRNA expression from baseline to end of treatment in patients with psychotic depression. Higher baseline expression of miR-126-3p and miR-106a-5p in psychotic depression patients compared to controls, with no between-group differences after ECT	[74]
Citalopram (12 weeks)	Whole blood	Major depressive episode patients ($n = 34$) and healthy controls ($n = 33$)	Identification of 414 miRNAs that may regulate one or several modules associated with clinical improvement	[75]
Naturalistic pharmacological treatment (8 weeks)	PBMCs	Several major depressive episode patients ($n = 16$)	Differential expression of miRNAs among patients that showed clinical improvement to treatment (seven upregulated and one downregulated)	[25]

ECT electroconvulsive therapy, *LCL* lymphoblastoid cell line, *MADRS-S* Montgomery-Åsberg Depression Rating Scale, self-rating version, *MDD* major depressive disorder, *PBMC* peripheral blood mononuclear cell miRNA-detecting technologies develop and our understanding of the biology of these small molecules increases. While most studies have focused on candidate miRNAs detected by quantitative PCR (qPCR) or performed genome-wide investigations using microarray technology, the development of small RNA next-generation sequencing offers a more valuable and unbiased method that should be considered in future studies. Moreover, the field is likely to benefit significantly from exploring more sophisticated and innovative approaches to the study of miRNAs, such as the investigation of such molecules in tissue-specific extracellular vesicles [62], the study of miRNA-mediated transgenerational transmission of complex behaviors [63, 64], and also the pharmacological targeting of specific miRNAs [65, 66].

Of the several limitations of the currently available studies, the main ones rely on the use of different methods for the analysis of miRNAs (which significantly limits the comparison of results between studies) and the heterogeneity of patient populations and tissues investigated between studies. In addition, more in-depth analyses of the correlation between peripheral and central tissues focused on miRNAs should also be performed so that peripheral findings can be further interpreted as proxy of brain tissue.

Finally, given the complexity of epigenetic mechanisms and the known cross talk between miRNA-mediated mechanisms of gene regulation with others, future studies should focus on analyzing miRNA alterations in combination with genotype, DNA methylation, and expression levels to provide a more complete and informative screenshot of the epigenetics and gene versus environment interaction mechanisms in MDD.

9.7 Conclusions

The field of psychiatric epigenetics is rapidly evolving, and the role of miRNAs in regulating gene expression in neuropsychiatric disorders has been consistently suggested by a growing body of evidence over the last few years. Although existing findings show convincing alterations in MDD and its treatment (especially with miR-132, miR-451a, miR-34a-5p, and miR-1202), most of these have not been sufficiently replicated in different populations, and some seem to be detected in other neuropsychiatric disorders. In this sense, the field is likely to significantly benefit from the investigation of more homogenous patient populations, the use of unbiased genome-wide sequencing methods, and a consistent investigation of the correlation between central and peripheral tissues. Nonetheless, targeting miRNAs seems to represent a valuable and innovative approach that may overcome the limitations of currently available antidepressants and provide new and powerful insights into the neurobiology of MDD and its transmission across generations.

Acknowledgments The Translational Psychiatry Program (USA) is funded by the Department of Psychiatry and Behavioral Sciences, McGovern Medical School, the University of Texas Health Science Center at Houston (UTHealth). The Center of Excellence on Mood Disorders (USA) is funded by the Pat Rutherford Jr. Chair in Psychiatry, John S. Dunn Foundation, and Anne and Don Fizer Foundation Endowment for Depression Research. Translational Psychiatry Laboratory (Brazil) is one of the centers of the National Institute for Molecular Medicine (INCT-MM) and one of the members of the Center of Excellence in Applied Neurosciences of Santa Catarina (NENASC). Its research is supported by grants from CNPq, FAPESC, Instituto Cérebro e Mente, and UNESC.

J.Q. is a 1A CNPq Research Fellow.

References

1. American Psychiatric Association (2013) Diagnostic and statistical manual of mental disorders, 5th edn: DSM-5. American Psychiatric Publishing, Washington, DC. ISBN-10: 8123923791
2. World Health Organization (2017) Depression and other common mental disorders: global health estimates. http://apps.who.int/iris/bitstream/handle/10665/254610/WHO-MSD-MER-2017.2-eng.pdf;jsessionid=6EE653ED0F776698A04D5181E8AB50D0?sequence=1
3. Penninx BW, Milaneschi Y, Lamers F, Vogelzangs N (2013) Understanding the somatic consequences of depression: biological mechanisms and the role of depression symptom profile. BMC Med 11:129. https://doi.org/10.1186/1741-7015-11-129
4. Chesney E, Goodwin GM, Fazel S (2014) Risks of all-cause and suicide mortality in mental disorders: a meta-review. World Psychiatry 13:153–160
5. Geschwind DH, Flint J (2015) Genetics and genomics of psychiatric disease. Science 349:1489–1494
6. Lorant V, Deliege D, Eaton W, Robert A, Philippot P, Ansseau M (2003) Socioeconomic inequalities in depression: a meta-analysis. Am J Epidemiol 157:98–112
7. Stetler C, Miller GE (2011) Depression and hypothalamic-pituitary-adrenal activation: a quantitative summary of four decades of research. Psychosom Med 73:114–126
8. Benros ME, Waltoft BL, Nordentoft M, Ostergaard SD, Eaton WW, Krogh J et al (2013) Autoimmune diseases and severe infections as risk factors for mood disorders: a nationwide study. JAMA Psychiat 70:812–820
9. Otte C, Gold SM, Penninx BW, Pariante CM, Etkin A, Fava M et al (2016) Major depressive disorder. Nat Rev Dis Primers 2:16065. https://doi.org/10.1038/nrdp.2016.65
10. Molendijk ML, Spinhoven P, Polak M, Bus BA, Penninx BW, Elzinga BM (2014) Serum BDNF concentrations as peripheral manifestations of depression: evidence from a systematic review and meta-analyses on 179 associations (N=9484). Mol Psychiatry 19:791–800
11. Boku S, Nakagawa S, Toda H, Hishimoto A (2018) Neural basis of major depressive disorder: beyond monoamine hypothesis. Psychiatry Clin Neurosci 72:3–12
12. Hansen KF, Obrietan K (2013) MicroRNA as therapeutic targets for treatment of depression. Neuropsychiatr Dis Treat 9:1011–1012
13. Dalton VS, Kolshus E, McLoughlin DM (2014) Epigenetics and depression: return of the repressed. J Affect Disord 155:1–12
14. Hunsberger JG, Austin DR, Chen G, Manji HK (2009) MicroRNAs in mental health: from biological underpinnings to potential therapies. NeuroMolecular Med 11:173–182
15. Huang Y, Shen XJ, Zou Q, Wang SP, Tang SM, Zhang GZ (2011) Biological functions of microRNAs: a review. J Physiol Biochem 67:129–139
16. Ha M, Kim VN (2014) Regulation of microRNA biogenesis. Nat Rev Mol Cell Biol 15:509–524
17. Liu Y, Yang X, Zhao L, Zhang J, Li T, Ma X (2016) Increased miR-132 level is associated with visual memory dysfunction in patients with depression. Neuropsychiatr Dis Treat 12:2905–2911
18. Li YJ, Xu M, Gao ZH, Wang YQ, Yue Z, Zhang YX et al (2013) Alterations of serum levels of BDNF-related miRNAs in patients with depression. PLoS One 8:e63648. https://doi.org/10.1371/journal.pone.0063648

19. Garbett KA, Vereczkei A, Kalman S, Brown JA, Taylor WD, Faludi G et al (2015) Coordinated messenger RNA/microRNA changes in fibroblasts of patients with major depression. Biol Psychiatry 77:256–265
20. Camkurt MA, Acar S, Coskun S, Gunes M, Gunes S, Yilmaz MF et al (2015) Comparison of plasma MicroRNA levels in drug naive, first episode depressed patients and healthy controls. J Psychiatr Res 69:67–71
21. Kuang WH, Dong ZQ, Tian LT, Li J (2018) MicroRNA-451a, microRNA-34a-5p, and microRNA-221-3p as predictors of response to antidepressant treatment. Braz J Med Biol Res 51:e7212. https://doi.org/10.1590/1414-431x20187212
22. Wan Y, Liu Y, Wang X, Wu J, Liu K, Zhou J et al (2015) Identification of differential microRNAs in cerebrospinal fluid and serum of patients with major depressive disorder. PLoS One 10:e0121975. doi: 10.1371/journal.pone.0121975
23. Azevedo JA, Carter BS, Meng F, Turner DL, Dai M, Schatzberg AF et al (2016) The microRNA network is altered in anterior cingulate cortex of patients with unipolar and bipolar depression. J Psychiatr Res 82:58–67
24. Gururajan A, Naughton ME, Scott KA, O'Connor RM, Moloney G, Clarke G et al (2016) MicroRNAs as biomarkers for major depression: a role for let-7b and let-7c. Transl Psychiatry 6:e862. https://doi.org/10.1038/tp.2016.131
25. Belzeaux R, Bergon A, Jeanjean V, Loriod B, Formisano-Treziny C, Verrier L et al (2012) Responder and nonresponder patients exhibit different peripheral transcriptional signatures during major depressive episode. Transl Psychiatry 2:e185. https://doi.org/10.1038/tp.2012.112
26. Li J, Meng H, Cao W, Qiu T (2015) MiR-335 is involved in major depression disorder and antidepressant treatment through targeting GRM4. Neurosci Lett 606:167–172
27. Roy B, Dunbar M, Shelton RC, Dwivedi Y (2017) Identification of MicroRNA-124-3p as a putative epigenetic signature of major depressive disorder. Neuropsychopharmacology 42:864–875
28. He S, Liu X, Jiang K, Peng D, Hong W, Fang Y et al (2016) Alterations of microRNA-124 expression in peripheral blood mononuclear cells in pre- and post-treatment patients with major depressive disorder. J Psychiatr Res 78:65–71
29. Fan HM, Sun XY, Guo W, Zhong AF, Niu W, Zhao L et al (2014) Differential expression of microRNA in peripheral blood mononuclear cells as specific biomarker for major depressive disorder patients. J Psychiatr Res 59:45–52
30. Lopez JP, Fiori LM, Cruceanu C, Lin R, Labonte B, Cates HM et al (2017) MicroRNAs 146a/b-5 and 425-3p and 24-3p are markers of antidepressant response and regulate MAPK/Wnt-system genes. Nat Commun 8:15497. https://doi.org/10.1038/ncomms15497
31. Smalheiser NR, Lugli G, Rizavi HS, Torvik VI, Turecki G, Dwivedi Y (2012) MicroRNA expression is down-regulated and reorganized in prefrontal cortex of depressed suicide subjects. PLoS One 7:e33201. https://doi.org/10.1371/journal.pone.0033201
32. Hansen KF, Sakamoto K, Wayma n GA, Impey S, Obrietan K (2010) Transgenic miR132 alters neuronal spine density and impairs novel object recognition memory. PLoS One 5:e15497. https://doi.org/10.1371/journal.pone.0015497
33. Scott HL, Tamagnini F, Narduzzo KE, Howarth JL, Lee YB, Wong LF et al (2012) MicroRNA-132 regulates recognition memory and synaptic plasticity in the perirhinal cortex. Eur J Neurosci 36:2941–2948
34. Salta E, De Strooper B (2017) microRNA-132: a key noncoding RNA operating in the cellular phase of Alzheimer's disease. FASEB J 31:424–433
35. Lee ST, Chu K, Im WS, Yoon HJ, Im JY, Park JE et al (2011) Altered microRNA regulation in Huntington's disease models. Exp Neurol 227:172–179
36. Sadakierska-Chudy A, Frankowska M, Miszkiel J, Wydra K, Jastrzebska J, Filip M (2017) Prolonged induction of miR-212/132 and REST expression in rat striatum following cocaine self-administration. Mol Neurobiol 54:2241–2254

37. Quinn RK, James MH, Hawkins GE, Brown AL, Heathcote A, Smith DW et al (2018) Temporally specific miRNA expression patterns in the dorsal and ventral striatum of addiction-prone rats. Addict Biol 23:631–642
38. Walker RM, Rybka J, Anderson SM, Torrance HS, Boxall R, Sussmann JE et al (2015) Preliminary investigation of miRNA expression in individuals at high familial risk of bipolar disorder. J Psychiatr Res 62:48–55
39. Yu HC, Wu J, Zhang HX, Zhang GL, Sui J, Tong WW et al (2015) Alterations of miR-132 are novel diagnostic biomarkers in peripheral blood of schizophrenia patients. Prog Neuro-Psychopharmacol Biol Psychiatry 63:23–29
40. Villela D, Ramalho RF, Silva AR, Brentani H, Suemoto CK, Pasqualucci CA et al (2016) Differential DNA methylation of MicroRNA genes in temporal cortex from Alzheimer's disease individuals. Neural Plast 2016:2584940. https://doi.org/10.1155/2016/2584940
41. Pagan C, Goubran-Botros H, Delorme R, Benabou M, Lemiere N, Murray K et al (2017) Disruption of melatonin synthesis is associated with impaired 14-3-3 and miR-451 levels in patients with autism spectrum disorders. Sci Rep 7:2096. https://doi.org/10.1038/s41598-017-02152-x
42. Liu S, Zhang F, Wang X, Shugart YY, Zhao Y, Li X et al (2017) Diagnostic value of blood-derived microRNAs for schizophrenia: results of a meta-analysis and validation. Sci Rep 7:15328. https://doi.org/10.1038/s41598-017-15751-5
43. Cosin-Tomas M, Antonell A, Llado A, Alcolea D, Fortea J, Ezquerra M et al (2017) Plasma miR-34a-5p and miR-545-3p as early biomarkers of Alzheimer's disease: potential and limitations. Mol Neurobiol 54:5550–5562
44. Reynolds RH, Petersen MH, Willert CW, Heinrich M, Nymann N, Dall M et al (2018) Perturbations in the p53/miR-34a/SIRT1 pathway in the R6/2 Huntington's disease model. Mol Cell Neurosci 88:118–129
45. Willner P (2017) The chronic mild stress (CMS) model of depression: history, evaluation and usage. Neurobiol Stress 6:78–93
46. Su M, Hong J, Zhao Y, Liu S, Xue X (2015) MeCP2 controls hippocampal brain-derived neurotrophic factor expression via homeostatic interactions with microRNA132 in rats with depression. Mol Med Rep 12:5399–5406
47. Zhao Y, Wang S, Chu Z, Dang Y, Zhu J, Su X (2017) MicroRNA-101 in the ventrolateral orbital cortex (VLO) modulates depressive-like behaviors in rats and targets dual-specificity phosphatase 1 (DUSP1). Brain Res 1669:55–62
48. Higuchi F, Uchida S, Yamagata H, Abe-Higuchi N, Hobara T, Hara K et al (2016) Hippocampal MicroRNA-124 enhances chronic stress resilience in mice. J Neurosci 36:7253–7267
49. Aschrafi A, Verheijen JM, Gordebeke PM, Olde Loohuis NF, Menting K, Jager A et al (2016) MicroRNA-326 acts as a molecular switch in the regulation of midbrain urocortin 1 expression. J Psychiatry Neurosci 41:342–353
50. Yi LT, Li J, Liu BB, Luo L, Liu Q, Geng D (2014) BDNF-ERK-CREB signalling mediates the role of miR-132 in the regulation of the effects of oleanolic acid in male mice. J Psychiatry Neurosci 39:348–359
51. Zurawek D, Kusmider M, Faron-Gorecka A, Gruca P, Pabian P, Kolasa M et al (2016) Time-dependent miR-16 serum fluctuations together with reciprocal changes in the expression level of miR-16 in mesocortical circuit contribute to stress resilient phenotype in chronic mild stress—an animal model of depression. Eur Neuropsychopharmacol 26:23–36
52. Lopez JP, Kos A, Turecki G (2018) Major depression and its treatment: microRNAs as peripheral biomarkers of diagnosis and treatment response. Curr Opin Psychiatry 31:7–16
53. Belzeaux R, Lin R, Ju C, Chay MA, Fiori LM, Lutz PE et al (2018) Transcriptomic and epigenomic biomarkers of antidepressant response. J Affect Disord 233:36–44
54. Belzeaux R, Lin R, Turecki G (2017) Potential use of MicroRNA for monitoring therapeutic response to antidepressants. CNS Drugs 31:253–262
55. Lopez JP, Pereira F, Richard-Devantoy S, Berlim M, Chachamovich E, Fiori LM et al (2017) Co-variation of peripheral levels of miR-1202 and brain activity and connectivity during antidepressant treatment. Neuropsychopharmacology 42:2043–2051

56. Fiori LM, Lopez JP, Richard-Devantoy S, Berlim M, Chachamovich E, Jollant F et al (2017) Investigation of miR-1202, miR-135a, and miR-16 in major depressive disorder and antidepressant response. Int J Neuropsychopharmacol 20:619–623
57. Lopez JP, Lim R, Cruceanu C, Crapper L, Fasano C, Labonte B et al (2014) miR-1202 is a primate-specific and brain-enriched microRNA involved in major depression and antidepressant treatment. Nat Med 20:764–768
58. Oved K, Morag A, Pasmanik-Chor M, Oron-Karni V, Shomron N, Rehavi M et al (2012) Genome-wide miRNA expression profiling of human lymphoblastoid cell lines identifies tentative SSRI antidepressant response biomarkers. Pharmacogenomics 13:1129–1139
59. Enatescu VR, Papava I, Enatescu I, Antonescu M, Anghel A, Seclaman E et al (2016) Circulating plasma micro RNAs in patients with major depressive disorder treated with antidepressants: a pilot study. Psychiatry Investig 13:549–557
60. Bocchio-Chiavetto L, Maffioletti E, Bettinsoli P, Giovannini C, Bignotti S, Tardito D et al (2013) Blood microRNA changes in depressed patients during antidepressant treatment. Eur Neuropsychopharmacol 23:602–611
61. Oved K, Morag A, Pasmanik-Chor M, Rehavi M, Shomron N, Gurwitz D (2013) Genome-wide expression profiling of human lymphoblastoid cell lines implicates integrin beta-3 in the mode of action of antidepressants. Transl Psychiatry 3:e313. doi: 10.1038/tp.2013.86
62. Fries GR, Quevedo J (2018) Exosomal MicroRNAs as potential biomarkers in neuropsychiatric disorders. Methods Mol Biol 1733:79–85
63. Sharma A (2014) Bioinformatic analysis revealing association of exosomal mRNAs and proteins in epigenetic inheritance. J Theor Biol 357:143–149
64. Nagy C, Turecki G (2015) Transgenerational epigenetic inheritance: an open discussion. Epigenomics 7:781–790
65. Velagapudi SP, Gallo SM, Disney MD (2014) Sequence-based design of bioactive small molecules that target precursor microRNAs. Nat Chem Biol 10:291–297
66. Schmidt MF (2014) Drug target miRNAs: chances and challenges. Trends Biotechnol 32:578–585
67. Bostrom AE, Ciuculete DM, Attwood M, Krattinger R, Nikontovic L, Titova OE et al (2017) A MIR4646 associated methylation locus is hypomethylated in adolescent depression. J Affect Disord 220:117–128
68. Wang X, Sundquist K, Hedelius A, Palmer K, Memon AA, Sundquist J (2015) Circulating microRNA-144-5p is associated with depressive disorders. Clin Epigenetics 7:69. https://doi.org/10.1186/s13148-015-0099-8
69. Maffioletti E, Cattaneo A, Rosso R, Maina G, Maj C, Gennarelli M et al (2016) Peripheral whole blood microRNA alterations in major depression and bipolar disorder. J Affect Disord 200:250–258
70. Sun N, Lei L, Wang Y, Yang C, Liu Z, Li X et al (2016) Preliminary comparison of plasma notch-associated microRNA-34b and -34c levels in drug naive, first episode depressed patients and healthy controls. J Affect Disord 194:109–114
71. Song MF, Dong JZ, Wang YW, He J, Ju X, Zhang L et al (2015) CSF miR-16 is decreased in major depression patients and its neutralization in rats induces depression-like behaviors via a serotonin transmitter system. J Affect Disord 178:25–31
72. Smalheiser NR, Lugli G, Zhang H, Rizavi H, Cook EH, Dwivedi Y (2014) Expression of microRNAs and other small RNAs in prefrontal cortex in schizophrenia, bipolar disorder and depressed subjects. PLoS One 9:e86469. https://doi.org/10.1371/journal.pone.0086469
73. Maheu M, Lopez JP, Crapper L, Davoli MA, Turecki G, Mechawar N (2015) MicroRNA regulation of central glial cell line-derived neurotrophic factor (GDNF) signalling in depression. Transl Psychiatry 5:e511. doi: 10.1038/tp.2015.11
74. Kolshus E, Ryan KM, Blackshields G, Smyth P, Sheils O, McLoughlin DM (2017) Peripheral blood microRNA and VEGFA mRNA changes following electroconvulsive therapy: implications for psychotic depression. Acta Psychiatr Scand 136:594–606
75. Belzeaux R, Lin CW, Ding Y, Bergon A, Ibrahim EC, Turecki G et al (2016) Predisposition to treatment response in major depressive episode: a peripheral blood gene coexpression network analysis. J Psychiatr Res 81:119–126

Chapter 10
Proteomic Markers for Depression

Licia C. Silva-Costa, Pamela T. Carlson, Paul C. Guest, Valéria de Almeida, and Daniel Martins-de-Souza

Abstract Major depressive disorder is a multifactorial disease, with molecular mechanisms not fully understood. A breakthrough could be reached with a panel of diagnostic biomarkers, which could be helpful to stratify patients and guide physicians to a better therapeutic choice, reducing the time between diagnostic and remission. This review brings the most recent works in proteomic biomarkers and highlights several potential proteins that could compose a panel of biomarkers to diagnostic and response to medication. These proteins are related to immune, inflammatory, and coagulatory systems and may also be linked to energy metabolism, oxidative stress, cell communication, and oligodendrogenesis.

Keywords Major depressive disorder · Mass spectrometry · Antidepressants · Drug response

10.1 General Overview

With 322 million people (4.4% of the population) affected worldwide according to the World Health Organization, major depressive disorder (MDD) is a long-lasting and recurrent disorder and one of the leading causes of disability in the Western world, with a lifetime prevalence at almost 15% of the diagnosed patients [1, 2]. Difficulties in social and occupational function, suicidal thoughts, and decline in physical health may occur in 10–30% of MDD patients who do not respond to treatment, possibly due

L. C. Silva-Costa · P. T. Carlson · P. C. Guest · V. de Almeida
Laboratory of Neuroproteomics, Department of Biochemistry and Tissue Biology, Institute of Biology, University of Campinas (UNICAMP), Campinas, SP, Brazil

D. Martins-de-Souza (✉)
Laboratory of Neuroproteomics, Department of Biochemistry and Tissue Biology, Institute of Biology, University of Campinas (UNICAMP), Campinas, SP, Brazil

Instituto Nacional de Biomarcadores em Neuropsiquiatria (INBION), Conselho Nacional de Desenvolvimento Científico e Tecnológico (CNPq), São Paulo, Brazil
e-mail: dmsouza@unicamp.br

P. C. Guest (ed.), *Reviews on Biomarker Studies in Psychiatric and Neurodegenerative Disorders*, Advances in Experimental Medicine and Biology 1118,
https://doi.org/10.1007/978-3-030-05542-4_10

to the syndrome's heterogeneity, which can make it difficult to diagnose [3–5]. The boundaries of depressive disorders, whether they can be considered only symptoms or true syndromes, are unclear as their symptoms are sometimes varying and even opposing, such as weight loss or weight gain and insomnia or hypersomnia, and seem to have no established mechanism [6–8]. Aside from this, the delayed response observed with antidepressants can hinder early observation of good or bad outcomes [9–11].

MDD is a multifactorial disorder [12–14] with several neurobiological hypotheses. The monoaminergic hypothesis, coined in the mid-1960s, postulates that disturbances in the levels of monoamines (serotonin, noradrenaline, and dopamine) are responsible for depression [15, 16]. However, the monoaminergic hypothesis was not sufficient to explain all changes observed in depression. In this regard, symptoms can be managed with electroconvulsive therapy or pharmacological manipulations of glutamatergic system [17, 18]. Despite neurotransmitters, studies have shown the involvement of the immune system in the pathophysiology of depression [19, 20]. The activation of the neuroimmune system leads to reduction of neurotrophic factors such as the brain-derived neurotrophic factor (BDNF), hampering the neurogenesis, changes that are compatible with the observation of impairment in the cognitive processes related to the disease [21–23].

Successful treatment of depression is also challenged by its various subtypes, with different neurobiological, biochemical, genetic, and anatomical characteristics. The molecular mechanisms of these subtypes are still poorly understood [24–26]. Still, gene polymorphisms are only considered a risk factor for depression, not a way to diagnose the disease, and few possible candidate single-nucleotide polymorphisms (SNPs) for MDD were considered replicable [27, 28]. The concept of endophenotypes in psychiatry reaffirms that the heterogeneity of symptoms of psychiatric illnesses such as MDD is the result of a complex network of interactions between genes, proteins, and circuits of cells and also between individuals and their experiences [29]. In order to establish a relationship between genes and clinical phenotypes, endophytic characterization has allowed some insights related to the mechanisms of MDD, through proteomic, transcriptomic, neuroanatomical, neurological, behavioral, and cognitive measurements, which must be inheritable and correlated with the disease, as well as measurable between affected and unaffected individuals, among other criteria [30–32]. The construction of endophenotypes can be favored through the establishment of proteomic biomarkers when influenced by genetic factors, although not every biomarker is considered an endophenotype [33].

A PubMed search on depression biomarkers reveals to what degree different biomarker approaches are highlighted in this field. In the last 5 years, the use of proteomic approaches in the study of psychiatric disorders has grown considerably, which has led to the identification of a wide array of differentially expressed proteins of which some could be potential biomarkers. Proteomics methods can be performed in various ways, such as with a high-throughput discovery setup or targeted quantitation, as the study of proteins can be performed both individually and in combination with others. Frequently, techniques are based on size characterization or antibody/aptamer binding [34]. In this review we analyze different approaches, in terms of scientific merit, for the discovery of biomarkers for MDD.

10.2 Biomarker Characterization

Biomarkers are measurable and evaluable characteristics capable of indicating a disease, a normal biological process, or the treatment response [35]. Diagnostic biomarkers can help stratify patients with depression, predictive biomarkers can assess response or remission, and moderators can determine the likelihood of response or remission to a particular treatment [34, 36]. Furthermore, biomarkers can be useful to identify new molecular targets, aiming to improve the development of new drugs [37]. However, a large data of proteomic biomarkers has been proposed for MDD without being implemented in the clinic because of lack of sensitivity or specificity or because standardization norms do not exist or are not widely accepted [36, 38]. Thus, it is necessary to compose a panel with several biomarkers, so that one set of proteins can display the changes in different biological mechanisms (Fig. 10.1).

Considering that both diagnostic biomarkers and response biomarkers should serve the ultimate purpose of reducing the overall time between diagnosis and successful treatment, the type of sample to be used for proteomic analysis should be easily obtained and must reflect the health status of the organism [39]. Therefore, although it is possible to obtain proteomic biomarkers of disease and response in materials such as tissues and cells or through image analysis, for a patient being treated in a clinical

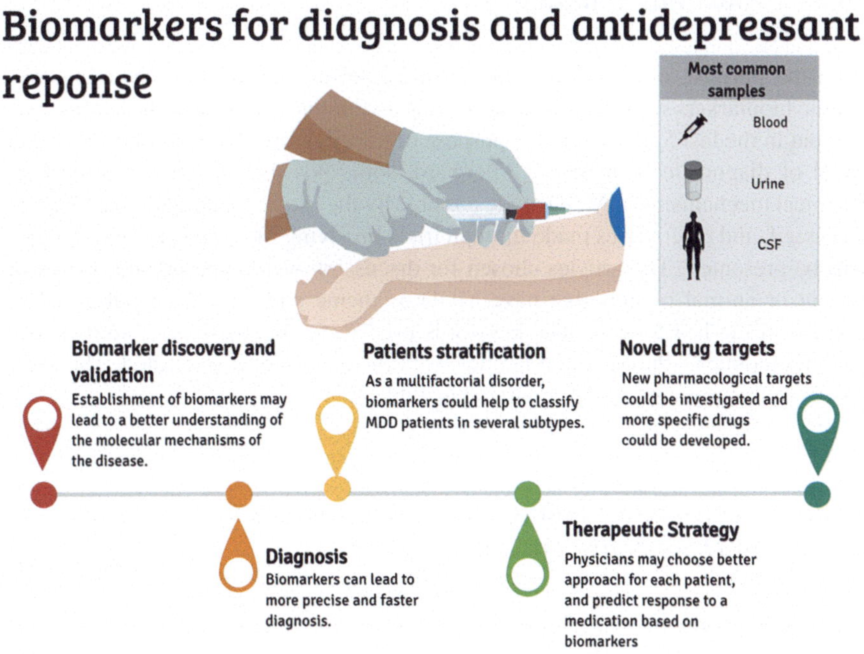

Fig. 10.1 Establishment of biomarkers to MDD can promote a breakthrough in therapeutic strategies

setting, samples such as saliva, urine, and blood are more suitable for this end [40]. Cerebrospinal fluid or fibroblasts could be obtained under more restrictive criteria, as the collection of this body fluid occurs in a relatively invasive manner [41].

Plasma is one of the most complex human samples with a dynamic range of proteins exceeding 10^{10} orders of magnitude [42] and containing proteomic sets of other tissues. It is also an abundant and easily obtained material. Its disadvantages for proteomic analysis include the presence of a large amount of albumin and heterogeneous glycoproteins, which make it difficult to observe other proteins [43, 44]. Plasma is composed of proteins secreted by tissues that act through the bloodstream or are the result of cellular damage; immunoglobulins; local or long-distance receptor ligands, such as hormones and cytokines; temporary passengers, such as lysosomal proteins; and aberrant secretions and proteins foreign to the organism originated from infections [42, 45]. Serum is a proteomic solution resulted from blood clotting, and because of the process of proteolysis, which may alter some proteins, one may prefer plasma to serum [46].

With the intention of delineating the molecular mechanisms of depression, other models have been employed, such as animal models, which can help to establish a trustworthy panel of biomarkers, and samples, such as postmortem tissue, which can be of limited use due to a patient's treatment and lifestyle [47, 48].

10.3 Proteomics Findings

With the purpose of elucidating the most interesting discoveries involving proteomic biomarkers related to depression, this text initially addresses the studies carried out in the last 5 years with the purpose of unveiling mechanisms and forming a panel of diagnostic biomarkers. Afterward, some possible insights into the biochemical mechanisms of MDD made possible by the use of an animal model will be discussed, and discoveries made through trials involving the response to treatments will be presented. The articles chosen for discussion were experimental works in patient or animal models that have MDD as theme and have been published in English in the last 5 years. The keywords used were "depression," "biomarkers," and "proteomics." PubMed and Scopus were the databases investigated. The main results of this research are summarized in Table 10.1.

10.3.1 Proteomics Findings Related to Diagnostic Biomarkers in Drug-Naive Patients

A relationship among the inflammatory, immune, and lipid systems and MDD has been found in some studies in drug-naive patients. These studies contribute to understand the proteome profile without the influence of drug effects. However, the

Table 10.1 Overview of recent potential biomarkers for MDD

Gene name	Characteristics of sample	Biological process
CP, EN-RAGE, FTH1, HPR, **IL-1**, IL-16, **MIF**, F3, **TNC** [49]	Blood from drug-naive patients	Inflammatory system
APOB, APOD, **CP, GC**, HRNR, PFN1 [50]	Blood from drug-naive patients	Immune and inflammatory systems/lipid metabolism
APOB, CP, GC [51]	Blood from drug-naive patients	Immune and inflammatory systems
CRP, **ITIH4**, **SAA1**, ANGPTL3 [52]	Blood from drug-naive patients	Immune and inflammatory systems
ASS1 [54]	Urine from drug-naive patients	Inflammatory system/urea cycle
CPE [57]	Postmortem pituitary tissues from BD and MDD patients	Carboxypeptidase activity
RBP-4, TTR [59]	Blood from BD and MDD patients	Retinoid metabolic process
END, B2RAN2 [60]	Blood from BD and MDD patients	Artery morphogenesis/ inflammatory system
CRP, **SAA1**, FX, PCI, TF, FVII, FV, TFPI, APC, F1+2 [61]	Blood from MDD suicide attempters	Inflammatory/coagulatory systems
PPP, **MIF, EN-RAGE**, IL-1RA, **TNC**, GROa, **vWF,** Prost, LH, AAT, UPA, CathD, HPN, MMP10, FABPA [66]	Blood from MDD remitted patients	Cell communication/ immune system/protein metabolism
CP, CC1QC, **ITIH4** [67]	Blood from MDD remitted patients	Inflammatory system
FGA [68]	Blood from MDD remitted patients	Inflammatory system
CCL11, IFN-γ [69]	Blood from MDD remitted/ nonremitted patients	Immune system
APOA4, CPB2, C7, CHEK1, ACTN1, **CRP**, THBS1, **FGA**, CFHR5, PYY2, F5, ARFIP1, CFHR2, MYH2 [70]	Blood from MDD remitted/ nonremitted patients	Immune system
vWF, SERPINA1, APOC3, A2M [73]	Blood from responders to CHM	Inflammatory system
VEGFC, Tie2, **BDNF** [74]	Blood from TRD patients	Inflammatory system
SAMP, C4BP [75]	Blood from TRD patients	Complement system
IGF-1, INS, CCL4, **BDNF** [76]	Blood from patients treated with antidepressants and ECT	Several biological processes
GRIA1, GRIA2, PRKCγ, PRKCβ, GRIN2B, SLC17A7, GNAQ, CAMK2α, PPP1R1A [77]	Hippocampus from VD animal model	Several processes involving regulation of the nervous system
NSF, ATP5A1, ACO2, STXBP1, DRP-2, SNAP-25 [78]	Hippocampus and frontal lobe from early-life stressed animals	Several biological processes

(continued)

Table 10.1 (continued)

Gene name	Characteristics of sample	Biological process
FGF-9, IL-4, **TNF-α**/mTOR, ERK1, PKCα, NSF, SYN1, PACN1, PSD95, NCDN, AATM, COMT, PPP3CC, PKCβ [79]	Serum/hippocampus and frontal cortex from animals treated with ketamine	Several biological processes
AK1, NDK B, HINT1, APT-2, GSTA4/GSTA6, GSTA4, RAN, Atp5h, Tagln3, Eif5a, SUMO2 [80]	Hippocampus from CMS animals/CMS animals treated with oleamide	Energy metabolism, oxidative stress, and cell communication
GFAP, VGLUT1, HOMER1, ATP1A2, UQCRFS1, UQCRC1, PRDX1, PRDX2 [81]	Synaptosomes from prefrontal cortex of CMS animals	Cell communication/energy metabolism
MMP9, **IL-1**, **CRP**, **TNF-α**/ MYPR, MBP, CN37 [82]	Serum and frontal cortex of stress-susceptible mice	Inflammatory system/ oligodendrogenesis

Notes: Bold gene names are recurrent potential biomarkers among the surveyed papers

number of studies performed with such groups of patients is small as samples from drug-naive patients are not easily available. Serum analysis performed by Stelzhammer et al. showed the involvement of the pro-inflammatory proteins ceruloplasmin (CP), Extracellular newly identified RAGE-binding protein (EN-RAGE), ferritin (FTH1), haptoglobin-related protein (HPR), interleukin-1 receptor antagonist (IL-1ra), interleukin-16 (IL-16), macrophage migration inhibitory factor (MIF), serotransferrin (F3), and tenascin-C (TNC), besides two other proteins linked to the oxidative stress process, in first onset patient samples [49]. Another study with drug-naive women identified six differentially expressed proteins - apolipoprotein B (APOB), apolipoprotein D (APOD), CP, vitamin D-binding protein (GC), hornerin (HRNR), and profilin 1 (PFN1) - that differentiated MDD patients from controls with up to 68% accuracy and which are related to the immune system, inflammatory system, and lipid metabolism [50]. Similar results were found using the combination of heart rate (HR) and plasma proteins in a study conducted by Kim et al. involving proteomics and a machine-learning approach [51]. Using these techniques, three proteins (in combination with HR) were identified as possible biomarkers for MDD, APOB, CP, and GC, which modulate the immune and inflammatory systems [51]. In another study, using a iTRAQ-based proteomics approach, serum samples from drug-free patients presented significant increases in immune- and lipid-related proteins such as C-reactive protein (CRP), inter-alpha-trypsin inhibitor heavy chain H4 (ITIH4), serum amyloid A-1 protein (SAA1), and angiopoietin-related protein 3 (ANGPTL3) [52]. Another study was conducted by Wu et al. using urine samples, in which the enzyme involved in the urea cycle and participating in the nitric oxide metabolic pathway, argininosuccinate synthase 1 (ASS1) [53], showed potential as an MDD biomarker [54].

Efforts have been made to find biomarkers that help psychiatrists differentiate between bipolar disorder (BD) and MDD that share changes in the hypothalamic-pituitary-adrenal (HPA) axis [55] and symptoms, such as oscillation of energy levels and mood disturbances, making it difficult to accurately diagnose each disease for proper treatment [56]. Stelzhammer et al. performed the first analysis using LC-MS^E to find differences in protein expression between postmortem pituitaries of BD and MDD patients and controls. In this study, proteins related to intracellular transport and remodeling of cytoskeletal pathways were found to be altered in MDD patients [57]. Furthermore, reduced levels of carboxypeptidase E (CPE) in MDD patient pituitaries suggest that prohormone conversion may be altered, although most hormones did not exhibit altered levels in MDD [57]. Retinol-binding protein 4 (RBP-4) and transthyretin (TTR) form a complex that is responsible for vitamin A transport and may be involved in mood disorders [58]. Their differential expression have been found from blood samples and may become distinct between BD I (a subtype of BD) and MDD [59]. Two other proteins, highly similar to vanin-1 protein (B2RAN2) and endoglin (END), were differentially expressed in plasma MDD drug-naive patients when compared to BD patients [60].

Suicide attempts are a relevant concern and may affect some MDD patients. One study investigated the plasma of drug-naive depressed attempters (MDD-SA), depressed suicide non-attempters (MDD-NA), and healthy controls using 2-DE-MALDI-TOF/TOF-MS and iTRAQ-LC-MS/M platforms and found alterations in the CRP, SAA1, coagulation factor X (FX), and protein C inhibitor (PCI) in drug-naive MDD-SA group compared with the other groups [61]. In a validation phase using enzyme-linked immunoadsorbent assay (ELISA), six proteins were found differentially expressed relative to both MDD-NA and health control subjects: tissue factor (TF), coagulation factor II (FVII), coagulation factor V (FV), tissue factor pathway inhibitor (TFPI), activated protein C (APC), and prothrombin fragment 1 + 2 (F1+2) [61]. Therefore, these results revealed differentially expressed inflammatory and coagulatory proteins. This indicates that further investigations are needed to understand potential mechanisms that may predispose to suicide [61].

10.3.2 Late-Life Depression

The term "late-life depression" often refers to depressive episodes occurring at around 60 years of age or later: the first episode may occur in early years (early-onset depression) or during aging (late-onset depression). It can be described as a heterogeneous and complex neuropsychiatric syndrome, triggered by another serious disease, by comorbidities, or may be the result of drug use [62].

The investigation of the differential expression of peripheral proteins through immunoassays in the blood plasma of elderly patients led to the association between geriatric depression and inflammatory processes and reduction of neurotrophic support, as well as proteostasis markers and nutrient detection [63]. Some of these same markers suggest that homeostatic dysregulation present in patients with geriatric depression can accelerate the aging process, which can increase the risk of Alzheimer's and dementia [64]. Another important investigation correlated higher SASP index (a set of proteins secreted by different senescent fibroblasts, responsible for inducing the aging of nearby cells) in patients with depression and may give clues about the molecular mechanisms that lead to LLD [65]. However, there are still few studies which have attempted to investigate the molecular mechanisms and obtain a consistent biological signature of LLD.

10.3.3 Markers Related to Response

Biomarkers for antidepressant response represent a strategy to personalize therapy through the characterization of a panel of proteins to guide physicians in the best antidepressant choice for each patient. Serum proteins from patients with current MDD (MDDc) and remitted MDD (MDDr) were compared in one large cohort study from Bot et al. Proteins related to cell communication, signal transduction processes, immune response, and protein metabolism were found to be differentially expressed in serum from MDDc patients compared to that from controls [66]. Those proteins were pancreatic polypeptide (PPP), MIF, EN-RAGE, IL-1 receptor antagonist (IL-1RA) and TNC, growth-regulated alpha protein (GROa), and von Willebrand factor (vWF), a marker involved in homeostasis. When MDDc and MDDr groups were compared, ten analytes were found to be altered: prostasin (Prost), luteinizing hormone (LH), alpha-1 antitrypsin (AAT), urokinase-type plasminogen activator (UPA) receptor, cathepsin D (CathD), hepsin (HPN), matrix metalloproteinase-10 (MMP10), IL-1RA, vWF, and fatty acid-binding protein adipocyte (FABPA) [66].

Regulation of inflammatory processes seems to be associated with a good response to antidepressant treatment. Comparative proteomic analysis of a small cohort of MDD subjects before and after treatment found a decrease of three proteins in serum after remission: CP, complement component 1q (CC1QC; a component of the classic activation), and ITIH4 [67]. Plasma fibrinogen alpha (FGA) levels were the object of investigation between drug responders and nonresponders in another work, without taking into account the type of antidepressant used. High levels of fibrinogen at baseline were associated with a poor response to antidepressants due possibly to an elevated inflammatory status [68].

Investigations of the association between antidepressant treatment response and the immune system have been carried out before and after treatment. In the study of Gadad et al., two proteins demonstrated significant association with treatment response. After 12 weeks of treatment, responders presented higher levels of eotaxin-1 (CCL11), and interferon-gamma (IFN-γ) levels were reduced in nonresponders. Eotaxin may be linked to increased levels of mammalian target of rapamycin (mTOR) protein, related to synaptogenesis. Higher levels of IFN-γ at baseline in nonresponders may be implicated in a higher activation of the kynurenine pathway, which has been associated with MDD, and thus implicating a worse response to medication [69]. Similarly, Turck et al. conducted a proteome profiling of plasma from MDD patients at baseline and after 6 weeks of treatment. Differences at baseline between responders and nonresponders showed significant differences between 29 proteins that could compose a panel of response prediction. Apolipoprotein A-IV (APOA4), carboxypeptidase B2 (CPB2), complement component C7 (C7), and serine/threonine-protein kinase Chk1 (CHEK1) were over twofold lower in responders, and alpha-actinin-1 (ACTN1), CRP, thrombospondin-1 (THBS1), FGA, complement factor H-related protein 5 (CFHR5), FV, arfaptin-1 (ARFIP1), and complement factor H-related protein 2 (CFHR2) were over twofold higher in responders [70]. After 6 weeks of treatment, 18 proteins were observed to have differences of more than twofold in responders when compared with the baseline levels. Among these proteins, putative peptide YY-2 (PYY2) and CHEK1 were increased, and mysosin-2 (MYH2) was decreased [70].

Integrin and RAS signaling pathways are involved in processes of synaptic signaling and remodeling in the central nervous system and can be potential biomarker candidates for response to antidepressants. Martins-de-Souza et al. compared blood mononuclear cell proteomes at baseline and after 6 weeks of antidepressant treatment, and proteins of integrin signaling pathways were found differentially expressed between responders and nonresponders [71]. Integrin, RAS, and fibrinogens are related to platelet activation with consequent inflammatory response, and N-glycan profiles may distinguish responders from nonresponders, since N-glycosylation has a relevant role in platelet activation [72]. Alterations of coagulation and complement cascades, lipid metabolism, platelet degranulation, and activation pathways were associated with a depressive status, and their changes were associated with response to Chinese herbal medicine (CHM). Proteins such as vWF, epididymis secretory sperm-binding protein (SERPINA1), apolipoprotein C-III (APOC3), and alpha-2-macroglobulin (A2M) were found at different levels in responding patients treated with CHM [73].

Although many antidepressant drugs are available, there is still a significant proportion of patients who do not respond well to them. Treatment-resistant depression (TRD) is the focus of few proteomic studies. Study from Pisoni et al. have demonstrated the role of altered vascular endothelial growth factor-C (VEGFC), angiopoietin-1 receptor (Tie2), and BDNF levels in impaired neurogenesis and neuroplasticity [74]. In another study, differentially expressed proteins identified in TRD patients showed altered complement activation, coagulation, and lipid transport processes, with more severely affected patients presenting altered serum amyloid P (SAMP) component and the C4b-binding protein (C4BP) [75]. Electroconvulsive therapy (ECT) is used to treat TRD patients with higher remission rates than antidepressant drugs, with molecular changes occurring after ECT treatment combined with antidepressants. Although there was an increase of insulin-like growth factor I (IGF-1) and C-peptide (INS) and decrease of MIP-1 beta (CCL4) and BDNF, only levels of C-peptide seem related to symptom improvements [76]. Additional studies are needed to unravel how these molecules are related to the therapeutic role of ECT.

10.3.4 Animal Models

Animal models have been used to investigate the biological mechanisms that lead to depression or even to investigate different responses to drugs. Like other psychiatric disorders, animal models to study depression present limitations in terms of mimicking clinical findings observed in humans. However, animal models have been widely used due to the possibility of investigating biochemical changes in tissues that are not accessible in humans, such as the brain.

The blockade of the bilateral carotid artery has been used to mimic vascular depression (VD). These animals present hippocampal damage and changes in animal weight and behavior similar to depressive symptoms [77]. Hippocampal proteomic analysis showed multiple changes in pathways related to neural plasticity (calcium signaling pathways and neurotransmission), energy, and amino acid metabolism, similar to patients with vascular depression [77]. Nine proteins related to these changes, glutamate receptor 1 (GRIA1), glutamate receptor 2 (GRIA2), protein kinase Cγ (PRKCγ), protein kinase Cβ (PRKCβ), glutamate receptor ionotropic NMDA 2B (GRIN2B), vesicular glutamate transporter 1 (SLC17A7), guanine nucleotide-binding protein G(q) subunit alpha (GNAQ), calcium-/calmodulin-dependent protein kinase type II subunit alpha (CAMK2α), and protein phosphatase 1 regulatory subunit 1A (PPP1R1A), were validated by Western blotting and showed reduced expression in the model compared to the sham group [77].

In a classical model of depression and maternal deprivation, changes were found through synaptosome analysis of hippocampal and whole frontal lobe. This study showed alterations in proteins related to energy metabolism and structural protein disturbances, which suggest that early-life stress may affect cytoskeletal dynamics of synapses. Specifically, vesicle-fusing ATPase (NSF), ATP synthase alpha (ATP5A1), aconitate hydratase (ACO2), syntaxin-binding protein 1 (STXBP1), dystrophin-related protein 2 (DRP-2), and synaptosomal-associated protein 25 (SNAP-25) were found to be differentially expressed [78].

Studies in rats investigated the proteomic and biochemical alterations of potential antidepressants. Wesseling et al. showed that administration of ketamine, an antagonist of the NMDA receptor, promoted changes in the frontal cortex, hippocampus, and serum of rodents. Only three serum proteins were affected, fibroblast growth factor 9 (FGF-9), interleukin-4 (IL-4), and tumor necrosis factor alpha (TNF-α), which suggested only minor changes in the peripheral system [79]. Proteins such as the mTOR, extracellular signal-regulated kinase 1 (ERK1), protein kinase Cα (PKCα), vesicle-fusing ATPase (NSF), synapsin (SYN1), syndapin-1 (PACN1), postsynaptic density protein 95 (PSD95), and neurochondrin (NCDN) were found to be differentially expressed in the hippocampus [79]. Increased levels of mitochondrial aspartate aminotransferase (AATM) occurred in the frontal cortex and hippocampus, and lower levels of catechol-O-methyltransferase (COMT) were detected in the hippocampus [79]. In addition, altered levels of calcineurin (PPP3CC) and protein kinase Cβ (PKCβ) were found in the frontal cortex and in the hippocampus after treatment with ketamine [79]. Similarly, rats submitted to chronic mild stress (CMS), a MDD animal model, were treated with oleamide. CMS affected the adenylate kinase isoenzyme protein 1 (AK1), nucleoside diphosphate kinase B (NDK B), histidine triad nucleotide-binding protein 1 (HINT1), acyl-protein thioesterase 2 (APT-2), and glutathione S-transferase A 4 (GSTA4) [80]. Rats subjected to CMS and treated with oleamide presented changes in the levels of glutathione S-transferase A6 (GSTA6), glutathione S-transferase A4 (GSTA4), GTP-binding protein Ran (RAN), ATP synthase subunit d, mitochondrial (Atp5h), transgelin-3 (Tagln3), eukaryotic translation initiation factor 5A-1 (Eif5a), and small ubiquitin-related modifier 2 (SUMO2) [80]. Oleamide was associated with increased sucrose intake, showing an antidepressant-like effect, affecting proteins involved in processes such as energy metabolism, oxidative stress, and cell communication [80].

Using the CMS model, a study showed that stress-resilient and stress-susceptible rats presented changes in proteins associated with cell interactions and glutamatergic signaling. Glial fibrillary acidic protein (GFAP) and vesicular glutamate transporter 1 (VGLUT1) proteins were found to be decreased in the stress-susceptible group, and proteins linked to ion regulation such as homer protein homolog 1 (HOMER1) and sodium-/potassium-transporting ATPase subunit alpha-2 (ATP1A2) were found upregulated in resilient rats [81]. Resilient rats have also shown higher levels of antioxidant proteins such as the mitochondrial proteins cytochrome b-c1 complex subunit Rieske (UQCRFS1) and cytochrome b-c1 complex subunit 1 (UQCRC1), related to cytochrome b-c1 complex subunits, and they had lower levels of peroxiredoxins, PRDX1 and PRDX2s [81]. Another animal model study investigated the serum and frontal cortex of stress-susceptible (SS) and stress-resilient (SR) mice. This study found that 20 proteins were differentially regulated in the serum and frontal cortex of these groups [82]. Changes in proteins related to inflammatory system such as matrix metalloproteinase-9 (MMP9), IL-1, CRP, and TNF-α were found in serum samples [82]. Resilient animals presented higher levels of myelin-associated proteins in the prefrontal cortex, including myelin proteolipid protein (MYPR), myelin basic protein (MBP), and 2,3-cyclic nucleotide 3 phosphodiesterase (CN37), suggesting an effect on the oligodendrogenesis process [82].

10.4 Conclusions

Despite many hypotheses that try to explain MDD etiology, there is a lack of defined molecular mechanisms that attempt to relate all main characteristics of this disorder. Although diagnostic systems such as DSM-V can systemize depressive symptoms, there are overlaps with other psychiatric disorders with respect to both symptoms and biomolecular pathways [83, 84]. Proteomics is a powerful set of techniques that enable the investigation of these biochemical characteristics aiming to propose a panel of diagnosis and response biomarkers.

Proteins such as CP, EN-RAGE, MIF, TNC, IL-1, GC, CRP, and SAA1 emerge several times in different studies with the purpose of identifying a molecular profile of MDD. FGA and vWF proteins are recurrent in studies of markers related to treatment response. Changes in the levels of BDNF seem to be related to resistance to treatments. Identification of a trustable and specific panel of markers to MDD can lead to changes for patients, since biomarkers can improve diagnostic capacity and reveal new drug targets, which can be used to develop newer and better drugs. However, there is a missing link between those findings and their validation in clinical practice.

The wide variety of differential expressed proteins may be due the complexity of interactions related to MDD, and there is a need for exploration of possible connections between these pathways. The development of a mathematical model exploring these connections and related symptoms could be useful in this sense. Although the focus of this chapter is proteomics, a field in which much further research is needed, there is also a need for translational and multi-omics research. Given the complexity of MDD, there is a call for proteomics assays that also interact with genomics, transcriptomics, metabolomics, epigenomics, lipidomics, and even metagenomics, giving rise to omics-based biomarkers.

Acknowledgments The authors thank FAPESP (Sao Paulo Research Foundation, grants 2013/08711-3, 2017/18242-1 2017/25588-1 and 2018/03422-7) and Serrapilheira Institute (grant number Serra-1709-16349) for funding. We also thank CAPES (Coordination for the Improvement of Higher Education Personnel) for the scholarship 88887.179832/2018-00. The authors thank Prof. Brett Vern Carlson (Technological Institute of Aeronautics (ITA)) for assistance with the manuscript.

References

1. Bromet E, Andrade LH, Hwang I, Sampson NA, Alonso J, de Girolamo G et al (2011) Cross-national epidemiology of DSM-IV major depressive episode. BMC Med 9:90. https://doi.org/10.1186/1741-7015-9-90
2. World Health Organization (2017) Depression and other common mental disorders: global health estimates. World Heal Organ. 1–24. doi: CC BY-NC-SA 3.0 IGO
3. Lamers F, Vogelzangs N, Merikangas KR, de Jonge P, Beekman ATF, Penninx BWJH (2013) Evidence for a differential role of HPA-axis function, inflammation and metabolic syndrome in melancholic versus atypical depression. Mol Psychiatry 18:692–699. https://doi.org/10.1038/mp.2012.144
4. KS A-H (2012) Treatment-resistant depression: therapeutic trends, challenges, and future directions. Patient Prefer Adherence 6:369–388. https://doi.org/10.2147/PPA.S29716
5. Souery D, Amsterdam J, de Montigny C, Lecrubier Y, Montgomery S, Lipp O et al (1999) Treatment resistant depression: methodological overview and operational criteria. Eur Neuropsychopharmacol 9:83–91
6. Paris J (2014) The mistreatment of major depressive disorder. Can J Psychiatry 59:148–151. https://doi.org/10.1177/070674371405900306
7. Belmaker RH, Agam G (2008) Major depressive disorder. N Engl J Med 358:55–68. https://doi.org/10.1056/NEJMra073096
8. Striegel-Moore R (2011) Developing an evidence-based classification of eating disorders: scientific findings for DSM-5, 1st edn. American Psychiatric Publishing, Washington, DC. ISBN:9780890426661
9. Stassen HH, Angst J, Delini-Stula A (1997) Delayed onset of action of antidepressant drugs? Survey of recent results. Eur Psychiatry 12:166–176
10. Fava M, Davidson KG (1996) Definition and epidemiology of treatment-resistant depression. Psychiatr Clin North Am 19:179–200
11. Duric V, Duman RS (2013) Depression and treatment response: dynamic interplay of signaling pathways and altered neural processes. Cell Mol Life Sci 70:39–53
12. Post RM (1992) Transduction of psychosocial stress into the neurobiology of recurrent affective disorder. Am J Psychiatry 149:999–1010
13. Vialou V, Feng J, Robison AJ, Nestler EJ (2013) Epigenetic mechanisms of depression and antidepressant action. Annu Rev Pharmacol Toxicol 53:59–87. https://doi.org/10.1146/annurev-pharmtox-010611-134540

14. Lopizzo N, Bocchio Chiavetto L, Cattane N, Plazzotta G, Tarazi FI, Pariante CM et al (2015) Gene-environment interaction in major depression: focus on experience-dependent biological systems. Front Psychiatry 6:68. https://doi.org/10.3389/fpsyt.2015.00068
15. López-Muñoz F, Alamo C (2009) Monoaminergic neurotransmission: the history of the discovery of antidepressants from 1950s until today. Curr Pharm Des 15:1563–1586
16. Schildkraut JJ (1965) The catecholamine hypothesis of affective disorders: a review of supporting evidence. Am J Psychiatry 122:509–522
17. Paul IA, Skolnick P (2003) Glutamate and depression. Ann N Y Acad Sci 1003:250–272
18. Sanacora G, Treccani G, Popoli M (2012) Towards a glutamate hypothesis of depression: an emerging frontier of neuropsychopharmacology for mood disorders. Neuropharmacology 62:63–77
19. Schiepers OJG, Wichers MC, Maes M (2005) Cytokines and major depression. Prog Neuropsychopharmacol Biol Psychiatry 29:201–217
20. Maes M (1995) Evidence for an immune response in major depression: a review and hypothesis. Prog Neuropsychopharmacol Biol Psychiatry 19:11–38
21. Tong L, Balazs R, Soiampornkul R, Thangnipon W, Cotman CW (2008) Interleukin-1β impairs brain derived neurotrophic factor-induced signal transduction. Neurobiol Aging 29:1380–1393
22. Hayley S (2014) The neuroimmune-neuroplasticity interface and brain pathology. Front Cell Neurosci 8:419. https://doi.org/10.3389/fncel.2014.00419
23. Pittenger C, Duman RS (2008) Stress, depression, and neuroplasticity: a convergence of mechanisms. Neuropsychopharmacology 33:88–109
24. Rush AJ (2007) The varied clinical presentations of major depressive disorder. J Clin Psychiatry 68(Suppl 8):4–10
25. Ostergaard SD, Jensen SOW, Bech P (2011) The heterogeneity of the depressive syndrome: when numbers get serious. Acta Psychiatr Scand 124:495–496686
26. WangPS,InselTR(2010)NIMH-fundedpragmatictrials:movingon.Neuropsychopharmacology 35:2489–2490
27. CONVERGE consortium (2015) Sparse whole-genome sequencing identifies two loci for major depressive disorder. Nature 523:588–591
28. Bosker FJ, Hartman CA, Nolte IM, Prins BP, Terpstra P, Posthuma D et al (2011) Poor replication of candidate genes for major depressive disorder using genome-wide association data. Mol Psychiatry 16:516–532
29. Gottesman II, Gould TD (2003) The endophenotype concept in psychiatry: etymology and strategic intentions. Am J Psychiatry 160:636–645
30. Glahn DC, Knowles EEM, McKay DR, Sprooten E, Raventós H, Blangero J et al (2014) Arguments for the sake of endophenotypes: examining common misconceptions about the use of endophenotypes in psychiatric genetics. Am J Med Genet B Neuropsychiatr Genet 165B:122–130
31. Hasler G, Drevets WC, Manji HK, Charney DS (2004) Discovering endophenotypes for major depression. Neuropsychopharmacology 29:1765–1781
32. Goldstein BL, Klein DN (2014) A review of selected candidate endophenotypes for depression. Clin Psychol Rev 34:417–427
33. Lenzenweger MF (2013) Endophenotype, intermediate phenotype, biomarker: definitions, concept comparisons, clarifications. Depress Anxiety 30:185–189
34. Gadad BS, Jha MK, Czysz A, Furman JL, Mayes TL, Emslie MP et al (2018) Peripheral biomarkers of major depression and antidepressant treatment response: current knowledge and future outlooks. J Affect Disord 233:3–14
35. Biomarkers Definitions Working Group (2001) Biomarkers and surrogate endpoints: preferred definitions and conceptual framework. Clin Pharmacol Ther 69:89–95
36. Strawbridge R, Young AH, Cleare AJ (2017) Biomarkers for depression: recent insights, current challenges and future prospects. Neuropsychiatr Dis Treat 13:1245–1262
37. Macaluso M, Preskorn SH (2012) How biomarkers will change psychiatry: from clinical trials to practice. Part I: introduction. J Psychiatr Pract 18:118–121

38. Jentsch MC, Van Buel EM, Bosker FJ, Gladkevich AV, Klein HC, Oude Voshaar RC et al (2015) Biomarker approaches in major depressive disorder evaluated in the context of current hypotheses. Biomark Med 9:277–297
39. Lim MD, Dickherber A, Compton CC (2011) Before you analyze a human specimen, think quality, variability, and bias. Anal Chem 83:8–13
40. Elliott P, Peakman TC (2008) The UK Biobank sample handling and storage protocol for the collection, processing and archiving of human blood and urine. Int J Epidemiol 37:234–244
41. Su S-Y, Hogrefe-Phi CE, Asara JM, Turck CW, Golub MS (2016) Peripheral fibroblast metabolic pathway alterations in juvenile rhesus monkeys undergoing long-term fluoxetine administration. Eur Neuropsychopharmacol 26:1110–1118
42. Anderson NL, Anderson NG (2002) The human plasma proteome: history, character, and diagnostic prospects. Mol Cell Proteomics 1:845–867
43. Urbas L, Brne P, Gabor B, Barut M, Strlic M, Petric TC et al (2009) Depletion of high-abundance proteins from human plasma using a combination of an affinity and pseudo-affinity column. J Chromatogr A 1216:2689–2694
44. Anderson NL, Leigh Anderson N, Anderson NG (2003) The human plasma proteome: history, character, and diagnostic prospects: fig 3. Mol Cell Proteomics 2:50–50
45. Putnam FW (2012) The plasma proteins: structure, function, and genetic control, 2nd edn. Elsevier, New York. ISBN:9780323138086
46. Jambunathan K, Galande AK (2014) Sample collection in clinical proteomics—proteolytic activity profile of serum and plasma. Proteomics Clin Appl 8:299–307
47. Martins-de-Souza D, Harris LW, Guest PC, Turck CW, Bahn S (2010) The role of proteomics in depression research. Eur Arch Psychiatry Clin Neurosci 260:499–506
48. Licznerski P, Duman RS (2013) Remodeling of axo-spinous synapses in the pathophysiology and treatment of depression. Neuroscience 251:33–50
49. Stelzhammer V, Haenisch F, Chan MK, Cooper JD, Steiner J, Steeb H et al (2014) Proteomic changes in serum of first onset, antidepressant drug-naïve major depression patients. Int J Neuropsychopharmacol 17:1599–1608
50. Lee MY, Kim EY, Kim SH, Cho K-C, Ha K, Kim KP et al (2016) Discovery of serum protein biomarkers in drug-free patients with major depressive disorder. Prog Neuropsychopharmacol Biol Psychiatry 69:60–68
51. Kim EY, Lee MY, Kim SH, Ha K, Kim KP, Ahn YM (2017) Diagnosis of major depressive disorder by combining multimodal information from heart rate dynamics and serum proteomics using machine-learning algorithm. Prog Neuropsychopharmacol Biol Psychiatry 76:65–71
52. Wang Q, Su X, Jiang X, Dong X, Fan Y, Zhang J et al (2016) iTRAQ technology-based identification of human peripheral serum proteins associated with depression. Neuroscience 330:291–325
53. Haines RJ, Pendleton LC, Eichler DC (2011) Argininosuccinate synthase: at the center of arginine metabolism. Int J Biochem Mol Biol 2:8–23
54. Wu D, Peng Y, Zhou J, Yang Y-T, Rao C-L, Bai S-J et al (2015) Identification and validation of argininosuccinate synthase as a candidate urinary biomarker for major depressive disorder. Clin Chim Acta 451:142–148
55. Grunze H (2011) The clinical side of bipolar disorders. Pharmacopsychiatry 44(Suppl 1):S43–S48.
56. Preece RL, Han SYS, Bahn S (2018) Proteomic approaches to identify blood-based biomarkers for depression and bipolar disorders. Expert Rev Proteomics 15:325–340
57. Stelzhammer V, Alsaif M, Chan MK, Rahmoune H, Steeb H, Guest PC et al (2015) Distinct proteomic profiles in post-mortem pituitary glands from bipolar disorder and major depressive disorder patients. J Psychiatr Res 60:40–48
58. Fleming CE, Nunes AF, Sousa MM (2009) Transthyretin: more than meets the eye. Prog Neurobiol 89:266–276
59. Frye MA, Nassan M, Jenkins GD, Kung S, Veldic M, Palmer BA et al (2015) Feasibility of investigating differential proteomic expression in depression: implications for biomarker development in mood disorders. Transl Psychiatry 5:e689. https://doi.org/10.1038/tp.2015.185

60. Ren J, Zhao G, Sun X, Liu H, Jiang P, Chen J et al (2017) Identification of plasma biomarkers for distinguishing bipolar depression from major depressive disorder by iTRAQ-coupled LC-MS/MS and bioinformatics analysis. Psychoneuroendocrinology 86:17–24
61. Yang Y, Chen J, Liu C, Fang L, Liu Z, Guo J et al (2016) The extrinsic coagulation pathway: a biomarker for suicidal behavior in major depressive disorder. Sci Rep 6:32882. https://doi.org/10.1038/srep32882
62. Aziz R, Steffens DC (2013) What are the causes of late-life depression? Psychiatr Clin North Am 36:497–516
63. Diniz BS, Lin C-W, Sibille E, Tseng G, Lotrich F, Aizenstein HJ et al (2016) Circulating biosignatures of late-life depression (LLD): towards a comprehensive, data-driven approach to understanding LLD pathophysiology. J Psychiatr Res 82:1–7
64. Diniz BS, Sibille E, Ding Y, Tseng G, Aizenstein HJ, Lotrich F et al (2015) Plasma biosignature and brain pathology related to persistent cognitive impairment in late-life depression. Mol Psychiatry 20:594–601
65. Diniz BS, Reynolds CF 3rd, Sibille E, Lin C-W, Tseng G, Lotrich F et al (2017) Enhanced molecular aging in late-life depression: the senescent-associated secretory phenotype. Am J Geriatr Psychiatry 25:64–72
66. Bot M, Chan MK, Jansen R, Lamers F, Vogelzangs N, Steiner J et al (2015) Serum proteomic profiling of major depressive disorder. Transl Psychiatry 5:e599. https://doi.org/10.1038/tp.2015.88
67. Lee J, Joo E-J, Lim H-J, Park J-M, Lee KY, Park A et al (2015) Proteomic analysis of serum from patients with major depressive disorder to compare their depressive and remission statuses. Psychiatry Investig 12:249–259
68. Martins-de-Souza D, Maccarrone G, Ising M, Kloiber S, Lucae S, Holsboer F et al (2014) Plasma fibrinogen: now also an antidepressant response marker? Transl Psychiatry 4:e352. https://doi.org/10.1038/tp.2013.129
69. Gadad BS, Jha MK, Grannemann BD, Mayes TL, Trivedi MH (2017) Proteomics profiling reveals inflammatory biomarkers of antidepressant treatment response: findings from the CO-MED trial. J Psychiatr Res 94:1–6
70. Turck CW, Guest PC, Maccarrone G, Ising M, Kloiber S, Lucae S et al (2017) Proteomic differences in blood plasma associated with antidepressant treatment response. Front Mol Neurosci 10:272. https://doi.org/10.3389/fnmol.2017.00272
71. Martins-de-Souza D, Maccarrone G, Ising M, Kloiber S, Lucae S, Holsboer F et al (2014) Blood mononuclear cell proteome suggests integrin and Ras signaling as critical pathways for antidepressant treatment response. Biol Psychiatry 76:e15–e17. https://doi.org/10.1016/j.biopsych.2014.01.022
72. Park DI, Štambuk J, Razdorov G, Pučić-Baković M, Martins-de-Souza D, Lauc G et al (2018) Blood plasma/IgG N-glycome biosignatures associated with major depressive disorder symptom severity and the antidepressant response. Sci Rep 8:179. https://doi.org/10.1038/s41598-017-17500-0
73. Chen C, Hu Y, Dong X-Z, Zhou X-J, Mu L-H, Liu P (2018) Proteomic analysis of the antidepressant effects of Shen-Zhi-Ling in depressed patients: identification of proteins associated with platelet activation and lipid metabolism. Cell Mol Neurobiol 38:1123–1135
74. Pisoni A, Strawbridge R, Hodsoll J, Powell TR, Breen G, Hatch S et al (2018) Growth factor proteins and treatment-resistant depression: a place on the path to precision. Front Psychiatry 9:386. https://doi.org/10.3389/fpsyt.2018.00386
75. Ruland T, Chan MK, Stocki P, Grosse L, Rothermundt M, Cooper JD et al (2016) Molecular serum signature of treatment resistant depression. Psychopharmacology 233:3051–3059
76. Stelzhammer V, Guest PC, Rothermundt M, Sondermann C, Michael N, Schwarz E et al (2013) Electroconvulsive therapy exerts mainly acute molecular changes in serum of major depressive disorder patients. Eur Neuropsychopharmacol 23:1199–1207
77. Zhao H, Du H, Liu M, Gao S, Li N, Chao Y et al (2018) Integrative proteomics-metabolomics strategy for pathological mechanism of vascular depression mouse model. J Proteome Res 17:656–669

78. Mallei A, Failler M, Corna S, Racagni G, Mathé AA, Popoli M (2014) Synaptoproteomic analysis of a rat gene-environment model of depression reveals involvement of energy metabolism and cellular remodeling pathways. Int J Neuropsychopharmacol 18. https://doi.org/10.1093/ijnp/pyu067
79. Wesseling H, Rahmoune H, Tricklebank M, Guest PC, Bahn S (2014) A targeted multiplexed proteomic investigation identifies ketamine-induced changes in immune markers in rat serum and expression changes in protein kinases/phosphatases in rat brain. J Proteome Res 14:411–421
80. Ge L, Zhu M-M, Yang J-Y, Wang F, Zhang R, Zhang J-H et al (2015) Differential proteomic analysis of the anti-depressive effects of oleamide in a rat chronic mild stress model of depression. Pharmacol Biochem Behav 131:77–86
81. Palmfeldt J, Henningsen K, Eriksen SA, Müller HK, Wiborg O (2016) Protein biomarkers of susceptibility and resilience to stress in a rat model of depression. Mol Cell Neurosci 74:87–95
82. Stelzhammer V, Ozcan S, Gottschalk MG, Steeb H, Hodes GE, Guest PC et al (2015) Central and peripheral changes underlying susceptibility and resistance to social defeat stress—a proteomic profiling study. Diagnostics in Neuropsychiatry 1:1–7. https://doi.org/10.1016/j.dineu.2015.08.001
83. Gottschalk MG, Wesseling H, Guest PC, Bahn S (2014) Proteomic enrichment analysis of psychotic and affective disorders reveals common signatures in presynaptic glutamatergic signaling and energy metabolism. Int J Neuropsychopharmacol 18. https://doi.org/10.1093/ijnp/pyu019
84. Goodwin GM (2015) The overlap between anxiety, depression, and obsessive-compulsive disorder. Dialogues Clin Neurosci 17:249–260

Chapter 11
Advances in Biomarker Studies in Autism Spectrum Disorders

Liming Shen, Yuxi Zhao, Huajie Zhang, Chengyun Feng, Yan Gao, Danqing Zhao, Sijian Xia, Qi Hong, Javed Iqbal, Xu Kun Liu, and Fang Yao

Abstract Autism spectrum disorder (ASD) is a neurological and developmental condition that begins early in childhood and lasts throughout life. The epidemiology of ASD is continuously increasing all over the world with huge social and economical burdens. As the etiology of autism is not completely understood, there is still no medication available for the treatment of this disorder. However, some behavioral interventions are available to improve the core and associated symptoms of autism, particularly when initiated at an early stage. Thus, there is an increasing demand for finding biomarkers for ASD. Although diagnostic biomarkers have not yet been established, research efforts have been carried out in neuroimaging and biological analyses including genomics and gene testing, proteomics, metabolomics, transcriptomics, and studies of the immune system, inflammation, and microRNAs. Here, we will review the current progress in these fields and focus on new methods, developments, research strategies, and studies of blood-based biomarkers.

Keywords Autism spectrum disorder · Diagnosis · Prediction · Biomarker · Proteomics

---

Liming Shen, and Yuxi Zhao contributed equally to this work.

L. Shen (✉) · Y. Zhao · H. Zhang · S. Xia · J. Iqbal · X. K. Liu · F. Yao
College of Life Science and Oceanography, Shenzhen University, Shenzhen, People's Republic of China
e-mail: slm@szu.edu.cn

C. Feng · Y. Gao · Q. Hong
Maternal and Child Health Hospital of Baoan, Shenzhen, People's Republic of China

D. Zhao
Department of Obstetrics and Gynecology, Affiliated Hospital of Guizhou Medical University, Guiyang, People's Republic of China

P. C. Guest (ed.), *Reviews on Biomarker Studies in Psychiatric and Neurodegenerative Disorders*, Advances in Experimental Medicine and Biology 1118,
https://doi.org/10.1007/978-3-030-05542-4_11

11.1 Introduction

Autism spectrum disorder (ASD) is a group of neurodevelopmental disorders characterized by impaired social communication and interaction with restricted and repetitive behavior, interests, or activities [1]. The data of past two decades shows that the prevalence of ASD has continued to rise [2]. The Centers for Disease Control and Prevention (CDC) has reported that the prevalence of ASD has increased in recent years and approximately 15 per 1000 children have been diagnosed with ASD in the United States [3]. Similar prevalence ratios have been reported in the latest worldwide surveys [4–8]. About 45% of individuals with autism have an intellectual disability [9], and 29–47% have regression using retrospective or categorical measures [10, 11]. Given these effects, ASD is a serious social problem and an increasing global burden with severe implications for public health services.

The etiology of ASD is complex and the exact causes are still unknown. Previous studies suggest that autism can be considered as a multifactorial disease in which both genetic and environmental factors are involved. Immune dysregulation, inflammation, oxidative stress, mitochondrial dysfunction, and environmental toxicant exposures have been reported in a large number of studies on associated physiological abnormalities in ASD [12–14]. Several possible risk factors for ASD have been investigated, which include advanced parental age, birth complications, prematurity, low birth weight, and assisted conception [15].

To date, there are no practical therapeutic approaches for ASD, and the drugs being used have limited efficacies. However, studies suggest that substantial improvements can be achieved by intensive behavioral intervention initiated prior to 24 months of age as neural plasticity is increased and challenging behaviors are less prominent [15]. Prior to the onset of behavioral abnormalities, behavioral interventions could conceivably minimize their severity or even result in prevention of a full-blown autism [16]. Thus, early diagnosis and prediction is essential for ASD. However, ASD diagnosis is limited as it is currently based on behavioral signs, symptoms, clinical observations, and behavioral evaluations. Without any biological determination, early diagnostics are difficult and subjective. The mean age of clinical diagnosis has found stable at 4–5 years with no evidence of decline [17].

As a result, the availability of reliable biomarkers for use in the prognosis and diagnosis of these patients remains an unmet clinical need [18]. Multiple approaches including neuroimaging, gene testing, transcriptomics, proteomics, and metabolomics are being used to discover new biomarker panels for ASD, and progress has been made in these fields. Other methods, such as electroencephalography (EEG) [19], eye tracking for visual orienting [20, 21], or pupillary light reflex [22], may have potential. There is now growing interest in the identification of molecular biomarkers that could be implemented easily in clinical practice through conventional laboratory medicine, following the routine collection of bodily fluids such as blood, urine, or saliva [18]. In this chapter, we have summarized the current progress in ASD biomarker studies and highlighted recent findings and emerging methodologies that could improve the timeliness of diagnosis in years to come.

11.2 Neuroimaging

Advances in neuroimaging methods, including functional magnetic resonance imaging (fMRI) [23, 24] and functional near-infrared spectroscopy (fNIRS) [25], have become important tools in investigating ASD. Cognitive processes can be characterized easily as biomarkers, including semantic memory processing, language processes in middle temporal gyrus, and visual perception in the inferior temporal gyrus, as well as integrate information from different senses and link these with different brain regions [26]. A recent study showed that a functional connectivity MRI-based machine learning algorithm applied at 6 months of age had 81% sensitivity and 100% specificity for diagnosis of ASD [27]. In addition, a deep-learning algorithm that primarily uses surface area information from MRI analyses of the brains of 6–12-month-old individuals predicted the diagnosis of autism in high-risk children at 24 months of age, with a positive predictive value of 81% and a sensitivity of 88% [28]. These findings suggest that early brain changes occur during the period in which autism-like behavior is first emerging and may be used to identify infants who would later meet the criteria for ASD.

11.3 Genetic Susceptibility and Genetic Testing

The etiology of ASD is complex. Early twin studies have suggested that autism has high heritability (more than 80%) [29], but when considered as a model with additive genetic makeup, heritability was estimated to be about 54% [30, 31]. Taking into account the high rate of inheritance of ASD, a significant amount of research has been aimed at identifying the genetic basis of the pathology development and identifying genetic markers that evaluate the risk of disease [32, 33]. By utilizing chromosomal microarray (CMA), whole-exome sequencing (WES), whole-genome sequencing (WGS), targeted gene sequencing, and copy number variation (CNV) research, hundreds of genes associated with autism risk have been identified [34]. The SFARI Gene web portal (https://gene.sfari.org/) seamlessly integrates different kinds of genetic data that are being generated by research studies, linking information on autism candidate genes within its original "Human Gene" module with corresponding data from a diverse array of supplementary data modules. ASD risk genes are then scored using a set of annotation rules developed in consultation with an external advisory board and classified into specific categories based on the evidence supporting their link to autism [35]. Many new susceptibility genes associated with ASD have now been discovered. These include CHD8, DYRK1A, SCN2A, ARID1B, ANK2, GRIN2B, SYNGAP1, ADNP, TBR1, POGZ, KATNAL2, SHANK2, and SHANK3. These findings implicate a large functional network of genes involved in transcriptional control, chromatin remodeling in the nucleus, protein synthesis, and the formation and function of synapses [36–42].

Recent research highlights the ASD-related variants in noncoding region [43–45], which show that precise coordination of gene activity is crucial for brain development and function. Moreover, 0.77% rare inherited cis-regulatory structural variants contribute to ASD [46]. The burden analysis has identified noncoding indels as a potential contributor [43], while polygenic risk further implicates conservation across vertebrate species. Of note, by disrupting regulatory elements to a degree greater than SNVs while occurring far more often than SVs, indels might represent a strong effect for coding regions. In the light of evolution, human-accelerated regions (HARs) represent conserved genomic loci with elevated divergence in humans, and rare de novo CNVs involving HARs can contribute to simplex ASD, and rare bi-allelic mutations in neurally active HARs can confer risk to ASD in as many as 5% of individuals from a consanguineous population [45]. Epigenomic profiling and in vitro analyses showed functional effects of candidate mutations in several HARs that interact with promoters of dosage-sensitive neurodevelopmental genes, including CUX1, PTBP2, GPC4, and MEF2C [45]. This suggests that disturbance in the regulation of gene expression but not mutation itself is important [47].

High genetic heterogeneity is one of the most prominent features of ASD, which leads to challenges in obtaining and interpreting genetic testing in clinical settings. Nevertheless, advances in techniques and methodologies in genetics and genomics provide opportunities for genetic research and discovery of biomarkers. Genetic testing of children with ASD is now standard in clinical settings through guidelines from the American College of Medical Genetics and Genomics (ACMGG) [48, 49]. ACMGG guidelines recommend for all children with ASD without a recognizable genetic diagnosis, along with fragile X testing for boys, single gene sequencing including MECP2 (methyl-CpG-binding protein 2 gene) for girls and PTEN for those with macrocephaly. Consideration of metabolic screening, brain MRI, and X-linked disability gene panel is also recommended in cases where medical history, physical exam, and/or family history supports it [48]. At present, utilization of high-throughput sequencing, including targeted gene panels and whole-exome sequencing, is becoming increasingly common in the clinical evaluation of children with ASD [49]. A recent review proposed an approach designed for a child who has been diagnosed with ASD by DSM-5 criteria, which highlights the incorporation of new testing methodologies for determining a molecular defect, including ASD-associated CNV on a microarray and ASD-associated changes identified through WES [50].

In terms of genetic testing, there are different strategies and techniques, and the estimated diagnostic yield is different for these. The current estimated diagnostic yield is relatively low as it has been estimated that a specific genetic etiology can be determined in about 15–20% of individuals with an ASD [49, 51]. However, this diagnostic yield may be increased with the development of new techniques and technologies. For example, development of targeted gene panels using next-generation sequencing (NGS) is an attractive strategy for the clinical evaluation of children with ASD, and some of these panels contain as many as 2000 genes (GeneDx, Gaithersburg autism/ID panel) [49]. Meanwhile, if the list of genes associated with ASD risk grows, the gene panel can be updated. Informed by better natural history data, it may become appropriate to screen for highly penetrant ASD variants in newborns in order to allow the introduction of behavioral interventions in the first year of life when the brain has its highest level of neuronal plasticity [50].

11.4 Blood Protein-Based Biomarkers

Despite the large number of genetic and environmental factors underlying autism, these factors can be predicted to converge upon a relatively limited number of intracellular biochemical pathways and neurodevelopmental mechanisms [18]. As mentioned above, many of the products encoded by ASD-related genes have similar functions or cause disease through some common signal transduction pathways, including neuronal development and axonal guidance, synaptic function, and chromatin remodeling [52]. Some mechanisms may be used to explain this phenomenon. For example, a master transcriptional repressor CHD8 (chromodomain helicase DNA-binding 8) has been found to play a critical role in ASD. It acts by remodeling chromatin structure and recruiting histone 1 to target genes and other ASD-risk genes. Thus, it functions at the center of a complex network of ASD genes and may prove to be an important underlying common mechanism for ASD, at least for a subset of affected individuals [53, 54]. Moreover, according to the Central Rule of Genetics, genetic levels indicate the probability of disease occurrence, and proteins are the undertakers of biological activities. Therefore, changes in the proteins can more directly reflect the occurrence and development of the disease. Furthermore, most of the disease biomarkers and drug targets are proteins. Taken together, ASD can be investigated from a protein perspective to understand the pathogenesis and obtain diagnostic biomarkers or therapeutic targets.

Blood collection has qualities of low invasiveness, ease of use, and cost-effectiveness. Therefore, blood-based biomarkers would be more amenable to large-scale screening and diagnosis in clinical applications. Blood proteins are the main components of blood and are often used for clinical diagnosis or screening. However, because of the existence of the blood-brain barrier (BBB), it is difficult to find the ASD protein markers from peripheral blood, especially disease-specific protein markers, which have actually originated from the brain. A recent study found that 75% of patients with ASD had reduced expression of barrier-forming "tight junction" (TJ) components (claudin-1, occluding, tricellulin) in the intestine and 66% had increased pore-forming claudins (claudin-2, claudin-10, claudin-15). This suggests increasing intestinal and BBB permeability in at least some subgroups of individuals with ASD [55]. Interestingly, alterations of the intestinal barrier in patients with ASD have been reported in a previous study [56]. Zonulin regulates tight junctions between enterocytes and is a physiological modulator controlling intestinal permeability. This protein was found to be increased in patients with ASD compared with healthy controls [57]. In addition, levels of zonulin and the Childhood Autism Rating Scale score have a positive correlation [57]. Therefore, changes in proteins or genes associated with intestinal and BBB permeability, as well as pathological and metabolic changes, and proteins and metabolites from blood may represent ASD biomarkers. Meanwhile, finding the commonalities and/or differences between peripheral samples and the CNS (central nervous system) includes peripheral signaling that is parallel with that seen in the CNS and could also uncover new or provide more possibilities to clarify the pathophysiology of disease as well as future candidates for biomarkers [58].

11.4.1 Proteomics

Proteomics is a large-scale study of protein expression in cells and tissues. It is a powerful tool for studying the biology of clinical conditions and may better reflect dynamic physiological processes compared to other methods [59]. Proteomics can also represent distinct primary structures of proteins (or "proteoforms") resulting from alternatively spliced RNA transcripts, genetic variations like coding SNP or mutations, and posttranslational modifications [60–62].

Several groups have conducted proteome analyses of ASD samples, including postmortem brain tissues [63, 64], serum [65–69], plasma [70–72], lymphocytes [73], neonatal blood [74, 75], urine [76, 77], and saliva [78, 79]. The results of these studies are summarized in Table 11.1. For brain tissues, Junaid et al. analyzed postmortem brain tissues from patients with autism using two-dimensional gel electrophoresis (2-DE) [63]. By selected reaction monitoring mass spectrometry (SRM-MS) analysis, Broek et al. analyzed postmortem brain tissues from the prefrontal cortex and cerebellum of ASD individuals and healthy subjects [64]. They found that proteins associated with myelination, synaptic vesicle regulation, and energy metabolism were dysregulated in ASD brain tissues. In addition, Wei et al. carried out a quantitative proteomic profiling study of cortical brain tissue from BTBR T + Itpr3tf (BTBR) mice, a model that displays an autism-like phenotype [80]. Since ASD is a neurodevelopmental disorder, the results of the brain tissue study provided information on the discovery of biomarkers for autism in peripheral tissues or blood, especially CNS-specific proteins. Furthermore, the same proteomic methods can be used to simultaneously analyze brain tissue and blood as a strategy for ultimately identifying blood diagnostic biomarkers.

Blood proteomics studies holistic changes in blood protein composition and quantity as well as posttranslational modification status, and these are helping to identify disease-associated markers and therapeutic targets and connect these to genetic risk [81–83]. Studies on biomarkers for the diagnosis of ASD blood proteins have also been reported and were mostly based on traditional proteomics methods (Table 11.1). In the last 5–10 years, liquid chromatography coupled to tandem mass spectrometry (LC-MS/MS) has become the technology of choice for high-throughput characterization of proteins [84]. For biomarker discovery, a suitable strategy for large-scale quantitative proteomics is to employ isobaric labeling such as with tandem mass tags (TMT) [85] or isobaric tags for relative and absolute quantitation (iTRAQ) [86]. Recently, we carried out iTRAQ-based plasma analysis to compare protein profiles from children with autism, compared to healthy control children, and 24 differentially expressed proteins were identified [72]. These proteins were found to be involved in different pathways that have previously been linked to the pathophysiology of ASDs. This supports the hypothesis that focal adhesions, acting cytoskeleton, cell adhesion, motility and migration, synaptogenesis, and the complement system are involved in the pathogenesis of autism. Moreover, it highlighted the important role of platelet function. Using enzyme-linked immunoadsorbent assay (ELISA) and ROC (receiver operating

Table 11.1 ASD-related proteomic studies reported in the literature

Studies	Sample types	Method	Proteins implicated	Validation	Pathways/Processes
Junaid et al. (2004) [63]	Brain (gray matter)	2D-PAGE	Glyoxalase I↓[a]	WB LC-MS/MS	Cell growth and differentiation
Corbett et al. (2007) [65]	Serum	LC-MS/MS	↑[d]: Apolipoprotein B-100, Complement C1q, FHR1, Fibronectin		Complement system Cellular apoptosis
Castagnola et al. (2008) [132]	Saliva	HPLC-ESI-IT-MS	[e]: Acidic proline-rich proteins, Histatin 1 Statherin		Calcium phosphates
Taurines et al. (2010) [66]	Serum	MALDI-TOF-MS	Three peaks with m/z ratios: 4.40, 5.15, 10.38 kDa		
Shen et al. (2011) [73]	B-lymphocytes	Antibody chips	IKK-α↑, EIF4G1↓, TYK2↓, Protein kinase C iota type ↓	WB RT-qPCR	Immune reactions
Schwarz et al. (2011) [133]	Serum	Immunoassay	Males↑: Chromogranin A, Connective tissue growth factor, ENA-78, Erythropoietin, Factor-VII, Fatty acid binding protein, Granulocyte colony-stimulating factor, IL-3, IL-4, IL-5, IL-10, IL-1B, IL-12p40, IL-12p70, IL-18, Intercellular adhesion molecule 1, Neuronal cell adhesion molecule, Sortilin 1, Stem cell factor, Tissue factor, Thrombopoietin, Tissue factor, Tumor necrosis factor-α Males↓: GOT1, Tenascin-C (TENA) Females↑: 31, 32-PI, Brain-derived neurotrophic factor, Free androgen index, IL-7, IL-1B, IL-12p40, Insulin, Luteinizing hormone, NMDA receptor regulated 1, TENA, Tissue factor. Females↓: Apo-A1, Apolipoprotein-CIII, Endothelin-1, Eotaxin-3, GOT1, Growth hormone, Immunoglobulin M, sRAGE		Inflammation Growth factors anabolism
Momeni et al. (2012) [134]	Serum	SELDI TOF MS MALDI-TOF/TOF MS ESI-FTICR MS	↑[d]: Complement protein C3 (C3)		Complement system

(continued)

Table 11.1 (continued)

Studies	Sample types	Method	Proteins implicated	Validation	Pathways/Processes
Broek et al. (2014) [64]	Brain (Prefrontal cortex, cerebellum)	SRM LC-MS/MS	Prefrontal cortex: ↑: Glial fibrillary acidic protein, Myelin-associated glycoprotein, Myelin basic protein, Myelin-oligodendrocyte glycoprotein, Myelin proteolipid protein, Synapsin-2 ↓: Creatine kinase B-type, Protein kinase C casein kinase substrate 1, Synaptotagmin-1,Syntaxin-1A, Vimentin Cerebellum: ↑: Creatine kinase B-type, Glial fibrillary acidic protein, Synapsin-2, Synaptotagmin-1, Syntaxin-1A ↓: Myelin-associated glycoprotein, Myelin basic protein, Myelin-oligodendrocyte glycoprotein, Myelin proteolipid protein, Protein kinase C casein kinase substrate 1, Vimentin		Astrocyte maturation Energy production Axon myelination Synaptic vesicle release
Ngounou Wetie et al. (2014) [69]	Serum	PAGE LC-MS/MS	↑: Apolipoproteins, Apolipoprotein A-I, Paraoxanase 1	WB	Oxidative stress responses Cholesterol metabolism
Steeb et al. (2014) [68]	Serum	Immunoassay, LC-MS/MS	Males↑: BMP6, CHGA, CTGF, EPO, ICAM1, IL-12p70, IL-16, IL-3, PAP, SHBG, TENA, TF, TNF Males↓: RGPD4 Females↑: APOE, APOC2, ARMC3, CLC4K, FETUB, GLCE, PTPA, RN149, TLE1, TRIPB, ZC3HE Females↓: ADIPO, APOA1, CHGA, EPO, IgA, IL-3, MRRP1, PAP, SHBG, TENA		Lipid metabolism Cell growth Inflammation Nucleic acid processing
Ngounou Wetie et al. (2015) [79]	Saliva	2D-PAGE LC-MS/MS	↑: FRAT1, Growth hormone regulated TBC protein 1, Integrin alpha6 subunit, Kinesin family member 14, MRP14, Mucin-16, Parotid secretory protein, Prolactin-inducible protein precursor ↓: Alpha-amylase, CREB-binding protein, Cystatin D, p532, Plasminogen, Transferrin, Zn alpha2 glycoprotein, Zymogen granule protein 16		Immune response Inflammation Lipid metabolism Oxidative stress
Ngounou Wetie et al. (2015) [78]	Saliva	LC-MS/MS	↑: DMBT1, LTF, Ig kappa chain C region, Ig gamma-1 chain C region, Ig lambda-2 chain C regions, Neutrophil elastase, PIP, Polymeric immunoglobulin receptor ↓: Histatin-1, Statherin, Acidic proline-rich phosphoprotein		Immune response (mucosal) Inflammation

Suganya et al. (2015) [76]	Urine	2D-PAGE MALDI-TOF-MS	↑: IGHG1, Kininogen-1, MASP2	ELISA	Nervous and neuroendocrine functions Immune response
Yang L et al. (2016) [77][b]	Urine	iTRAQ labeling LC-MALDI-MS/MS	alpha 1-acid glycoprotein, prostaglandin-H2 D-isomerase, kininogen-1 isoform 2, leucine-rich alpha-2-glycoprotein 1, immunoglobulin fragment Fab New lambda light chain, vitelline membrane outer layer 1 homolog, isoform CRA_b, lithostathine-1-alpha, alpha-2-glycoprotein 1, zinc, isoform CRA_a, collagen alpha-1(XII) chain long isoform, inter-alpha-trypsin inhibitor heavy chain H4 isoform 1		Organ Injury and abnormalities Inflammatory response Renal and urological disease Gastro-intestinal disease
Cortelazzo et al. (2016) [70][c]	Plasma	2D-PAGE LC-MS/MS	↑: Alpha-2-macroglobulin, Alpha-1-antitrypsin, Haptoglobin, Fibrinogen beta chain, Fibrinogen gamma chain, IGHA1, IGHG ↓: Apolipoprotein A-I, Apolipoprotein A-IV, Apolipoprotein J, Prealbumin, Serum transferrin		Inflammatory response Lipid peroxidation
Feng et al. (2017) [71]	Plasma	2D-Oxyblot MALDI-TOF	Complement component C8 alpha chain[f], Ig kappa chain C[f]	1D-Oxyblot	Immune system Oxidative stress
Singh S et al. (2017) [135]	Serum	Immunoassay	Thyroid-stimulating hormone↓ Interleukin-8↑	MSD platform	Neutrophil activation Thyroid structure and metabolism
Shen et al. (2018) [72]	Plasma	iTRAQ labeling, Nano-LC-MS/MS	↑: Alpha-1-antitrypsin, Vitronectin Angiotensinogen, Apolipoprotein E, C3, C5, Ehd3, Fibronectin, Fibulin-1, IGFALS ↓: Actin, cytoplasmic 2, Alpha-actinin-1, Alpha-enolase, Beta-parvin, Calmodulin, Calreticulin, Ehd3, Fermitin family homolog 3, Integrin Alpha-IIb, MAPRE2, Talin-1, Thrombospondin-1, Vcp, Vinculin	ELISA	Immune system Inflammatory Focal adhesions Acting cytoskeleton Cell adhesion Cell motility and migration Synaptogenesis Complement system Platelet function

(continued)

Table 11.1 (continued)

Studies	Sample types	Method	Proteins implicated	Validation	Pathways/Processes
Yang et al. (2018) [136]	Serum	MB-WCX MALDI-TOF MS LC-ESI-MS/MS	↑[d]:Alpha-fetoprotein precursor, Apolipoprotein C-I precursor, Carboxypeptidase B2, Fatty acid binding protein 1, Fibrinogen alpha chain, Plasma serine protease inhibitor- precursor, Platelet factor4, Trace amine-associated receptor 6	ELISA	Immune response Extracellular matrix remodelling

[a]Expression changes in ASD, compared with the control: ↑: upregulated; ↓: downregulated
[b]Top ten proteins based on multiplicative differences in abundance between autism and control subjects
[c]Lipid peroxidation-derived aldehyde was also evaluated in this study
[d]Identified peptides associated with proteins
[e]Phosphorylation level decreased in autistic patients
[f]Oxidatively modified (protein carbonylation)

Abbreviations: *2D-Oxyblot* 2-DE (two-dimensional gel electrophoresis) plus Western blot analysis(WB), 31, 32-PI Des31,32-proinsulin, *ADIPO* Adiponectin, *APOA1* Apolipoprotein A1, *APOC2* Apolipoprotein C2, *APOE* Apolipoprotein E, *ARMC3* Armadillo repeat containing 3, *BMP6* Bone morphogenic protein-6, *C5* Complement C5, *CHGA* Chromogranin A, *CLC4K* C-type lectin domain family 4 member K, *CTGF* Connective tissue growth factor, *DMBT1* Deleted in malignant brain tumors 1, *Ehd3* EH domain-containing containing protein 3, *EIF4G1* Eukaryotic translation initiation factor 4 gamma 1, *EPO* Erythropoietin, *FETUB* Fetuin B, *FRAT1* Frequently rearranged in advanced T-cell lymphomas 1, *GLCE* Glucuronic acid epimerase, *GOT1* Serum glutamic oxaloacetic transaminase, *ICAM1* Intracellular adhesion molecule-1, *IgA* Immunoglobulin A, *IGFALS* Insulin-like growth factor-binding protein complex acid labile subunit, *IGHA1* Ig alpha-1 chain C region, *IGHG* Ig gamma heavy chain, *IKKA* I-kappa-B kinase 1, *IL* Interleukin, *LTF* lactotr-ansferrin, *MAPRE2* Microtubule-associated protein RP/EB family member 2, *MASP2* Mannan-binding lectin serine protease-2 isoform-2 precursor, *MB-WCX* Magnetic beads cation-exchange chromatography, *MRP14* Migration inhibitory factor-related protein 14, *MRRP1* Mito-chondrial RNase P protein 1, *PAGE* Polyacrylamide gel electrophoresis, *PAP* Prostatic acid phosphatase, *PIP* Prolactin-inducible protein, *PTPA* PP 2A activator, reg subunit 4, *RT-qPCR* quantitative reverse transcription polymerase chain reaction, *RGPD4* RANBP2-like and GRIP domain containing 4, *RN149* Ring finger protein 149, *SELDI TOF MS* Surface-enhanced laser desorption/ionization time-of-flight mass spectrometry, *SHBG* Sex hormone binding globulin, *sRAGE* receptor for advanced glycosylation end product, *TF* Tissue factor, *TLE1* Transducin-like enhancer of split 1, *TNF* Tumour necrosis factor-alpha, *TRIPB* Thyroid hormone receptor interactor 11, *TYK2* Non-receptor tyrosine-protein kinase TYK2, *Vcp* Transitional endoplasmic reticulum ATPase, *ZC3HE* Zinc finger CCCH domain-containing protein 14 characteristic) analysis, combinations of five proteins (C3, C5, GC, ITGA2B, and TLN1) were used to distinguish children with autism from healthy controls with a high AUC (area under the ROC curve) value [72]. Interestingly, these three pathways (focal adhesion, cell adhesion molecules, and leukocyte transendothelial migration) were related to a Han Chinese cohort in a recent study [87]. Together with other blood proteomics studies (Table 11.1), the results support the view that ASD-related proteins are mainly concentrated in some pathways. Indeed, these proteins may be related to the cause of ASD or are a consequence or merely an epiphenomenon.

Labeling samples with isobaric tags is an optimal workflow for comparative analysis of medium-sized projects. However, to support progress in the field of psychiatric issues including proteomic analyses of autism samples, further advances in proteomic profiling techniques are required. For example, if hundreds of samples need to be analyzed simultaneously, the issues of data incompleteness and batch effects can become apparent across sets of multiplexed samples. Currently, two alternative mass spectrometric strategies are used in the large number of samples. These include the classical label-free data-dependent acquisition (DDA) proteomics workflow [88] and sequential window acquisition of all theoretical mass spectra (SWATH-MS) [89].

Additionally, validation of biomarker candidates can be addressed by transferring the candidates found in the explorative studies to targeted assays based on SRM-MS and parallel reaction monitoring (PRM) [90]. The so-called targeted proteomics is a bridge between discovery and validation. Such assays typically require smaller sample volumes, less sample preparation, and significantly shorter analysis times and often provide absolute quantification with high precision [91]. An advantage of targeted proteomics is the possibility for multiplexing, allowing for analysis of a number of compounds in one analysis. Therefore, they enable evaluation of large panels of biomarker candidates in large clinical studies and can be further employed in clinical routine work [90, 91]. As mentioned above, SRM-MS has been used for targeted analysis of proteins associated with different pathways in accordance with literature evidence in brain tissues from ASD subjects and healthy controls [64].

11.4.2 Ultrasensitive Techniques in Blood

Apart from the above discussion, it is also worth mentioning that CNS-specific proteins may be present at very low concentrations in the blood and thus may be difficult to quantify reliably using standard immunochemical technologies such as ELISA. Recent technical breakthroughs in the field of ultrasensitive assays have started to change this scenario [92]. These technologies include single-molecule array (Simoa) technology, immunomagnetic reduction (IMR) [92], and immunoprecipitation (IP) coupled with mass spectrometry [93]. Furthermore, these methods have high sensitivity and accuracy and have been used in the study of neurological diseases such as Alzheimer's disease (AD) [92, 93]. For example, the average sensitivity improvement of the Simoa immunoassays versus conventional ELISA was >1200-fold, with coefficients of variation of <10% [94]. Thus, these ultrasensitive technologies may be useful for measuring peripheral CNS-specific proteins or protein biomarkers at low concentrations in the bloodstream in studies of ASD and other neurological and neurodevelopmental conditions.

11.5 Transcriptomics

Transcriptomic studies are essential as a link between measuring protein levels and analyzing genetic information. Several studies using transcriptome profiling analysis of peripheral blood lymphocytes (PBL) [95] and the lymphoblast cell line (LCL) [96–99] from patients with ASD and unaffected controls have revealed a number of transcripts that are differentially expressed in the individuals with autism compared to controls. This highlights the point that transcriptomic signatures from peripheral blood that could be used as a surrogate for understanding the genetics of ASD and as potentials of biomarkers. Recently, Ansel et al. reviewed gene expression changes in ASD detected in multiple independent studies in various tissues [100]. It is worth noting that some differentially expressed genes were found in common between the brain and peripheral blood or LCL cells (Table 11.2). The biological processes associated with these genes included response to cytokines, defense response, blood vessel development, cellular response to cytokine stimulus, angiogenesis, cytokine-mediated signaling pathway, positive regulation of cell communication, and regulation of response to stimulus. The affected biological pathways included cytokine-cytokine receptor interaction, chemokine signaling pathway, leukocyte transendothelial migration, taurine and hypotaurine metabolism, and the Wnt signaling pathway. Interestingly, in ASD subjects, the genes associated with cytokines were upregulated, whereas those associated with synaptic signaling (i.e., GAD1 and GRIA3) were downregulated. Likewise, other dysregulated genes were found in common between intestinal biopsy samples and peripheral blood or LCL cells (Table 11.2). These genes were mainly involved in several biological processes and pathways including cell differentiation, response to cytokines, immune system development, and cytokine-cytokine receptor interaction.

Clearly, detection of the gene changes presented in the two or more types of samples can enhance detection specificity. Given the key role of brain and intestinal tissues in the pathogenesis of ASD, simultaneous detection of these genes in the brain and blood/LCL cells, or in the intestinal biopsy samples and peripheral blood or LCL cells, might lead to a new biomarker approach for ASD diagnosis.

Table 11.2 The commonly dysregulated genes in brain and peripheral blood or LCL cells

Gene symbols	Blood/LCL studies	Brain studies
SLC9A9	Talebizadeh et al. (2014) ↑[96]	Garbett et al. (2008) ↑[137]
NDUFB5	Talebizadeh et al. (2014) ↑[96]	Anitha et al. (2012) ↓[139]
CMKOR1	Nishimura et al. (2007) ↑[97]	Garbett et al. (2008) ↑[137] Voineagu and Eapen (2013) ↑[138]
HIST1H3H	Nishimura et al. (2007) ↓[97]	Chow et al. (2012) ↑[140]
MeCP2	Kuwano et al. (2011) ↑[95]	James et al. (2014) ↓[141] Zhubi et al. (2014)↓[142]
ABHD3	Kong et al. (2013) ↑[143]	Garbett et al. (2008) ↑[137]
CTNNB1	Kong et al. (2013) ↑[143]	Chow et al. (2012) ↓[140]
SYCE1	Kong et al. (2013) ↓[143]	Chow et al. (2012) ↓[140]
FOSL1	Ivanov et al. (2015) ↓[144]	Chow et al. (2012) ↑[140]
KIF1B	Talebizadeh et al. (2014) ↑[96]	Garbett et al. (2008) ↑[137]
PITPNC1	Nishimura et al. (2007) ↑[97]	Garbett et al. (2008) ↑[137]
CX3CR1	Gregg et al. (2008) ↑[145] Enstrom et al. (2009) ↑[146]	Ziats and Rennert (2013) ↑[147]
STOM	Glatt et al. (2012) ↓[118]	Garbett et al. (2008) ↑[137]
TAP1	Glatt et al. (2012) ↓[118]	Garbett et al. (2008) ↑[137]
PARP9	Glatt et al. (2012) ↓[118]	Garbett et al. (2008) ↑[137]
GAD1	Chien et al. (2013) ↓[99]	Zhubi et al. (2014) ↓[142]
ANXA1	Chien et al. (2013) ↓[99]	Garbett et al. (2008) ↑[137]
CHI3L1	Chien et al. (2013) ↓[99]	Garbett et al. (2008) ↑[137]
WWTR1	Chien et al. (2013) ↓[99]	Garbett et al. (2008) ↑[137]
CXCL10	Chien et al. (2013) ↑[99]	Chow et al. (2012) ↑[140]
CXCR4	Chien et al. (2013) ↑[99]	Chow et al. (2012) ↑[140]
DNASE1L3	Chien et al. (2013) ↓[99]	Chow et al. (2012) ↓[140]
GRIA3	Chien et al. (2013) ↓[99]	Chow et al. (2012) ↓[140]
SERPINA1	Chien et al. (2013) ↓[99]	Chow et al. (2012) ↑[140]
TNFRSF19	Chien et al. (2013) ↓[99]	Chow et al. (2012) ↓[140]

Expression changes in ASD, compared with the control: ↑: upregulated; ↓: downregulated

11.6 Metabolomics

Another "-omics" technology being used for biomarker discovery is metabonomics. It is considered as one of the fastest developing workflows in biomarker research [101]. Profiles of small-molecular-weight substances present in different sample types (cells, tissue, and body fluids) are known as metabolites. Nuclear magnetic resonance (NMR) is a particularly powerful tool for metabolite analysis, and MS is also a sensitive platform to identify and quantify these in complex biological systems [101]. Metabolomic analysis of human biofluids provides another sensitive approach to identify metabolite profiles potentially usable as biomarkers for ASD. Researchers have discovered many potential biomarkers from blood and urine [102–106] (Table 11.3). Among these, metabolites associated with the tricarboxylic

Table 11.3 ASD-related metabolomics studies reported in the literature

Study	Sample types	Method	Number of dysregulated Implicated listed	Pathway/ Processes
Yap et al. (2010) [104]	Urine	^{1}H-NMR	Increased: dimethylamine, N-acetyl glycoprotein, platelet serotonin. Decreased: Urinary hippurate, phenylacetylglutamine, amino acids aspartate, glutamine, γ-aminobutyrate	Abnormalities of gut microbiota Gastrointestinal dysfunction
Emond et al. (2010) [150]	Urine	GC-MS	Increased: Succinate glycolate, Z5, Z37 Decreased: Z18, Z41, Z26, Z21, X16, hippurate, palmitate 3-hydroxyphenylacetatevanillylhydracrylate, 3-hydroxy-hippurate, p-hydroxy mandelate, 1H-indole-3-acetate phosphate, stearate, 3-methyladipate	Microbial pathways associated with gut bacteria
Ming et al. (2012) [105]	Urine	Untargeted metabolomics analysis	Increased: Trans-urocanate, glutarylcarnitine, 3-methylglutarylcarnitine, 2-(4-hydroxyphenyl) propionate, taurocholenate sulphate Decreased: Glycine, N-acetylglycine, serine, threonine, β-alanine, alanine, histidine, taurine, gamma-glutamylleucine, gamma-glutamyltyrosine, gamma-glutamylthreonine, 3-(3-hydroxyphenyl) propionate, 5-aminovalerate, carnosine, urate	Amino acid metabolism, Oxidative stress, Mammalian microbial co-metabolism
Kuwabara et al. (2013) [149]	Plasma	CE-TOF-MS	Increased: Glutaric acid, Serine, glucuronic acid, taurine, arginine Decreased: Citric acid, lactic acid, uric acid, creatinine, cystine 2-Hydroxyvaleric acid.	Oxidative stress, Mitochondrial dysfunction

West et al. (2014) [102]	Plasma	MS	Increased: Aspartic acid, serine, glutamic acid, DHEA sulfate, glutaric acid, 5-hydroxynorvaline, 5-aminovaleric acid lactam, succinin acid, 2-hydroxyvaleric acid, 3-aminoisobutyric acid Decreased: Homocitrulline, 2-hydroxyvaleric acid, cystine, heptadecanoic acid, isoleucine, creatinine, methylhexadecanoic acid, 4-hydroxyphenyllactic acid, citric acid, myristic, lactic acid	Oxidative stress Energy production Mitochondrial disease or dysfunction
Diémé et al. (2015) [106]	Urine	^{1}H–^{13}C NMR- LC–HRMS	Increased: Dihydroxy-1H-indole glucuronide I, indoxyl sulfate, N-α-acetyl-l-arginine, N-acetylasparagine, glucuronic acid, indoxyl, valine, norvaline, 5-aminopentanoic acid, α-N-phenylacetyl-l-glutaminep-cresol sulphate Decreased: Methylguanidine, guanidine succinic acid, desaminotyrosine, dihydrouracil	Gut microbiota dysfunction
Wang et al. (2016) [103]	serum	UPLC/Q-TOF MS/MS	Increased: Pregnanetriol, 9,10-epoxyoctadecenoic acid, sphingosine 1-phosphate, LPA(18:2(9Z,12Z)/0:0) LysoPE (0:0/16:0), LysoPE (18:0/0:0), LysoPE(0:0/20:2(11Z,14Z)), LysoPC(18:3(6Z,9Z,12Z)) LysoPC(20:3(5Z,8Z,11Z)), phytosphingosine Decreased: L-acetylcarnitine, decanoylcarnitine, uric acid, arachidonic acid, docosahexaenoic acid, docosapentaenoic acid, adrenic acid	Fatty acid β-oxidation Mitochondrial dysfunction

(continued)

Table 11.3 (continued)

Study	Sample types	Method	Number of dysregulated Implicated listed	Pathway/ Processes
Gevi et al. (2016) [151]	Urine	HILIC UHPLC-MS	Increased: Xanthurenic acid, quinolinic acid, indolyl-3-acetic acid, indolyl lactate, inosine, hypoxanthine, xanthosine. Decreased: Melatonin, N-acetyl-5-methoxytryptamine, kynurenic acid.	Tryptophan and purine metabolic pathways
El-Ansary et al. (2018) [152]	Plasma	Metabolism-Associated Markers	Increased: Lactic acid, pyruvic acid Decreased: MRC complex I complex-I/ coenzyme Q10, MRC complex-I / caspase-7, glutathione / glutathione-S-transferase	Mitochondrial dysfunction
Lv et al. (2018) [153]	Plasma	MS/MS	Decreased: Free carnitine, glutaryl carnitine, long chain acyl carnitines	Fat metabolism Mitochondrial dysfunction

“X” compounds are TMS derivatives, “Z” compounds are TMSO derivatives. “MRC” compounds are mitochondrial respiratory chain. Increased or decreased: compared with the control acid cycle, saturated fatty acid metabolism, and oxidative stress suggest that mitochondrial dysfunction may be a risk factor for autism. In addition, Diémé et al. analyzed the samples of children with ASD and healthy controls using NMR and a liquid chromatography (LC)-MS-based approach [106]. The results showed that the metabolites with the most significant differences between autism and control children were indoxyl sulfate, N-α-acetyl-L-arginine, methyl guanidine, and phenylacetylglutamine. Interestingly, indoxyl and indoxyl sulfate, which are produced by tryptophan metabolism in gut bacteria, were reproducibly found in urine investigations, which suggest these as potential biomarkers [107].

11.7 Immune System and Cytokines

Currently, many studies have suggested that the immune system plays an important role in the pathophysiology of ASD [108], even though this system has a less pronounced genetic component. Therefore, the effects on the immune system are most likely caused by environmental factors. Maternal immune activation (MIA), resulting from either genetic or environmental causes, has been highlighted as a factor that can increase the risk of ASD [109–112]. It may act at early periods of fetal brain development and potentially alter gene expression regulation leading to effects on shared pathways for ASD-related phenotypes. A study showed that many genes which are dysregulated in early fetal brain development by MIA overlap with known ASD-associated genes [113].

Multiple studies have confirmed cytokine, chemokine, and growth factor abnormalities in ASD, but the results have been inconsistent. A recent meta-analysis found elevations in the plasma or serum concentrations of cytokines IL-1b, IL-6, and IL-8, interferon (IFN)-c, eotaxin, and monocyte chemotactic protein-1 (MCP-1) in individuals with ASD, as well as a reduction in the concentration of transforming growth factor (TGF)-β1. Among these, IL-1 and IL-6 were of high interest due to repeatability of the results concerning associated behavioral abnormalities [114]. Interestingly, activation of microglia and astrocytes in multiple brain regions, as well as elevated levels of IFN-c, IL-1b, IL-6, MCP-1, TGF-β1, and tumor necrosis factor (TNF)-α, has also been observed in postmortem brain tissues of ASD patients [115–117]. These findings suggest that some cytokines may serve as potential biomarkers for ASD and that immunomodulatory therapies may be used as a treatment strategy. In addition to the immune cells of the brain (i.e., microglia), peripheral blood immune cells have also been extensively studied. These may be used to investigate systemic neurochemical changes in neurodevelopmental diseases. For example, it has been proposed that infection-induced changes in the mother's circulation, including expression of pro-inflammatory cytokines and activation of Th17 cells, could affect the immune status of the fetus [109]. These changes, in combination with genetic susceptibilities of the fetus and subsequent immune challenges, could increase the risk for ASD [109]. Moreover, it is interesting that three genes (STOM, LMNB1, and VWF) were found to be dysregulated at the transcriptional level in

PBMCs from ASD subjects [118], and these were also reported as differentially expressed proteins in the plasma of children with autism in our previous proteomic studies [72]. Thus, it can be proposed that simultaneous detection of these genes in the plasma and PBMCs could increase the specificity of detection.

11.8 MicroRNAs and Exosomes

MicroRNAs (miRNAs) are small endogenous noncoding regulatory RNAs (typically 21–23 nucleotides long) that function as posttranscriptional regulators of gene expression [119] and thereby play an important role in CNS development and function [120]. Recently, Hicks et al. reviewed the expression patterns of miRNAs altered in ASD include the brain, blood, saliva, and olfactory precursor cells [121]. This study encapsulated 219 target miRNAs from 12 human studies of ASD and identified 27 miRNAs that were dysregulated in two or more investigations. These miRNAs target brain-expressed genes related to neurodevelopment and which have been implicated in ASD. Among these, three miRNAs (miR-23a-3p, miR-146a-5p, and miR-106b-5p) showed consistent dysregulation across three or more studies [121]. The dysregulated miRNAs in the blood [122–124] and saliva [125] may serve as potential biomarkers for ASD given the ease of accessibility of these biofluids.

Exosomes are approximately 100 nm extracellular vesicles secreted from most cell types including neurons, which contain proteins, mRNAs, noncoding RNAs, and DNA [126]. Exosomes obtained from peripheral blood have attracted attention in various disease research areas, including AD [127, 128] and depression [129]. Since exosomes can cross the BBB from both directions, circulating exosomes in blood are expected to reveal the pathophysiology of brain diseases [130]. With this in mind, several methods to extract neuron-derived exosomes (NDE) from peripheral blood have been developed [131].

11.9 Conclusions

As the etiology and pathogenesis have not yet been elucidated, ASD lacks effective prevention and treatment drugs. Research shows that early interventions can improve the prognosis of children with ASD. Therefore, early diagnosis is essential for ASD. The identification of biomarkers for ASD has made some progress in imaging and molecular analyses. However, due to the heterogeneity of the disease, the utility of these biomarkers still faces difficulties and challenges. Nevertheless, with the development of genomic, proteomic, metabolomic, transcriptomic, and microRNA profiling methods, as well as the emergence of some new technologies and applications, more research should be focused on identifying reliable, noninvasive, and inexpensive biomarkers in accessible tissues and body fluids such as the blood and blood cells. In this way, the genes, proteins, mRNA, miRNA, and exosomes could be useful sources of biomarkers. More attention should be given to the consistent changes between the CNS and peripheral blood or blood cells, focusing on those changes in CNS-specific or CNS-derived proteins and genes. It is also important to use biomarker panels, high-throughput multiplexed targeted technologies, and integrated applications of multiple methods for both discovery and validation of findings. Eventually, the study of ASD biomarkers may lead to a better understanding of the underlying mechanisms of ASD and identify new biomarkers of disease and therapeutic targets.

Acknowledgments The authors would like to acknowledge the National Natural Science Foundation of China (grant no. 31870825) and Shenzhen Bureau of Science, Technology and Information(nos.JCYJ20150402100258220,JCYJ20150529164656093,JCYJ20170412110026229) for funds to support this work.

References

1. American Psychiatric Association (2013) Diagnostic and statistical manual of mental disorders, 5th edn. American Psychiatric Publishing, Washington, DC. ISBN-10: 0890425558
2. Lai M-C, Lombardo MV, Baron-Cohen S (2014) Autism. Lancet 383:896–910
3. Baio J, Wiggins L, Christensen DL, Maenner MJ, Daniels J, Warren Z et al (2018) Prevalence of autism spectrum disorder among children aged 8 years—autism and developmental disabilities monitoring network, 11 sites, United States, 2014. MMWR Surveill Summ 67:1–23
4. Baron-Cohen S, Scott FJ, Allison C, Williams J, Bolton P, Matthews FE et al (2009) Prevalence of autism-spectrum conditions: UK school-based population study. Br J Psychiatry 194:500–509
5. Saemundsen E, Magnússon P, Georgsdóttir I, Egilsson E, Rafnsson V (2013) Prevalence of autism spectrum disorders in an Icelandic birth cohort. BMJ Open3(6). https://doi.org/10.1136/bmjopen-2013-002748
6. Russell G, Rodgers LR, Ukoumunne OC, Ford T (2014) Prevalence of parent-reported ASD and ADHD in the UK: findings from the Millennium Cohort Study. J Autism Dev Disord 44:31–40
7. Baxter AJ, Brugha TS, Erskine HE, Scheurer RW, Vos T, Scott JG (2015) The epidemiology and global burden of autism spectrum disorders. Psychol Med 45:601–613
8. Jin Z, Yang Y, Liu S, Huang H, Jin X (2018) Prevalence of DSM-5 autism spectrum disorder among school-based children aged 3-12 years in Shanghai, China. J Autism Dev Disord 48:2434–2443
9. Fombonne E (2009) Epidemiology of pervasive developmental disorders. Pediatr Res 65:591–598
10. Barger BD, Campbell JM, McDonough JD (2013) Prevalence and onset of regression within autism spectrum disorders: a meta-analytic review. J Autism Dev Disord 43:817–828
11. Ozonoff S, Gangi D, Hanzel EP, Hill A, Hill MM, Miller M et al (2018) Onset patterns in autism: variation across informants, methods, and timing. Autism Res 11:788–797
12. Rossignol DA, Frye RE (2012) A review of research trends in physiological abnormalities in autism spectrum disorders: immune dysregulation, inflammation, oxidative stress, mitochondrial dysfunction and environmental toxicant exposures. Mol Psychiatry 17:389–401
13. Frustaci A, Neri M, Cesario A, Adams JB, Domenici E, Dalla Bernardina B et al (2012) Oxidative stress-related biomarkers in autism: systematic review and meta-analyses. Free Radic Biol Med 52:2128–2141

14. Kaur K, Chauhan V, Gu F, Chauhan A (2014) Bisphenol A induces oxidative stress and mitochondrial dysfunction in lymphoblasts from children with autism and unaffected siblings. Free Radic Biol Med 76:25–33
15. Zachor DA (2012) Autism spectrum disorders—a syndrome on the rise: risk factors and advances in early detection and intervention. Harefuah 151:162–164 189
16. Dawson G (2008) Early behavioral intervention, brain plasticity, and the prevention of autism spectrum disorder. Dev Psychopathol 20:775–803
17. Brett D, Warnell F, McConachie H, Parr JR (2016) Factors affecting age at ASD diagnosis in UK: no evidence that diagnosis age has decreased between 2004 and 2014. J Autism Dev Disord 46:1974–1984
18. Ruggeri B, Sarkans U, Schumann G, Persico AM (2014) Biomarkers in autism spectrum disorder: the old and the new. Psychopharmacology 231:1201–1216
19. Heunis T, Aldrich C, Peters JM, Jeste SS, Sahin M, Scheffer C et al (2018) Recurrence quantification analysis of resting state EEG signals in autism spectrum disorder—a systematic methodological exploration of technical and demographic confounders in the search for biomarkers. BMC Med 16:101. https://doi.org/10.1186/s12916-018-1086-7
20. Jones W, Klin A (2013) Attention to eyes is present but in decline in 2–6-month-old infants later diagnosed with autism. Nature 504:427–431
21. Constantino JN, Kennon-McGill S, Weichselbaum C, Marrus N, Haider A, Glowinski AL et al (2017) Infant viewing of social scenes is under genetic control and is atypical in autism. Nature 547:340–344
22. Nyström P, Gliga T, Nilsson Jobs E, Gredebäck G, Charman T, Johnson MH et al (2018) Enhanced pupillary light reflex in infancy is associated with autism diagnosis in toddlerhood. Nat Commun 9(1):1678. https://doi.org/10.1038/s41467-018-03985-4
23. Bhaumik R, Pradhan A, Das S, Bhaumik DK (2018) Predicting autism spectrum disorder using domain-adaptive cross-site evaluation. Neuroinformatics 16:197–205
24. Dickie EW, Ameis SH, Shahab S, Calarco N, Smith DE, Miranda D et al (2018) Personalized intrinsic network topography mapping and functional connectivity deficits in Autism Spectrum disorder. Biol Psychiatry 84:278–286
25. Liu T, Liu X, Yi L, Zhu C, Markey PS, Pelowski M (2017) Assessing autism at its social and developmental roots: a review of Autism Spectrum Disorder studies using functional near-infrared spectroscopy. NeuroImage 185:955–967
26. Onitsuka T, Shenton ME, Salisbury DF, Dickey CC, Kasai K, Toner SK et al (2004) Middle and inferior temporal gyrus gray matter volume abnormalities in chronic schizophrenia: an MRI study. Am J Psychiatry 161:1603–1611
27. Emerson RW, Adams C, Nishino T, Hazlett HC, Wolff JJ, Zwaigenbaum L et al (2017) Functional neuroimaging of high-risk 6-month-old infants predicts a diagnosis of autism at 24 months of age. Sci Transl Med 9(393). https://doi.org/10.1126/scitranslmed.aag2882
28. Hazlett HC, Gu H, Munsell BC, Kim SH, Styner M, Wolff JJ et al (2017) Early brain development in infants at high risk for autism spectrum disorder. Nature 542:348–351
29. Ronald A, Hoekstra RA (2011) Autism spectrum disorders and autistic traits: a decade of new twin studies. Am J Med Genet B Neuropsychiatr Genet 156:255–274
30. Sandin S, Lichtenstein P, Kuja-Halkola R, Larsson H, Hultman CM, Reichenberg A (2014) The familial risk of autism. JAMA 311:1770–1777
31. Gaugler T, Klei L, Sanders SJ, Bodea CA, Goldberg AP, Lee AB et al (2014) Most genetic risk for autism resides with common variation. Nat Genet 46:881–885
32. de la Torre-Ubieta L, Won H, Stein JL, Geschwind DH (2016) Advancing the understanding of autism disease mechanisms through genetics. Nat Med 22:345–361
33. Vorstman JAS, Parr JR, Moreno-De-Luca D, Anney RJL, Nurnberger JI Jr, Hallmayer JF (2017) Autism genetics: opportunities and challenges for clinical translation. Nat Rev Genet 18:362–376
34. Geschwind DH, State MW (2015) Gene hunting in autism spectrum disorder: on the path to precision medicine. Lancet Neurol 14:1109–1120

35. Abrahams BS, Arking DE, Campbell DB, Mefford HC, Morrow EM, Weiss LA et al (2013) SFARI Gene 2.0: a community-driven knowledgebase for the autism spectrum disorders (ASDs). Mol Autism 4:36. doi: 10.1186/2040-2392-4-36
36. Gilman SR, Iossifov I, Levy D, Ronemus M, Wigler M, Vitkup D (2011) Rare de novo variants associated with autism implicate a large functional network of genes involved in formation and function of synapses. Neuron 70:898–907
37. Sanders SJ, Murtha MT, Gupta AR, Murdoch JD, Raubeson MJ, Willsey AJ et al (2012) De novo mutations revealed by whole-exome sequencing are strongly associated with autism. Nature 485:237–241
38. Iossifov I, O'Roak BJ, Sanders SJ, Ronemus M, Krumm N, Levy D et al (2014) The contribution of de novo coding mutations to autism spectrum disorder. Nature 515:216–221
39. Takata A, Ionita-Laza I, Gogos JA, Xu B, Karayiorgou M (2016) De novo synonymous mutations in regulatory elements contribute to the genetic etiology of autism and schizophrenia. Neuron 89:940–947
40. Kosmicki JA, Samocha KE, Howrigan DP, Sanders SJ, Slowikowski K, Lek M et al (2017) Refining the role of de novo protein-truncating variants in neurodevelopmental disorders by using population reference samples. Nat Genet 49:504–510
41. Wilfert AB, Sulovari A, Turner TN, Coe BP, Eichler EE (2017) Recurrent de novo mutations in neurodevelopmental disorders: properties and clinical implications. Genome Med 9(1):101. https://doi.org/10.1186/s13073-017-0498-x
42. Takata A, Miyake N, Tsurusaki Y, Fukai R, Miyatake S, Koshimizu E et al (2018) Integrative analyses of de novo mutations provide deeper biological insights into autism spectrum disorder. Cell Rep 22:734–747
43. Werling DM, Brand H, An J-Y, Stone MR2, Zhu L, Glessner JT et al (2018) An analytical framework for whole-genome sequence association studies and its implications for autism spectrum disorder. Nat Genet 50:727–736
44. Turner TN, Hormozdiari F, Duyzend MH, McClymont SA, Hook PW, Iossifov I et al (2016) Genome sequencing of autism-affected families reveals disruption of putative noncoding regulatory DNA. Am J Hum Genet 98:58–74
45. Doan RN, Bae B-I, Cubelos B, Chang C, Hossain AA, Al-Saad S et al (2016) Mutations in human accelerated regions disrupt cognition and social behavior. Cell 167:341–354.e12
46. Brandler WM, Antaki D, Gujral M, Kleiber ML, Whitney J, Maile MS et al (2018) Paternally inherited cis-regulatory structural variants are associated with autism. Science 360:327–331
47. Qin L, Ma K, Wang Z-J, Hu Z, Matas E, Wei J et al (2018) Social deficits in Shank3-deficient mouse models of autism are rescued by histone deacetylase (HDAC) inhibition. Nat Neurosci 21:564–575
48. Schaefer GB, Mendelsohn NJ, Professional Practice and Guidelines Committee (2013) Clinical genetics evaluation in identifying the etiology of autism spectrum disorders: 2013 guideline revisions. Genet Med 15:399–407
49. Kalsner L, Twachtman-Bassett J, Tokarski K, Stanley C, Dumont-Mathieu T, Cotney J et al (2018) Genetic testing including targeted gene panel in a diverse clinical population of children with autism spectrum disorder: findings and implications. Mol Genet Genomic Med 6:171–185
50. Fernandez BA, Scherer SW (2017) Syndromic autism spectrum disorders: moving from a clinically defined to a molecularly defined approach. Dialogues Clin Neurosci 19:353–371
51. Jiang Y-H, Wang Y, Xiu X, Choy KW, Pursley AN, Cheung SW (2014) Genetic diagnosis of autism spectrum disorders: the opportunity and challenge in the genomics era. Crit Rev Clin Lab Sci 51:249–262
52. Won H, Mah W, Kim E (2013) Autism spectrum disorder causes, mechanisms, and treatments: focus on neuronal synapses. Front Mol Neurosci 6(19). https://doi.org/10.3389/fnmol.2013.00019
53. Wang P, Mokhtari R, Pedrosa E, Kirschenbaum M, Bayrak C, Zheng D et al (2017) CRISPR/Cas9-mediated heterozygous knockout of the autism gene CHD8 and characterization of its

transcriptional networks in cerebral organoids derived from iPS cells. Mol Autism 8(11). https://doi.org/10.1186/s13229-017-0124-1
54. Wilkinson B, Grepo N, Thompson BL, Kim J, Wang K, Evgrafov OV et al (2015) The autism-associated gene chromodomain helicase DNA-binding protein 8 (CHD8) regulates non-coding RNAs and autism-related genes. Transl Psychiatry 5:e568. https://doi.org/10.1038/tp.2015.62
55. Fiorentino M, Sapone A, Senger S, Camhi SS, Kadzielski SM, Buie TM et al (2016) Blood–brain barrier and intestinal epithelial barrier alterations in autism spectrum disorders. Mol Autism;7:49. https://doi.org/10.1186/s13229-016-0110-z
56. De Magistris L, Familiari V, Pascotto A, Sapone A, Frolli A, Iardino P et al (2010) Alterations of the intestinal barrier in patients with autism spectrum disorders and in their first-degree relatives. J Pediatr Gastroenterol Nutr 51:418–424
57. Esnafoglu E, Cırrık S, Ayyıldız SN, Erdil A, Ertürk EY, Daglı A et al (2017) Increased serum zonulin levels as an intestinal permeability marker in autistic subjects. J Pediatr 188:240–244
58. Hayashi-Takagi A, Vawter MP, Iwamoto K (2014) Peripheral biomarkers revisited: integrative profiling of peripheral samples for psychiatric research. Biol Psychiatry 75:920–928
59. Graves PR, Haystead TAJ (2002) Molecular biologist's guide to proteomics. Microbiol Mol Biol Rev 66:39–63
60. Smith LM, Kelleher NL, Consortium for Top Down Proteomics (2013) Proteoform: a single term describing protein complexity. Nat Methods 10:186–187
61. Wilhelm M, Schlegl J, Hahne H, Gholami AM, Lieberenz M, Savitski MM et al (2014) Mass-spectrometry-based draft of the human proteome. Nature 509:582–587
62. Aebersold R, Agar JN, Amster IJ, Baker MS, Bertozzi CR, Boja ES et al (2018) How many human proteoforms are there? Nat Chem Biol 14:206–214
63. Junaid MA, Kowal D, Barua M, Pullarkat PS, Sklower Brooks S et al (2004) Proteomic studies identified a single nucleotide polymorphism in glyoxalase I as autism susceptibility factor. Am J Med Genet A 131:11–17
64. Broek JA, Guest PC, Rahmoune H, Bahn S (2014) Proteomic analysis of post mortem brain tissue from autism patients: evidence for opposite changes in prefrontal cortex and cerebellum in synaptic connectivity-related proteins. Mol Autism 5:41. https://doi.org/10.1186/2040-2392-5-41
65. Corbett BA, Kantor AB, Schulman H, Walker WL, Lit L, Ashwood P et al (2007) A proteomic study of serum from children with autism showing differential expression of apolipoproteins and complement proteins. Mol Psychiatry 12:292–306
66. Taurines R, Dudley E, Conner AC, Grassl J, Jans T, Guderian F et al (2010) Serum protein profiling and proteomics in autistic spectrum disorder using magnetic bead-assisted mass spectrometry. Eur Arch Psychiatry Clin Neurosci 260:249–255
67. Ramsey JM, Guest PC, Broek JA, Glennon JC, Rommelse N, Franke B et al (2013) Identification of an age-dependent biomarker signature in children and adolescents with autism spectrum disorders. Mol Autism 4:27. https://doi.org/10.1186/2040-2392-4-27
68. Steeb H, Ramsey JM, Guest PC, Stocki P, Cooper JD, Rahmoune H et al (2014) Serum proteomic analysis identifies sex-specific differences in lipid metabolism and inflammation profiles in adults diagnosed with Asperger syndrome. Mol Autism 5:4. https://doi.org/10.1186/2040-2392-5-4
69. Ngounou Wetie AG, Wormwood K, Thome J, Dudley E, Taurines R, Gerlach M et al (2014) A pilot proteomic study of protein markers in autism spectrum disorder. Electrophoresis 35:2046–2054
70. Cortelazzo A, De Felice C, Guerranti R, Signorini C, Leoncini S, Zollo G et al (2016) Expression and oxidative modifications of plasma proteins in autism spectrum disorders: interplay between inflammatory response and lipid peroxidation. Proteomics Clin Appl 10:1103–1112
71. Feng C, Chen Y, Pan J, Yang A, Niu L, Min J et al (2017) Redox proteomic identification of carbonylated proteins in autism plasma: insight into oxidative stress and its related biomarkers in autism. Clin Proteomics 14:2. https://doi.org/10.1186/s12014-017-9138-0

72. Shen L, Zhang K, Feng C, Chen Y, Li S, Iqbal J et al (2018) iTRAQ-based proteomic analysis reveals protein profile in plasma from children with autism. Proteomics Clin Appl 12(3):e1700085. https://doi.org/10.1002/prca.201700085
73. Shen C, Zhao X, Ju W, Huo LR, Yan W, Zou JH et al (2011) A proteomic investigation of B lymphocytes in an autistic family: a pilot study of exposure to natural rubber latex (NRL) may lead to autism. J Mol Neurosci 43:443–452
74. Zerbo O, Yoshida C, Grether JK, Van de Water J, Ashwood P, Delorenze GN et al (2014) Neonatal cytokines and chemokines and risk of Autism Spectrum Disorder: the Early Markers for Autism (EMA) study: a case-control study. J Neuroinflammation 11:113. https://doi.org/10.1186/1742-2094-11-113
75. Krakowiak P, Goines PE, Tancredi DJ, Ashwood P, Hansen RL, Hertz-Picciotto I et al (2017) Neonatal cytokine profiles associated with autism spectrum disorder. Biol Psychiatry 81:442–451
76. Suganya V, Geetha A, Sujatha S (2015) Urine proteome analysis to evaluate protein biomarkers in children with autism. Clin Chim Acta 450:210–219
77. Yang L, Rudser K, Golnik A, Wey A, Higgins LA, Gourley GR (2016) Urine protein biomarker candidates for autism. J Proteomics Bioinform S14:004. https://doi.org/10.4172/jpb.S14-004
78. Wetie AGN, Wormwood KL, Russell S, Ryan JP, Darie CC, Woods AG (2015) A pilot proteomic analysis of salivary biomarkers in autism spectrum disorder. Autism Res 8:338–350
79. Wetie AGN, Wormwood KL, Charette L, Ryan JP, Woods AG, Darie CC (2015) Comparative two-dimensional polyacrylamide gel electrophoresis of the salivary proteome of children with autism spectrum disorder. J Cell Mol Med 19:2664–2678
80. Wei H, Ma Y, Liu J, Ding C, Hu F, Yu L (2016) Proteomic analysis of cortical brain tissue from the BTBR mouse model of autism: evidence for changes in STOP and myelin-related proteins. Neuroscience 312:26–34
81. Wu L, Candille SI, Choi Y, Xie D, Jiang L, Li-Pook-Than J et al (2013) Variation and genetic control of protein abundance in humans. Nature 499:79–82
82. Suhre K, Arnold M, Bhagwat AM, Cotton RJ, Engelke R, Raffler J et al (2017) Connecting genetic risk to disease end points through the human blood plasma proteome. Nat Commun 8:14357
83. Sun BB, Maranville JC, Peters JE, Stacey D, Staley JR, Blackshaw J et al (2018) Genomic atlas of the human plasma proteome. Nature 558:73–79
84. Aebersold R, Mann M (2016) Mass-spectrometric exploration of proteome structure and function. Nature 537:347–355
85. Thompson A, Schäfer J, Kuhn K, Kienle S, Schwarz J, Schmidt G et al (2003) Tandem mass tags: a novel quantification strategy for comparative analysis of complex protein mixtures by MS/MS. Anal Chem 75:1895–1904
86. Unwin RD, Pierce A, Watson RB, Sternberg DW, Whetton AD (2005) Quantitative proteomic analysis using isobaric protein tags enables rapid comparison of changes in transcript and protein levels in transformed cells. Mol Cell Proteomics 4:924–935
87. Skafidas E, Testa R, Zantomio D, Chana G, Everall IP, Pantelis C (2014) Predicting the diagnosis of autism spectrum disorder using gene pathway analysis. Mol Psychiatry 19:504–510
88. Geyer PE, Kulak NA, Pichler G, Holdt LM, Teupser D, Mann M (2016) Plasma proteome profiling to assess human health and disease. Cell Syst 2:185–195
89. Gillet LC, Navarro P, Tate S, Röst H, Selevsek N, Reiter L et al (2012) Targeted data extraction of the MS/MS spectra generated by data-independent acquisition: a new concept for consistent and accurate proteome analysis. Mol Cell Proteomics 11:O111.016717
90. Ebhardt HA, Root A, Sander C, Aebersold R (2015) Applications of targeted proteomics in systems biology and translational medicine. Proteomics 15:3193–3208
91. Harlan R, Zhang H (2014) Targeted proteomics: a bridge between discovery and validation. Expert Rev Proteomics 11:657–661

92. Andreasson U, Blennow K, Zetterberg H (2016) Update on ultrasensitive technologies to facilitate research on blood biomarkers for central nervous system disorders. Alzheimers Dement (Amst) 3:98–102
93. Grasso G (2018) Mass spectrometry is a multifaceted weapon to be used in the battle against Alzheimer's disease: amyloid beta peptides and beyond. Mass Spectrom Rev 38(1):34–48
94. Wilson DH, Rissin DM, Kan CW, Fournier DR, Piech T, Campbell TG et al (2016) The Simoa HD-1 analyzer: a novel fully automated digital immunoassay analyzer with single-molecule sensitivity and multiplexing. J Lab Autom 21:533–547
95. Kuwano Y, Kamio Y, Kawai T, Katsuura S, Inada N, Takaki A et al (2011) Autism-associated gene expression in peripheral leucocytes commonly observed between subjects with autism and healthy women having autistic children. PLoS One 6(9):e24723. https://doi.org/10.1371/journal.pone.0024723
96. Talebizadeh Z, Aldenderfer R, Wen Chen X (2014) A proof-of-concept study: exon-level expression profiling and alternative splicing in autism using lymphoblastoid cell lines. Psychiatr Genet 24:1–9
97. Nishimura Y, Martin CL, Vazquez-Lopez A, Spence SJ, Alvarez-Retuerto AI, Sigman M et al (2007) Genome-wide expression profiling of lymphoblastoid cell lines distinguishes different forms of autism and reveals shared pathways. Hum Mol Genet 16:1682–1698
98. Hu VW, Frank BC, Heine S, Lee NH, Quackenbush J (2006) Gene expression profiling of lymphoblastoid cell lines from monozygotic twins discordant in severity of autism reveals differential regulation of neurologically relevant genes. BMC Genomics 7:118. https://doi.org/10.1186/1471-2164-7-118
99. Chien W-H, Gau SS-F, Chen C-H, Tsai W-C, Wu Y-Y, Chen P-H et al (2013) Increased gene expression of FOXP1 in patients with autism spectrum disorders. Mol Autism 4:23. https://doi.org/10.1186/2040-2392-4-23
100. Ansel A, Rosenzweig JP, Zisman PD, Melamed M, Gesundheit B (2017) Variation in gene expression in autism spectrum disorders: an extensive review of transcriptomic studies. Front Neurosci 10:601. https://doi.org/10.3389/fnins.2016.00601
101. Liu Y, Qing H, Deng Y (2014) Biomarkers in Alzheimer's disease analysis by mass spectrometry-based proteomics. Int J Mol Sci 15:7865–7882
102. West PR, Amaral DG, Bais P, Smith AM, Egnash LA, Ross ME et al (2014) Metabolomics as a tool for discovery of biomarkers of autism spectrum disorder in the blood plasma of children. PLoS One 9:e112445. https://doi.org/10.1371/journal.pone.0112445
103. Wang H, Liang S, Wang M, Gao J, Sun C, Wang J et al (2016) Potential serum biomarkers from a metabolomics study of autism. J Psychiatry Neurosci 41:27–37
104. Yap IKS, Angley M, Veselkov KA, Holmes E, Lindon JC, Nicholson JK (2010) Urinary metabolic phenotyping differentiates children with autism from their unaffected siblings and age-matched controls. J Proteome Res 9:2996–3004
105. Ming X, Stein TP, Barnes V, Rhodes N, Guo L (2012) Metabolic perturbance in autism spectrum disorders: a metabolomics study. J Proteome Res 11:5856–5862
106. Diémé B, Mavel S, Blasco H, Tripi G, Bonnet-Brilhault F, Malvy J et al (2015) Metabolomics study of urine in autism spectrum disorders using a multiplatform analytical methodology. J Proteome Res 14:5273–5282
107. Gilbert JA, Blaser MJ, Caporaso JG, Jansson JK, Lynch SV, Knight R (2018) Current understanding of the human microbiome. Nat Med 24:392–400
108. Voineagu I, Wang X, Johnston P, Lowe JK, Tian Y, Horvath S et al (2011) Transcriptomic analysis of autistic brain reveals convergent molecular pathology. Nature 474:380–384
109. Estes ML, McAllister AK (2016) Maternal immune activation: implications for neuropsychiatric disorders. Science 353:772–777
110. Careaga M, Murai T, Bauman MD (2017) Maternal immune activation and autism spectrum disorder: from rodents to nonhuman and human primates. Biol Psychiatry 81:391–401

111. Chen S, Zhong X, Jiang L, Zheng X, Xiong Y, Ma S et al (2016) Maternal autoimmune diseases and the risk of autism spectrum disorders in offspring: a systematic review and meta-analysis. Behav Brain Res 296:61–69
112. Patel S, Masi A, Dale RC, Whitehouse AJO, Pokorski I, Alvares GA et al (2017) Social impairments in autism spectrum disorder are related to maternal immune history profile. Mol Psychiatry 23(8):1794–1797
113. Lombardo MV, Moon HM, Su J, Palmer TD, Courchesne E, Pramparo T (2018) Maternal immune activation dysregulation of the fetal brain transcriptome and relevance to the pathophysiology of autism spectrum disorder. Mol Psychiatry 23:1001–1013
114. Masi A, Quintana DS, Glozier N, Lloyd AR, Hickie IB, Guastella AJ (2015) Cytokine aberrations in autism spectrum disorder: a systematic review and meta-analysis. Mol Psychiatry 20:440–446
115. Vargas DL, Nascimbene C, Krishnan C, Zimmerman AW, Pardo CA (2005) Neuroglial activation and neuroinflammation in the brain of patients with autism. Ann Neurol 57:67–81
116. Li X, Chauhan A, Sheikh AM, Patil S, Chauhan V, Li X-M et al (2009) Elevated immune response in the brain of autistic patients. J Neuroimmunol 207:111–116
117. Morgan JT, Chana G, Pardo CA, Achim C, Semendeferi K, Buckwalter J et al (2010) Microglial activation and increased microglial density observed in the dorsolateral prefrontal cortex in autism. Biol Psychiatry 68:368–376
118. Glatt SJ, Tsuang MT, Winn M, Chandler SD, Collins M, Lopez L et al (2012) Blood-based gene expression signatures of infants and toddlers with autism. J Am Acad Child Adolesc Psychiatry 51(9):934–44.e2. https://doi.org/10.1016/j.jaac.2012.07.007
119. Felekkis K, Touvana E, Stefanou C, Deltas C (2010) microRNAs: a newly described class of encoded molecules that play a role in health and disease. Hippokratia 14:236–240
120. Cao D-D, Li L, Chan W-Y (2016) MicroRNAs: key regulators in the central nervous system and their implication in neurological diseases. Int J Mol Sci 17(6):pii: E842. https://doi.org/10.3390/ijms17060842
121. Hicks SD, Middleton FA (2016) A comparative review of microRNA expression patterns in autism spectrum disorder. Front Psych 7:176. https://doi.org/10.3389/fpsyt.2016.00176
122. Popov NT, Madjirova NP, Minkov IN, Vachev TI (2012) Micro RNA HSA-486-3P gene expression profiling in the whole blood of patients with autism. Biotechnol Biotechnol Equip 26:3385–3388
123. Mundalil Vasu M, Anitha A, Thanseem I, Suzuki K, Yamada K, Takahashi T et al (2014) Serum microRNA profiles in children with autism. Mol Autism 5:40. https://doi.org/10.1186/2040-2392-5-40
124. Huang F, Long Z, Chen Z, Li J, Hu Z, Qiu R et al (2015) Investigation of gene regulatory networks associated with autism spectrum disorder based on MiRNA expression in China. PLoS One 10(6):e0129052. https://doi.org/10.1371/journal.pone.0129052
125. Hicks SD, Ignacio C, Gentile K, Middleton FA (2016) Salivary miRNA profiles identify children with autism spectrum disorder, correlate with adaptive behavior, and implicate ASD candidate genes involved in neurodevelopment. BMC Pediatr 16:52. https://doi.org/10.1186/s12887-016-0586-x
126. Zhang J, Li S, Li L, Li M, Guo C, Yao J et al (2015) Exosome and exosomal microRNA: trafficking, sorting, and function. Genomics Proteomics Bioinformatics 13:17–24
127. Fiandaca MS, Kapogiannis D, Mapstone M, Boxer A, Eitan E, Schwartz JB et al (2015) Identification of preclinical Alzheimer's disease by a profile of pathogenic proteins in neurally derived blood exosomes: a case-control study. Alzheimers Dement 11:600–607.e1. https://doi.org/10.1016/j.jalz.2014.06.008
128. Goetzl EJ, Mustapic M, Kapogiannis D, Eitan E, Lobach IV, Goetzl L et al (2016) Cargo proteins of plasma astrocyte-derived exosomes in Alzheimer's disease. FASEB J 30:3853–3859
129. Tavakolizadeh J, Roshanaei K, Salmaninejad A, Yari R, Nahand JS, Sarkarizi HK et al (2018) MicroRNAs and exosomes in depression: potential diagnostic biomarkers. J Cell Biochem 119:3783–3797

130. András IE, Leda A, Contreras MG, Bertrand L, Park M, Skowronska M et al (2017) Extracellular vesicles of the blood-brain barrier: role in the HIV-1 associated amyloid beta pathology. Mol Cell Neurosci 79:12–22
131. Kuwano N, Kato TA, Mitsuhashi M, Sato-Kasai M, Shimokawa N, Hayakawa K et al (2018) Neuron-related blood inflammatory markers as an objective evaluation tool for major depressive disorder: an exploratory pilot case-control study. J Affect Disord 240:88–98
132. Castagnola M, Messana I, Inzitari R, Fanali C, Cabras T, Morelli A et al (2008) Hypophosphorylation of salivary peptidome as a clue to the molecular pathogenesis of autism spectrum disorders. J Proteome Res 7:5327–5332
133. Schwarz E, Guest PC, Rahmoune H, Wang L, Levin Y, Ingudomnukul E et al (2011) Sex-specific serum biomarker patterns in adults with Asperger's syndrome. Mol Psychiatry 16:1213–1220
134. Momeni N, Bergquist J, Brudin L, Behnia F, Sivberg B, Joghataei MT et al (2012) A novel blood-based biomarker for detection of autism spectrum disorders. Transl Psychiatry 2:e91. https://doi.org/10.1038/tp.2012.19
135. Singh S, Yazdani U, Gadad B, Zaman S, Hynan LS, Roatch N et al (2017) Serum thyroid-stimulating hormone and interleukin-8 levels in boys with autism spectrum disorder. J Neuroinflammation 14:113. https://doi.org/10.1186/s12974-017-0888-4
136. Yang J, Chen Y, Xiong X, Zhou X, Han L, Ni L et al (2018) Peptidome analysis reveals novel serum biomarkers for children with autism spectrum disorder in China. Proteomics Clin Appl 12(5):e1700164. https://doi.org/10.1002/prca.201700164
137. Garbett K, Ebert PJ, Mitchell A, Lintas C, Manzi B, Mirnics K et al (2008) Immune transcriptome alterations in the temporal cortex of subjects with autism. Neurobiol Dis 30:303–311
138. Voineagu I, Eapen V (2013) Converging pathways in autism spectrum disorders: interplay between synaptic dysfunction and immune responses. Front Hum Neurosci. https://doi.org/10.3389/fnhum.2013.00738
139. Anitha A, Nakamura K, Thanseem I, Yamada K, Iwayama Y, Toyota T et al (2012) Brain region-specific altered expression and association of mitochondria-related genes in autism. Mol Autism 3:12. https://doi.org/10.1186/2040-2392-3-12
140. Chow ML, Pramparo T, Winn ME, Barnes CC, Li HR, Weiss L et al (2012) Age-dependent brain gene expression and copy number anomalies in autism suggest distinct pathological processes at young versus mature ages. PLoS Genet 8(3):e1002592. https://doi.org/10.1371/journal.pgen.1002592
141. James SJ, Shpyleva S, Melnyk S, Pavliv O, Pogribny IP (2014) Elevated 5-hydroxymethylcytosine in the Engrailed-2 (EN-2) promoter is associated with increased gene expression and decreased MeCP2 binding in autism cerebellum. Transl Psychiatry 4:e460. https://doi.org/10.1038/tp.2014.87
142. Zhubi A, Chen Y, Dong E, Cook EH, Guidotti A, Grayson DR (2014) Increased binding of MeCP2 to the GAD1 and RELN promoters may be mediated by an enrichment of 5-hmC in autism spectrum disorder (ASD) cerebellum. Transl Psychiatry 4:e349. https://doi.org/10.1038/tp.2013.123
143. Kong S, Shimizu-Motohashi Y, Campbell M, Lee I, Collins C, Brewster S et al (2013) Peripheral blood gene expression signature differentiates children with autism from unaffected siblings. Neurogenetics 14:143–152
144. Ivanov H, Stoyanova V, Popov N, Bosheva M, Vachev T (2015) Blood-based gene expression in children with Autism spectrum disorder. Bio Discovery 17:e8966. doi: 10.7750/BioDiscovery.2015.17.2
145. Gregg JP, Lit L, Baron CA, Hertz-Picciotto I, Walker W, Davis RA et al (2008) Gene expression changes in children with autism. Genomics 91:22–29
146. Enstrom AM, Lit L, Onore CE, Gregg JP, Hansen RL, Pessah IN et al (2009) Altered gene expression and function of peripheral blood natural killer cells in children with autism. Brain Behav Immun 23:124–133
147. Ziats MN, Rennert OM (2013) Sex-biased gene expression in the developing brain: implications for autism spectrum disorders. Mol Autism 4:10. https://doi.org/10.1186/2040-2392-4-10

148. Ginsberg MR, Rubin RA, Falcone T, Ting AH, Natowicz MR (2012) Brain transcriptional and epigenetic associations with autism. PLoS One 7(9):e44736. https://doi.org/10.1371/journal.pone.0044736
149. Kuwabara H, Yamasue H, Koike S, Inoue H, Kawakubo Y, Kuroda M et al (2013) Altered metabolites in the plasma of autism spectrum disorder: a capillary electrophoresis time-of-flight mass spectroscopy study. PLoS One 8:e73814. doi: 10.1371/journal.pone.0073814
150. Emond P, Mavel S, Aïdoud N, Nadal-Desbarats L, Montigny F, Bonnet-Brilhault F et al (2013) GC-MS-based urine metabolic profiling of autism spectrum disorders. Anal Bioanal Chem 405:15:5291–5300
151. Gevi F, Zolla L, Gabriele S, Persico AM (2016) Urinary metabolomics of young Italian autistic children supports abnormal tryptophan and purine metabolism. Mol Autism 7:47. https://doi.org/10.1186/s13229-016-0109-5
152. El-Ansary AK, Bacha AGB, Al-Ayahdi LY (2011) Plasma fatty acids as diagnostic markers in autistic patients from Saudi Arabia. Lipids Health Dis 10:62. https://doi.org/10.1186/1476-511X-10-62
153. Lv Q-Q, You C, Zou X-B, Deng H-Z (2018) Acyl-carnitine, C5DC, and C26 as potential biomarkers for diagnosis of autism spectrum disorder in children. Psychiatry Res 267:277–280

Chapter 12
Proteomic Investigations of Autism Spectrum Disorder: Past Findings, Current Challenges, and Future Prospects

Joseph Abraham, Nicholas Szoko, and Marvin R. Natowicz

Abstract Proteomics is a powerful tool to study biological systems and is potentially useful in identifying biomarkers for clinical screening and diagnosis, for monitoring treatment, and for exploring pathogenetic mechanisms in autism. Unlike numerous other experimental approaches employed in autism research, there have been few proteomic-based analyses. Herein, we discuss the findings of studies regarding autism that utilized a proteomic approach and review key considerations in sample acquisition, processing, and analysis. Most proteomic studies on autism used blood or other peripheral tissues. Few studies used brain tissue, the main site of biological difference between persons with autism and others. The findings have varied and are not yet replicated. Some showed abnormalities of synaptic proteins or proteins of mitochondrial bioenergetics. Various abnormalities of proteins relating to immune processes and lipid metabolism have also been noted. Whether any of the proteomic differences between autism and control cases are primary or secondary phenomena is currently unclear. Consequently, no definitive biomarkers for autism have been identified, and the pathophysiological insights provided by proteomic studies to date are uncertain in the absence of replication. Based on this body of work and the challenges in using proteomics to study autism, we suggest considerations for future study design. These include attention to subject and specimen inclusion/

J. Abraham · N. Szoko
Cleveland Clinic Lerner College of Medicine of Case Western Reserve University, Cleveland, OH, USA

M. R. Natowicz (✉)
Cleveland Clinic Lerner College of Medicine of Case Western Reserve University, Cleveland, OH, USA

Pathology & Laboratory Medicine, Genomic Medicine, Neurology and Pediatrics Institutes, Cleveland Clinic, Cleveland, OH, USA
e-mail: NATOWIM@ccf.org

P. C. Guest (ed.), *Reviews on Biomarker Studies in Psychiatric and Neurodegenerative Disorders*, Advances in Experimental Medicine and Biology 1118,
https://doi.org/10.1007/978-3-030-05542-4_12

exclusion criteria, attention to the state of specimens prior to proteomic analysis, and use of a replicate set of specimens. We end by discussing especially promising applications of proteomics in the study of autism pathobiology.

Keywords Autism · ASD · Proteomic · Proteomics · Mass spectrometry · Neuroproteomics

12.1 Introduction

Autism spectrum disorder (ASD) is a heterogeneous group of neurodevelopmental conditions characterized by impairments in socialization and communication with repetitive or stereotyped behaviors or interests [1, 2]. In addition, persons with ASD frequently have medical, neurological, and/or psychiatric comorbidities [3, 4]. ASD is also common as recent data on children in the United States indicate a prevalence of ASD at about 1 in 59 to 1 in 42 children and a global prevalence in children of nearly 1% [5–7]. Consequently, ASD is associated with varied and considerable personal, family, and societal costs. For these reasons, determination of the underlying pathobiology of ASD and efforts directed toward ASD prevention, early diagnosis, and effective treatment are public health priorities.

The understanding of ASD pathogenesis has progressed substantially in recent years [8]. Many experimental approaches, ranging from histopathological analyses of postmortem brain tissue, biochemical analyses of brain and other tissues, genomic analyses of ASD cases, transcriptomic studies of various tissues, neuropsychological studies, structural and functional neuroimaging studies, and diverse studies of animal models of autism, have contributed to our current understanding of the pathogenesis of ASD. Atypical functional brain connectivity is a central finding of studies of the pathobiology, and the concept of autism as a collection of developmental disconnection syndromes is widely held [9]. Some of the best supported and non-mutually exclusive theories of autism pathogenesis at a molecular level include abnormalities of synapse formation and function and abnormalities of genes that regulate chromatin remodeling and transcription, although there is support for many other molecular processes [10–13]. Whether there is a single unifying pathophysiological model for the diverse causes of autism remains to be determined. Some important unifying models that incorporate genetic, biochemical, and electrophysiological findings in autism research have been proposed [14, 15].

Unlike other "omics" approaches such as genomics and transcriptomics, which have been highly productive approaches in advancing our knowledge of autism, there have been few proteomics-based analyses which have investigated the pathobiology of autism. In this review, we describe the challenges in using proteomic methodologies in autism research, summarize and critically assess proteomics-based studies of ASD, and discuss some potentially useful applications of proteomics in advancing the understanding of autism biology.

12.2 Potential Applications of Proteomics in Autism Research

There are multiple potential applications of proteomics in clinical and basic research related to autism. Proteomic analyses hold forth the potential for the use in screening persons who are at risk to develop autism or in defining targets for screening and, similarly, for potential use in the diagnosis of autism. Screening and early diagnosis of autism are important since early identification permits early intervention, often with therapeutic benefits [2, 16]. In addition, the development of effective screening and diagnostic biomarkers might result in substantial cost savings compared to the generally inefficient and slow process of screening and diagnosis that commonly occurs [17]. It might also enable a more accurate categorization of persons with autism since an understanding of the biology of autism and the development of more effective therapies might be facilitated by the delineation of reproducible and biomedically meaningful biomarker-defined subtypes of persons with ASD [2, 18]. In addition, proteomic analyses could also conceivably have use in monitoring impacts of treatment. Although such uses of proteomics in ASD are presently speculative, proteomic methods have recently gained acceptance in several clinical laboratory applications, and many other applications are being investigated, and the adoption of additional clinical laboratory uses of proteomics is anticipated [19–21].

Related to the clinical imperative of better delineation of clinical subtypes of autism, the other major potential application of proteomics in autism is in the study of the pathogenesis of autism. There are numerous known causes of autism, and there is evidence of derangement of many cellular process and metabolic pathways [11, 15]. It is presently unclear if there is a final common pathway or unifying pathophysiological mechanism for autism that encompasses its diverse etiologies. Proteomic investigation of specimens from persons with autism and of diverse models of autism is, therefore, the other major area of application of proteomics to autism.

12.3 Challenges in Using Proteomic Approaches in Autism Research

Apart from the complexities of the analytic dimensions of proteomics in studying autism, several autism-specific pre-analytic concerns need to be considered in any experimental study. These include:

1. How do the authors define autism?
2. How, if at all, is the etiological heterogeneity of autism considered?
3. Do the authors account for comorbidities?
4. What is the biological state of the subject(s) from which the samples were derived?
5. What tissue or cell type is used in the experimental analysis?
6. What was the tissue procurement process for the samples?

The concept of autism has undergone substantial change since its earliest formulation, with different neurobehavioral phenotypes described [22, 23]. Even today, the categorical diagnosis of ASD includes diverse clinical phenotypes and does not define a neurobiologically or etiologically well-defined, homogenous condition. Consequently, its use remains problematic and, sometimes, contested [24, 25]. Moreover, not all studies include information regarding the specific criteria used in establishing the diagnosis. There are multiple diagnostic criteria and tools used to establish the diagnosis. This information should be described for any study of ASD, not just those involving proteomics, as the criteria that are implemented will have impacts regarding which persons or samples will be used in a study [26].

In addition, ASD is etiologically heterogeneous. There are more than 100 known causes of ASD, most of which are uncommon or rare [8, 10, 27, 28]. Combining data from ASD cases having different etiologies can result in confounding of the results. Studies should specify both the etiologies of the ASD cases from which samples are derived and the basis for the designation of those etiologies. In those instances where a diagnosis is not known, pertinent negative diagnostic evaluations should be noted, if possible.

Medical, neurological, and psychiatric comorbidities are common in persons with ASD [3, 4, 29]. Comorbidities should be assessed when considering candidacy for a study since the comorbidities and the medications and other treatments of the comorbidities can impact the proteome. Consequently, inclusion and exclusion criteria should be well-defined prior to sample collection, and pertinent clinical information from evaluations of the individuals from whom samples are derived should be available as appropriate.

Inclusion and exclusion criteria should also consider the prandial state of the subjects from which the samples originated and, if postmortem samples are obtained, the cause of death. The prandial state should, if possible, be controlled for in studies using serum, plasma, or saliva to reduce another potential source of confounding factors. Not unexpectedly, the cause of death can also have significant impacts on tissue proteomes [30, 31].

What cell type or tissue to analyze is another significant issue in proteomic studies of autism. Autism is first and foremost a disorder of the brain, despite the significant number of comorbidities in non-CNS systems. Proteomic analyses of brain tissue are, therefore, likely to be more informative than analysis of other tissues in terms of understanding the pathogenesis of brain dysfunction in autism. However, relatively few postmortem autism brains are available for research purposes. Of those that are available, the number of samples accessible for research may be further reduced because of limitations imposed by the study with respect to age, cause of death, or other eligibility criteria. Moreover, even when adequate numbers of brain samples are available, proteomic analyses of brain tissue are inherently complicated by the dynamic changes in brain during neurodevelopment and the cellular heterogeneity of brain tissue [32–34].

Processing of samples for proteomic analysis is essential to consider, particularly in relation to the postmortem interval time. The duration between death and banking of samples can be especially relevant in proteomics because of time-dependent proteolysis and other modifications that can occur to some proteins [35–37]. Thus, matching cases and controls for the postmortem interval in human tissue studies is an important consideration. Issues related to storage of tissue and histological processing of specimens, if done, are also relevant for proteomic studies and are reviewed elsewhere [38, 39].

Study design and bioinformatic analysis decisions when evaluating mass spectrometry (MS) data can have profound impacts on the overall results [40–44]. Investigations using postmortem ASD brain tissue or samples from rare genetic subtypes of ASD are frequently limited by small sample size. Moreover, comparative MS-based methodologies typically require adjusting significance for multiple comparisons to ensure that differences that are observed are not due to chance from performing several statistical tests [45]. Advanced statistical methods, such as linear mixed models, can be used to address the particular difficulties faced when comparing a small number of cases and replicates in the context of vast amounts of proteomic data [46, 47].

In addition to technical replicates, MS-based proteomic findings should be validated with a second method. For example, interesting findings from a label-free unbiased MS analysis should be validated using an independent method such as selected reaction monitoring assays. Replication of the findings using a second set of biological samples further strengthens a study. Finally, biological relevance should be assessed for those interesting findings that were statistically significant. This can be challenging, typically requiring support from additional investigative approaches.

12.4 Proteomic Studies of Autism

To date, there have been few proteomic studies of autism [48]. Unbiased or global proteomic studies of ASD in humans using nonneural tissues include 12 studies of plasma or serum, an analysis of peripheral B lymphocytes, 3 studies of saliva, and 1 study of urine [49–65]. The characteristics and key findings of these studies are shown in Table 12.1.

Several interesting observations were observed in these analyses. Of studies using nonneural tissues, many noted altered levels of one or more proteins involved in inflammation or immune system regulation, including some acute phase reactants and interleukins [49, 52, 53, 55–64]. Abnormalities of the complement system were noted in several analyses [62, 63]. This is interesting in view of recent work demonstrating that the complement pathway can affect synaptic remodeling and has roles in neurodevelopment and some neurodegenerative processes [66, 67]. Varied abnormalities of immunologic markers or of immunologic function including, in some instances, immune abnormalities involving cerebrospinal fluid or brain tissue has been a recurring theme in ASD research [68, 69].

Several proteomic studies identified proteins involved in lipid metabolism as differentially expressed in ASD [49, 52, 57, 58, 60, 64]. These findings align with some prior metabolic investigations of ASD using other experimental methods [70, 71].

Table 12.1 Proteomic studies of autism spectrum disorder

References	Tissue	State of ASD tissue[a]	Subjects	ASD diagnostic criteria	ASD inclusion criteria	ASD exclusion criteria	Comorbidity characterization	Method[b]	Proteins implicated[c]	Validation[d]	Suggested pathways/ processes[e]
[72]	Brain (gray matter)	PMI: 22.5 h (17.8–26)	8 ASD (8M) 10 controls (7M/3F)	Variable, incomplete clinical histories	–	–	Reported (medical/ behavioral)	2D-PAGE	GLO1	WB LC-MS/ MS	–
[49]	Sera	Prandial state: fasting Others: no illness <72 h, no vaccines <2 weeks	69 ASD (58M/11F) 35 controls (29M/6F)	DSM-IV, confirmed by ADOS-G and ADI-R	ASD diagnosis by DSM-IV	PDD-NOS or Asperger; Fragile X; neuro/psych disorders; other medical illness; failure to complete research protocol	Reported indirectly through exclusion criteria	LC-MS/MS	**FHR1**, **C1Q**, **FN1**, *APOB-100*	–	–
[50]	Saliva	Prandial state: no food/ beverage <30 min Others: same collection time	27 ASD (20M/7F) 23 controls	DSM-IV	ASD (including Asperger and PDD-NOS) per DSM-IV criteria	Not reported	Not reported	LC-MS/MS	STATH, HTN1, PRP	–	–
[51]	Sera	Prandial state: not reported	16 ASD (16M) 16 controls (16M)	DSM-IV, confirmed by FSK/SCQ, ADI-R, ADOS	ASD by DSM-IV criteria, confirmed by other measures	Current/chronic physical illness; severe somatic or neuro disorder; schizophrenic psychoses	Reported (medical/ behavioral)	MALDI-TOF MS	–	–	–

[53]	B lymphocytes	Prandial state: not reported	2 ASD (1M/1F) 2 normal siblings (2F)	DSM-IV	–		Not reported	Immunoassay, WB, RT-qPCR	**IKKA**, *TYK2*, *PRKC1*, *EIF4G1*	WB RT-qPCR	–
[52]	Sera	Prandial state: not reported	45 ASD (22M/23F) 50 controls (26M/24F)	DSM-IV-TR	ASD diagnosis by DSM-IV-TR	Diagnosis revision during study	Not reported	Immunoassay	Males: **IL12B**, **FABP2**, **IL3**, **EPO**, **F3**, **IL5**, **CSF3**, **IL1B**, **CHGA**, **NCAM**, **TNC**, **TNF**, **CXCL5**, **IL18**, **F7**, **CTGF**, **IL4**, **THPO**, **KITLG**, **SORT1**, **IL10**, **IL12A**, **ICAM1**, *GOT1* Females: **NARG1**, **IL12A**, **FAI**, **IL1B**, **LH**, **IL7**, **TF**, **BDNF**, *GOT1*, *APOC3*, *IGHM*, *AGER*, *APOA1*, *TNC*, *CCL26*, *EDN1*, *GH1*	–	Inflammation Growth factors/ anabolism
[54]	Sera	Prandial state: not reported	32 ASD (23M/5F) 31 controls (14M/16F)	DSM-IV	ASD diagnosis by DSM-IV	Infection/illness less than 2 weeks before examination; collection errors	Not reported beyond exclusion of schizophrenia	MALDI-TOF MS	–	–	–

(continued)

Table 12.1 (continued)

References	Tissue	State of ASD tissue[a]	Subjects	ASD diagnostic criteria	ASD inclusion criteria	ASD exclusion criteria	Comorbidity characterization	Method[b]	Proteins implicated[c]	Validation[d]	Suggested pathways/ processes[e]
[73]	Brain (BA10 and cerebellum)	PMI: 19 h (13–26.25)	16 ASD (13 M/3F) 18 controls (13 M/5F)	ADI-R, CARS	ASD by ADI-R and CARS	Genetic disorders/ mutations by "gene analysis"	Not reported	SRM LC-MS/ MS	BA10: **GFAP**, **MAG**, **MBP**, **MOG** **PLPP1**, **SYN2**,*VIME*, *CKB*, *STX1A*, *SYT1*, *PACSIN1* Cerebellum: **GFAP**, **CKB**, **STX1A**, **SYN2**, **SYT1**, *VIME*, *MAG*, *MBP*, *MOG*, *PLP1*, *PACSIN1*	–	Energy metabolism Synaptic function Myelination
[57]	Sera	Prandial state: not reported	7 ASD (7 M) 7 controls (7 M)	DSM-IV, confirmed by FSK/SCQ, ADI-R, ADOS	ASD by DSM-IV criteria	Current/chronic physical illness; severe neuro disorder; schizophrenic psychoses	Reported (medical and behavioral) in prior paper on the same patients	PAGE, LC-MS/ MS	**APOA1**, **APOA4**, **PON1**	WB	–
[58]	Sera	Prandial state: not reported Others: BMI, exercise level, alcohol/ smoking status, OCPs	30 (14 M/16F) 29 controls (13 M/16F)	DSM-IV-TR	ASD diagnosis by DSM-IV-TR	Family history of psych or medical illness, medication use, tobacco/ marijuana use	Not reported	Immunoassay, LC-MS/MS	Males: **SHBG**, **EPO**, **CHGA**, **PAP**, **BMP6**, **IL3**, **TENA**, **TNF**, **TF**, **CTGF**, **IL16**, **IL12A**, **ICAM1**, *RGPD4* Females: *SHBG*, *ADIPO*, *APOA1*, **APOE**, **APOC2**, **FETUB**, **GLCE**, *EPO*, *CHGA*, *PAP*, **PTPA**, **TLE1**, **RN149**, **TRIPB**, *IL3*, *TENA*, *IGA*, **CLC4K**, *MRRP1*, **ZC3HE**, **ARMC3**	–	Lipid metabolism Cell growth Inflammation

[56]	Saliva	Prandial state: not reported	6 ASD (6 M) 6 controls	DSM-IV-TR	ASD diagnosis by DSM-IV-TR	Not reported	Reported (medical and behavioral)	2D-PAGE, LC-MS/MS	**FRAT1**, **KIF14**, **ITGA6**, **GRTP1**, **PSP**, **PIP**, **MUC16**, **MRP14**, *AMY1A*, *CREBBP*, *HERC1*, *TF*, *AZGP1*, *ZG16*, *CST5*, *PLG*	–	Oxidative stress Lipid metabolism Immune response Inflammation
[55]	Saliva	Prandial state: not reported	6 ASD (6M) 6 controls	DSM-IV-TR	ASD diagnosis by DSM-IV-TR	Not reported	Reported (medical and behavioral)	LC-MS/MS	**PIP**, **LTF**, **IGKC**, **IGHG1**, **IGLC2**, **ELANE**, **PIGR**, **DMBT1**, *HTN1*, *STATH*, *PRH1*	–	Immune response (mucosal) Inflammation
[59]	Urine	20–30 ml first void urine	30 ASD (19M/11F) 30 controls (18M/12F)	CHAT	ASD diagnosis by CHAT	Not reported	Minimal (antibiotic use, GI complaints, gluten sensitivity)	2D-PAGE, MALDI-TOF MS	**KNG1**, **IGHG1**, **MASP2**	ELISA	–
[60]	Sera	Prandial state: fasting	30 ASD (24M/6F) 30 controls (20M/10F)	DSM-V	ASD diagnosis by DSM-V	Not reported (one subject with celiac, none with other comorbidities)	No comorbidities beyond one patient with celiac	2D-PAGE, LC-MS/MS	**A2M**, **SERPINA3**, **HP**, **A1AT**, *TF*, *TTR*, *RBP4*, *AHSG*, *APOJ*	WB	Acute-phase response
[61]	Plasma	Prandial state: fasting Others: no medications or dietary supplements	15 ASD (12M/3F) 15 controls (12M/3F)	DSM-IV	ASD diagnosis by DSM-IV	No physical or neurological conditions or family history of ASD	Not expanded upon (no medications)	2D-PAGE, followed by Western blot for protein carbonylation and then protein identification by MALDI-TOF	**C8A**, **IGKC**	Chemical assay	Immune system

(continued)

Table 12.1 (continued)

References	Tissue	State of ASD tissue[a]	Subjects	ASD diagnostic criteria	ASD inclusion criteria	ASD exclusion criteria	Comorbidity characterization	Method[b]	Proteins implicated[c]	Validation[d]	Suggested pathways/ processes[e]
[62]	Sera	Prandial state: not reported	65 ASD 65 controls	DSM-V	ASD diagnosis by DSM-V	History of tuberous sclerosis complex, Rett syndrome, Prader-Willi syndrome, Angelman syndrome, or Fragile X syndrome. Infection in prior 2 weeks	Not reported	Lectin microarrays, lectin-magnetic peptide conjugate-assisted LC-MS/MS	**74 proteins differentially glycosylated**	WB, antibody microarray	Complement cascade, aberrant cellular regulation of response-to-stimulus
[63]	Plasma	Prandial state: fasting	30 ASD (24M/6F) 30 controls (24M/6F)	DSM-IV	ASD diagnosis by DSM-IV	Not reported	Not reported	iTRAQ	**APOE**, **SERPINA1**, **FBLN1**, **FN1**, **C3**, **C5**, **AGT**, **VTN** **SERPIN4A**, **IGFALS**, *ACTN1*, *CALM1*, *CALR*, *ACTG1*, *PARVB*, *MAPRE2*, *ENO1*, *ITGA2B*, *FERMT3*, *EHD3*, *TLN1*, *VCL*, *VCP*, *THBS1*	ELISA	Focal adhesions, cytoskeleton, motility and migration, synaptogenesis, complement system, platelet function
[65]	Sera	Prandial state: fasting	32 ASD (28M, 4F) 20 controls (20M)	DSM-V	ASD diagnosis by DSM-V	History of tuberous sclerosis complex, Rett syndrome, Prader-Willi syndrome, Angelman syndrome, or Fragile X syndrome	Not reported	MALDI-TOF MS	8 unidentified peaks	–	–

[64]	Sera	Prandial state: fasting	68 ASD (56M, 12F) 80 controls (60M, 20F)	DSM-IV	ASD diagnosis by DSM-IV	Children with an infectious disease, an organic disease of the nervous system, or a psychiatric disorder	Not reported beyond exclusion criteria	MALDI-TOF-MAS	**SERPINA5**, **PF4**, **FAPB1**, **APOC1**, **AFP**, **CPB2**, **TAAR6**, **FGA**	ELISA	Platelet function, lipid metabolism

ASD autism spectrum disorder, *AS* Asperger syndrome, *PMI* postmortem interval, *M* males, *F* females, *DSM (TR)* Diagnostic and Statistical Manual of Mental Disorders (Text Revision), *ADOS (G)* autism diagnostic observation schedule (Generic), *ADI-R* autism diagnostic interview, revised, *CHAT* checklist for autism in toddlers, *SCQ* social communication questionnaire, *FSK* German version of SCQ, *CARS* childhood autism rating scale, *PDD-NOS* pervasive developmental disorder, not otherwise specified, *iTRAQ* isobaric tags for relative and absolute quantitation, *PAGE* polyacrylamide gel electrophoresis, *LC-MS/MS* liquid chromatography tandem mass spectrometry, *MALDI-TOF MS* matrix-assisted laser desorption/ionization time-of-flight mass spectrometry, *SRM* selected reaction monitoring, *WB* Western blot, *RT-qPCR* quantitative reverse transcription polymerase chain reaction, *ELISA* enzyme-linked immunosorbent assay

[a]PMI reported as median (interquartile range)[b]Method used for global proteomic analysis[c]Proteins in **bold** and *italics* are up- and downregulated in ASD vs. controls, respectively[d]Validation of the proteomic findings by use of immunoassay such as ELISA, Western blot (WB), or targeted MS[e]The indicated pathways or processes are those provided by the researchers of the various studies. The methodology for designating these pathways, when described, differs between studies. No pathways or processes are indicated when fewer than five differentially expressed proteins were observed

Nonetheless, it is unclear if the abnormalities of lipid metabolism noted in ASD tissues are primary or secondary occurrences in ASD pathophysiology.

Because the brain is central to the pathobiology of ASD, studies examining neural tissue will likely offer the most understanding into the mechanistic underpinnings of ASD. There has been one global proteomic analysis of human ASD brain and one targeted proteomic analysis of ASD brain [72, 73]. In an early proteomic study using 2D gel electrophoresis followed by MS analysis of a single polypeptide, a polymorphic form of glyoxalase 1 in ASD brain was observed. That isoform was found to have reduced enzyme activity, and its underlying DNA sequence variant was found to be more prevalent in ASD versus control cases [72]. A targeted proteomic analysis of postmortem prefrontal cortex and cerebellum from persons with ASD used LC-MS/MS, and this revealed several differentially expressed proteins. Most of these proteins were implicated in processes of energy metabolism, synaptic function, and myelination [73]. As noted above, abnormalities of synapse biology have been consistently noted in diverse studies of ASD pathobiology, but, as most synaptic proteins are specific to neural tissue, this finding has not been noted in proteomic studies of nonneural tissue, affirming a limitation in the use of nonneural tissues in proteomic studies of ASD pathogenesis.

Most proteomic studies of ASD to date have been limited by one or more of the challenges described above (Table 12.1). Characterizations of ASD cases and controls were often incomplete. The studies were variably powered to determine significant differences between cases and controls. In addition, the studies employed different protein extraction and separation methods and MS methodologies and, therefore, varied considerably in the comprehensiveness of detection of the proteome of their samples. Moreover, the use of a follow-up quantification technique to validate key proteomic results has become increasingly standard in contemporary proteomics, with follow-up targeted LC-MS/MS becoming a preferred approach [74]. However, the published ASD proteomic studies have varied in the use of a validation of the MS methodology. Finally, there has been a consistent absence of a biological replication set in the aforementioned studies.

12.5 Promising Approaches for Future Research

The future of ASD proteomics will require attention to important nontechnical ASD-specific issues and other nontechnical matters, in addition to technical issues of sample analysis. Important nontechnical issues, noted above, include subject inclusion/exclusion criteria, tissue selection, sample size and the powering of the study, the state and processing of tissues prior to analysis, the use of technical replicates, validation of the methodology, and replication of the findings using an independent set of biological samples. In addition, there are multiple methodologies available for mass spectrometric proteomic analysis, each with different strengths and applications, and each will likely be impacted by technological advancements over the next few years [42]. Methods for proteomic analyses in addition to MS also show significant promise [75].

Certain types of proteomics-based research in ASD hold, in our view, great promise. Application of proteomic methods to postmortem ASD brain is likely to be informative. Complexities relating to the cellular and tissue heterogeneity of brain can also be addressed to some extent. Technologies such as laser capture microdissection and fluorescence-activated cell sorting enable investigation of highly specific tissue or cell types, while various affinity-based biochemical approaches also allow cell type-specific proteomic analysis [32]. Matrix-assisted laser desorption/ionization mass spectrometry imaging permits region- and even cell-specific measurement of hundreds of proteins in an intact sample of interest in an anatomical context, such as in tissue slices, and has already provided insights in several neuroproteomic contexts [76, 77]. Proteomic analyses of subsets of proteins having specific biochemical modifications, such as the phosphoproteome, are another type of proteomic analysis that have potential in furthering our understanding of ASD biology, as has been the case for other types of clinical conditions and biological processes [78].

The use of differentiated pluripotent stem cells from ASD subjects is an important new approach to study the biology of autism. Another promising new tool involves the use of brain organoids, miniature in vitro "brain structures" that can be developed from cells derived from persons with autism and others, to learn about brain biology in autism. Both of these approaches have already generated interesting findings related to ASD [12, 79].Analysis of induced pluripotent stem cells from ASD cases found increased proliferation of neural progenitor cells and neuron numbers mediated by dysregulation of the beta-catenin/BRN2 transcriptional cascade, while RNA-seq analysis on cerebral organoids deficient in the ASD risk gene, CHD8, and comparison to other ASD organoid models affirmed the importance of DLX genes in ASD pathobiology [12, 80]. However, to our knowledge, proteomic analyses have not yet been used in these and related ASD model systems.

Proteomic and other studies of animal models of ASD have been informative, especially as these experimental systems allow for experimental manipulation that is not possible with human tissue [81–84]. Animal models of autism have included syndromic forms of autism attributable to single gene mutations or cytogenomic lesions. For example, using murine fmr1 knockout neuronal cultures, synaptic abnormalities of Fragile X syndrome were connected to differential cortical expression of proteins involved in the regulation of synaptic structure, neurotransmission, and dendritic mRNA transport [85]. Another study using a MeCP2-deficient murine model recapitulating Rett syndrome (RTT) found proteomic changes reflecting perturbations in RNA metabolism, proteostasis, monoamine metabolism, and cholesterol synthesis, which have been noted in RTT [86]. Caution is required when considering the generalizability of such models to human experience and to ASD specifically. Nonetheless, proteomic analyses of animal models of ASD will likely continue to provide insights into ASD pathogenesis.

Lastly, integration of multiple experimental methods, for example, a transcriptomic and proteomic combined analysis, has provided knowledge regarding dynamic networks of molecular interaction [42, 87]. To date, there have been no human studies of ASD that included such an integrated analysis that incorporates a proteomic approach. Coupling other omics methods with proteomics to studying human ASD tissues and various models of ASD will be powerful means for advancing our understanding of the complexities of ASD pathobiology.

12.6 Conclusions

There have been relatively few proteomics-based studies of autism thus far. In addition to technical considerations that are pertinent to any proteomic analysis, there are important ASD-related considerations that relate to the diagnosis of ASD, its etiological heterogeneity and varied comorbidities, and the brain as the central affected organ. Published findings from proteomics-based analyses of ASD have varied and require replication, including differential expression of proteins of lipid metabolism, inflammation and immune function, synaptic biology, and mitochondrial bioenergetics. Moving forward, multi-omics approaches that incorporate proteomic analysis of human brain samples, stem cells, brain organoids, and animal models of ASD are likely to be informative. Beyond elucidating ASD pathobiology, proteomic approaches could lead to the identification of novel biomarkers for improved diagnosis and therapeutic monitoring. In turn, this could result in discovery of new drug targets or therapeutic approaches to help improve the lives of individuals affected by ASD.

References

1. American Psychiatric Publishing (2013) Diagnostic and statistical manual of mental disorders: DSM-5, 5th edn. American Psychiatric Publishing, Washington DC, pp 50–59. isbn:8123923791
2. Lai MC, Lombardo MV, Baron-Cohen S (2014) Autism. Lancet 383(9920):896–910
3. Bauman ML (2010) Medical comorbidities in autism: challenges to diagnosis and treatment. Neurotherapeutics 7(3):320–327
4. Muskens JB, Velders FP, Staal WG (2017) Medical comorbidities in children and adolescents with autism spectrum disorders and attention deficit hyperactivity disorders: a systematic review. Eur Child Adolesc Psychiatry 26(9):1093–1103
5. Baio J, Wiggins L, Christensen DL, Maenner MJ, Daniels J, Warren Z et al (2018) Prevalence of autism spectrum disorder among children aged 8 years—autism and developmental disabilities monitoring network, 11 sites, United States, 2014. MMWR Surveill Summ 67(6):1–23
6. Xu G, Strathearn L, Liu B, Bao W (2018) Corrected prevalence of autism spectrum disorder among US children and adolescents. JAMA 319(5):505. https://doi.org/10.1001/jama.2018.0001
7. Baxter AJ, Brugha TS, Erskine HE, Scheurer RW, Vos T, Scott JG (2015) The epidemiology and global burden of autism spectrum disorders. Psychol Med 45(3):601–613
8. Willsey AJ, State MW (2015) Autism spectrum disorders: from genes to neurobiology. Curr Opin Neurobiol 30:92–99

9. Geschwind DH, Levitt P (2007) Autism spectrum disorders: developmental disconnection syndromes. Curr Opin Neurobiol 17(1):103–111
10. Amaral DG (2017) Examining the causes of Autism Cerebrum 2017. pii: cer-01-17. eCollection 2017 Jan-Feb
11. Kleijer KTE, Huguet G, Tastet J, Bourgeron T, Burbach JPH (2017) Anatomy and cell biology of autism spectrum disorder: lessons from human genetics. Adv Anat Embryol Cell Biol 224:1–25
12. Forsberg SL, Ilieva M, Maria Michel T (2018) Epigenetics and cerebral organoids: promising directions in autism spectrum disorders. Transl Psychiatry 8(1):14. https://doi.org/10.1038/s41398-017-0062-x
13. Andrews SV, Ellis SE, Bakulski KM, Sheppard B, Croen LA, Hertz-Picciotto I et al (2017) Cross-tissue integration of genetic and epigenetic data offers insight into autism spectrum disorder. Nat Commun 8(1):1011. https://doi.org/10.1038/s41467-017-00868-y
14. Courchesne E, Pramparo T, Gazestani VH, Lombardo MV, Pierce K, Lewis NE (2018) The ASD living biology: from cell proliferation to clinical phenotype. Mol Psychiatry. https://doi.org/10.1038/s41380-018-0056-y. [Epub ahead of print]
15. Mullins C, Fishell G, Tsien RW (2016) Unifying views of autism spectrum disorders: a consideration of autoregulatory feedback loops. Neuron 89(6):1131–1156
16. Murphy CM, Wilson CE, Robertson DM, Ecker C, Daly EM, Hammond N et al (2016) Autism spectrum disorder in adults: diagnosis, management, and health services development. Neuropsychiatr Dis Treat 12:1669–1686
17. Durkin MS, Elsabbagh M, Barbaro J, Gladstone M, Happe F, Hoekstra RA et al (2015) Autism screening and diagnosis in low resource settings: challenges and opportunities to enhance research and services worldwide. Autism Res 8(5):473–476
18. Masi A, DeMayo MM, Glozier N, Guastella AJ (2017) An overview of autism spectrum disorder, heterogeneity and treatment options. Neurosci Bull 33(2):183–193
19. Vrana JA, Theis JD, Dasari S, Mereuta OM, Dispenzieri A, Zeldenrust SR et al (2014) Clinical diagnosis and typing of systemic amyloidosis in subcutaneous fat aspirates by mass spectrometry-based proteomics. Haematologica 99(7):1239–1247
20. Belczacka I, Latosinska A, Metzger J, Marx D, Vlahou A, Mischak H et al (2018) Proteomics biomarkers for solid tumors: current status and future prospects. Mass Spectrom Rev. https://doi.org/10.1002/mas.21572. [Epub ahead of print]
21. Sabbagh B, Mindt S, Neumaier M, Findeisen P (2016) Clinical applications of MS-based protein quantification. Proteomics Clin Appl 10(4):323–345
22. Evans B (2013) How autism became autism: the radical transformation of a central concept of child development in Britain. Hist Human Sci 26(3):3–31
23. Verhoeff B (2013) Autism in flux: a history of the concept from Leo Kanner to DSM-5. Hist Psychiatry 24(4):442–458
24. London EB (2014) Categorical diagnosis: a fatal flaw for autism research? Trends Neurosci 37(12):683–686
25. Müller RA, Amaral DG (2017) Editorial: time to give up on autism spectrum disorder? Autism Res 10(1):10–14
26. Battaglia A (2007) On the selection of patients with developmental delay/mental retardation and autism spectrum disorders for genetic studies. Am J Med Genet 143A(8):789–790
27. Betancur C (2011) Etiological heterogeneity in autism spectrum disorders: more than 100 genetic and genomic disorders and still counting. Brain Res 1380:42–77
28. Sztainberg Y, Zoghbi HY (2016) Lessons learned from studying syndromic autism spectrum disorders. Nat Neurosci 19(11):1408–1417
29. Jones RM, Lord C (2013) Diagnosing autism in neurobiological research studies. Behav Brain Res 251:113–124
30. Sarkis GA, Mangaonkar MD, Moghieb A, Lelling B, Guertin M, Yadikar H et al (2017) The application of proteomics to traumatic brain and spinal cord injuries. Curr Neurol Neurosci Rep 17(3):23. https://doi.org/10.1007/s11910-017-0736-z

31. Shao G, Wang Y, Guan S, Burlingame AL, Lu F, Knox R et al (2017) Proteomic analysis of mouse cortex postsynaptic density following neonatal brain hypoxia-ischemia. Dev Neurosci 39(1–4):66–81
32. Kitchen RR, Rozowsky JS, Gerstein MB, Nairn AC (2014) Decoding neuroproteomics: integrating the genome, translatome and functional anatomy. Nat Neurosci 17(11):1491–1499
33. Hosp F, Mann M (2017) A primer on concepts and applications of proteomics in neuroscience. Neuron 96(3):558–571
34. Ramadan N, Ghazale H, El-Sayyad M, El-Haress M, Kobeissy FH (2017) Neuroproteomics studies: challenges and updates. Methods Mol Biol 1598:3–19
35. Fountoulakis M, Hardmeier R, Höger H, Lubec G (2001) Postmortem changes in the level of brain proteins. Exp Neurol 167(1):86–94
36. Crecelius A, Götz A, Arzberger T, Fröhlich T, Arnold GJ, Ferrer I et al (2008) Assessing quantitative post-mortem changes in the gray matter of the human frontal cortex proteome by 2-D DIGE. Proteomics 8(6):1276–1291
37. ElHajj Z, Cachot A, Müller T, Riederer IM, Riederer BM (2016) Effects of postmortem delays on protein composition and oxidation. Brain Res Bull 121:98–104
38. Banks RE (2008) Preanalytical influences in clinical proteomic studies: raising awareness of fundamental issues in sample banking. Clin Chem 54(1):6–7
39. Becker KF (2015) Using tissue samples for proteomic studies-critical considerations. Proteomics Clin Appl 9(3–4):257–267
40. Oberg AL, Vitek O (2009) Statistical design of quantitative mass spectrometry-based proteomic experiments. J Proteome Res 8(5):2144–2156
41. Schmidt A, Forne I, Imhof A (2014) Bioinformatic analysis of proteomics data. BMC Syst Biol 8(Suppl 2):S3. https://doi.org/10.1186/1752-0509-8-S2-S3
42. Aebersold R, Mann M (2016) Mass-spectrometric exploration of proteome structure and function. Nature 537(7620):347–355
43. Hu A, Noble WS, Wolf-Yadlin A (2016) Technical advances in proteomics: new developments in data-independent acquisition. F1000Res 5. pii: F1000 Faculty Rev-419. https://doi.org/10.12688/f1000research.7042.1
44. Ruderman D (2017) Designing successful proteomics experiments. Methods Mol Biol 1550:271–288
45. Listgarten J, Emili A (2005) Statistical and computational methods for comparative proteomic profiling using liquid chromatography-tandem mass spectrometry. Mol Cell Proteomics 4(4):419–434
46. Clough T, Thaminy S, Ragg S, Aebersold R, Vitek O (2012) Statistical protein quantification and significance analysis in label-free LC-MS experiments with complex designs. BMC Bioinformatics 13(Suppl 16):S6. https://doi.org/10.1186/1471-2105-13-S16-S6
47. Ritchie ME, Phipson B, Wu D, Hu Y, Law CW, Shi W et al (2015) Limma powers differential expression analyses for RNA-sequencing and microarray studies. Nucleic Acids Res 43(7):e47. https://doi.org/10.1093/nar/gkv007
48. Szoko N, McShane AJ, Natowicz MR (2017) Proteomic explorations of autism spectrum disorder. Autism Res 10(9):1460–1469
49. Corbett BA, Kantor AB, Schulman H, Walker WL, Lit L, Ashwood P et al (2007) A proteomic study of serum from children with autism showing differential expression of apolipoproteins and complement proteins. Mol Psychiatry 12(3):292–306
50. Castagnola M, Messana I, Inzitari R, Fanali C, Cabras T, Morelli A et al (2008) Hypophosphorylation of salivary peptidome as a clue to the molecular pathogenesis of autism spectrum disorders. J Proteome Res 7(12):5327–5332
51. Taurines R, Dudley E, Conner AC, Grassl J, Jans T, Guderian F et al (2010) Serum protein profiling and proteomics in autistic spectrum disorder using magnetic bead-assisted mass spectrometry. Eur Arch Psychiatry Clin Neurosci 260(3):249–255
52. Schwarz E, Guest PC, Rahmoune H, Wang L, Levin Y, Ingudomnukul E et al (2011) Sex-specific serum biomarker patterns in adults with Asperger's syndrome. Mol Psychiatry 16(12):1213–1220

53. Shen C, Zhao Xl JW, Zou XB, Huo LR, Yan W et al (2011) A proteomic investigation of B lymphocytes in an autistic family: a pilot study of exposure to natural rubber latex (NRL) may lead to autism. J Mol Neurosci 43(3):443–452
54. Momeni N, Bergquist J, Brudin L, Behnia F, Sivberg B, Joghataei MT et al (2012) A novel blood-based biomarker for detection of autism spectrum disorders. Transl Psychiatry 2:e91. https://doi.org/10.1038/tp.2012.19
55. Ngounou Wetie AG, Wormwood KL, Russell S, Ryan JP, Darie CC, Woods AG (2015) A pilot proteomic analysis of salivary biomarkers in autism spectrum disorder. Autism Res 8(3):338–350
56. Ngounou Wetie AG, Wormwood KL, Charette L, Ryan JP, Woods AG, Darie CC (2015) Comparative two-dimensional polyacrylamide gel electrophoresis of the salivary proteome of children with autism spectrum disorder. J Cell Mol Med 19(11):2664–2678
57. Ngounou Wetie AG, Wormwood K, Thome J, Dudley E, Taurines R, Gerlach M et al (2014) A pilot proteomic study of protein markers in autism spectrum disorder. Electrophoresis 35(14):2046–2054
58. Steeb H, Ramsey JM, Guest PC, Stocki P, Cooper JD, Rahmoune H et al (2014) Serum proteomic analysis identifies sex-specific differences in lipid metabolism and inflammation profiles in adults diagnosed with Asperger syndrome. Mol Autism 5(1):4. https://doi.org/10.1186/2040-2392-5-4
59. Suganya V, Geetha A, Sujatha S (2015) Urine proteome analysis to evaluate protein biomarkers in children with autism. Clin Chim Acta 450:210–219
60. Cortelazzo A, De Felice C, Guerranti R, Signorini C, Leoncini S, Zollo G et al (2016) Expression and oxidative modifications of plasma proteins in autism spectrum disorders: interplay between inflammatory response and lipid peroxidation. Proteomics Clin Appl 10(11):1103–1112
61. Feng C, Chen Y, Pan J, Yang A, Niu L, Min J et al (2017) Redox proteomic identification of carbonylated proteins in autism plasma: insight into oxidative stress and its related biomarkers in autism. Clin Proteomics 14:2. https://doi.org/10.1186/s12014-017-9138-0
62. Qin Y, Chen Y, Yang J, Wu F, Zhao L, Yang F et al (2017) Serum glycopattern and Maackia amurensis lectin-II binding glycoproteins in autism spectrum disorder. Sci Rep 7:46041. https://doi.org/10.1038/srep46041
63. Shen L, Zhang K, Feng C, Chen Y, Li S, Iqbal J et al (2018) Itraq-based proteomic analysis reveals protein profile in plasma from children with autism. Proteomics Clin 12(3):e1700085. https://doi.org/10.1002/prca.201700085
64. Yang J, Chen Y, Xiong X, Zhou X, Han L, Ni L et al (2018) Peptidome analysis reveals novel serum biomarkers for children with autism spectrum disorder in China. Proteomics Clin Appl 13:e1700164. https://doi.org/10.1002/prca.201700164. [Epub ahead of print]
65. Chen YN, Du HY, Shi ZY, He L, He YY, Wang D (2018) Serum proteomic profiling for autism using magnetic bead-assisted matrix-assisted laser desorption ionization time-of-flight mass spectrometry: a pilot study. World J Pediatr 14(3):233–237
66. Stephan AH, Barres BA, Stevens B (2012) The complement system: an unexpected role in synaptic pruning during development and disease. Annu Rev Neurosci 35:369–389
67. Presumey J, Bialas AR, Carroll MC (2017) Complement system in neural synapse elimination in development and disease. Adv Immunol 135:53–79
68. Mead J, Ashwood P (2015) Evidence supporting an altered immune response in ASD. Immunol Lett 163(1):49–55
69. Meltzer A, Van de Water J (2017) The role of the immune system in autism spectrum disorder. Neuropsychopharmacology 42(1):284–298
70. Tamiji J, Crawford DA (2010) The neurobiology of lipid metabolism in autism spectrum disorders. Neurosignals 18(2):98–112
71. Mazahery H, Stonehouse W, Delshad M, Kruger MC, Conlon CA, Beck KL et al (2017) Relationship between long chain n-3 polyunsaturated fatty acids and autism spectrum disorder: systematic review and meta-analysis of case-control and randomised controlled trials. Nutrients 9(2). pii: E155. https://doi.org/10.3390/nu9020155

72. Junaid MA, Kowal D, Barua M, Pullarkat PS, Sklower Brooks S, Pullarkat RK (2004) Proteomic studies identified a single nucleotide polymorphism in glyoxalase I as autism susceptibility factor. Am J Med Genet A 131(1):11–17
73. Broek JA, Guest PC, Rahmoune H, Bahn S (2014) Proteomic analysis of post mortem brain tissue from autism patients: evidence for opposite changes in prefrontal cortex and cerebellum in synaptic connectivity-related proteins. Mol Autism 5:41. https://doi.org/10.1186/2040-2392-5-41
74. Aebersold R, Burlingame AL, Bradshaw RA (2013) Western blots versus selected reaction monitoring assays: time to turn the tables? Mol Cell Proteomics 12(9):2381–2382
75. Cayer DM, Nazor KL, Schork NJ (2016) Mission critical: the need for proteomics in the era of next-generation sequencing and precision medicine. Hum Mol Genet 25(R2):R182–R189
76. Wang K, Huang C, Nice E (2014) Recent advances in proteomics: towards the human proteome. Biomed Chromatogr 28(6):848–857
77. Schubert KO, Weiland F, Baune BT, Hoffmann P (2016) The use of MALDI-MSI in the investigation of psychiatric and neurodegenerative disorders: a review. Proteomics 16(11–12):1747–1758
78. Rigbolt KTG, Blagoev B (2012) Quantitative phosphoproteomics to characterize signaling networks. Semin Cell Dev Biol 23(8):863–8671
79. Ilieva M, Fex Svenningsen Å, Thorsen M, Michel TM (2018) Psychiatry in a dish: stem cells and brain organoids modeling autism spectrum disorders. Biol Psychiatry 83(7):558–568
80. Wang P, Mokhtari R, Pedrosa E, Kirschenbaum M, Bayrak C, Zheng D et al (2017) Crispr/cas9-mediated heterozygous knockout of the autism gene CHD8 and characterization of its transcriptional networks in cerebral organoids derived from IPS cells. Mol Autism 8:11. https://doi.org/10.1186/s13229-017-0124-1
81. Daimon CM, Jasien JM, Wood WH, Zhang Y, Becker KG, Silverman JL et al (2015) Hippocampal transcriptomic and proteomic alterations in the BTBR mouse model of autism spectrum disorder. Front Physiol 6:324. https://doi.org/10.3389/fphys.2015.00324
82. Wei H, Ma Y, Liu J, Ding C, Hu F, Yu L (2016) Proteomic analysis of cortical brain tissue from the BTBRmouse model of autism: evidence for changes in STOP and myelin-related proteins. Neuroscience 312:26–34
83. Bidinosti M, Botta P, Krüttner S, Proenca CC, Stoehr N, Bernhard M et al (2016) Clk2 inhibition ameliorates autistic features associated with shank3 deficiency. Science 351(6278):1199–1203
84. Niere F, Namjoshi S, Song E, Dilly GA, Schoenhard G, Zemelman BV et al (2016) Analysis of proteins that rapidly change upon mechanistic/mammalian target of rapamycin complex 1 (mTORC1) repression identifies parkinson protein 7 (PARK7) as a novel protein aberrantly expressed in tuberous sclerosis complex (TSC). Mol Cell Proteomics 15(2):426–444
85. Liao L, Park SK, Xu T, Vanderklish P, Yates JR (2008) Quantitative proteomic analysis of primary neurons reveals diverse changes in synaptic protein content in fmr1 knockout mice. Proc Natl Acad Sci U S A 105(40):15281–15286
86. Pacheco NL, Heaven MR, Holt LM, Crossman DK, Boggio KJ, Shaffer SA et al (2017) RNA sequencing and proteomics approaches reveal novel deficits in the cortex of MECP2-deficient mice, a model for RETT syndrome. Mol Autism 8:56. https://doi.org/10.1186/s13229-017-0174-4
87. Sabidó E, Selevsek N, Aebersold R (2012) Mass spectrometry-based proteomics for systems biology. Curr Opin Biotechnol 23(4):591–597

Chapter 13
Role of the Gut Microbiome in Autism Spectrum Disorders

Joby Pulikkan, Agnisrota Mazumder, and Tony Grace

Abstract Autism spectrum disorder (ASD) is a severe neurodevelopmental or neuropsychiatric disorder with elusive etiology and obscure pathophysiology. Cognitive inabilities, impaired communication, repetitive behavior pattern, and restricted social interaction and communication lead to a debilitating situation in autism. The pattern of co-occurrence of medical comorbidities is most intriguing in autism, compared to any other neurodevelopmental disorders. They have an elevated comorbidity burden among which most frequently are seizures, psychiatric illness, and gastrointestinal disorders. The gut microbiota is believed to play a pivotal role in human health and disease through involvement in physiological homoeostasis, immunological development, glutathione metabolism, amino acid metabolism, etc., which in a reasonable way explain the role of gut-brain axis in autism. Branded as a neurodevelopmental disorder with psychiatric impairment and often misclassified as a mental disorder, many experts in the field think that a therapeutic solution to autism is unlikely to emerge. As the pathophysiology is still elusive, taking into account of the various symptoms that are concurrent in autism is important. Gastrointestinal problems that are seen associated with most of the autism cases suggest that it is not just a psychiatric disorder as many claim but have a physiological base, and alleviating the gastrointestinal problems could help alleviating the symptoms by bringing out the much needed overall improvement in the affected victims. A gut disorder akin to Crohn's disease is, sometimes, reported in autistic children, an extremely painful gastrointestinal disease which is named as autistic enterocolitis. This disturbed situation hypothesized to be initiated by dysbiosis or microbial imbalance could in turn perturb the coordination of microbiota-gut-brain axis which is important in human mental health as goes the popular dictum: "fix your gut, fix your brain."

J. Pulikkan · A. Mazumder
Department of Genomics, Central University of Kerala, Kasaragod, Kerala, India

T. Grace (✉)
Department of Genomics, Central University of Kerala, Kasaragod, Kerala, India

Division of Biology, Kansas State University, Manhattan, KS, USA
e-mail: tonygrace@cukerala.ac.in; tonygrac@ksu.edu

P. C. Guest (ed.), *Reviews on Biomarker Studies in Psychiatric and Neurodegenerative Disorders*, Advances in Experimental Medicine and Biology 1118,
https://doi.org/10.1007/978-3-030-05542-4_13

Keywords Autism spectrum disorder (ASD) · Gastrointestinal problems · Microbial dysbiosis · Gut-brain axis

13.1 Introduction

Life will not be possible as we know it today without microbes. Hitherto unknown and unpredicted, the role of gut microbiome in human health and disease is being explored eloquently with the help of next-generation techniques and allied improvements. Coined by Joshua Lederberg, the term microbiome signifies the natural community of microorganisms that factually share our body space and can be categorized as commensal, symbiotic, or pathogenic and have been all but ignored as determinants of health and disease [1]. Gone are the times when the role of microbes in human health was just focused on identifying and treating pathogens in patients, usually with antibiotics. Having colonized by trillions of microorganisms [2], these microbial communities outnumber the total number of human cells by a factor of 10 and contain 150 times as many genes as those encoded by the human genome [3], suggesting us to consider human gut microbiota as an organ or superorganism that coevolved with the human host to achieve a beneficial symbiotic relationship [4]. The term "superorganism" would refer to a collection of individuals which behave as a single unit with enhanced function [4]. The human host is immensely benefitted by the gut microbiota, attested by the fact that there is a plethora of interactions happening between them throughout the development [4] and its disruption could lead to manifold negative consequences [5]. The major role of these microbiota can be defined in terms of a metabolic one. A marvelous metabolic capacity equivalent to that of the liver is envisioned by the combined metabolic capacity of the microbiota, justifying their portrayal as an additional human organ [6] and revealing the exemplary host-microbiota mutualism in the gut. Their supreme role in the maintenance of host health includes facilitating digestion of otherwise indigestible dietary fibers into readily absorbable short-chain fatty acids (SCFAs), synthesis of vitamins and other beneficial metabolites, immune system regulation, detoxification of harmful substances, and enhanced resistance against colonization by pathogenic microorganisms [5]. The defense against pathogen colonization is achieved by nutrient competition and production of antimicrobial activities [7]. Their role in fortification of intestinal epithelial barrier and their ability to induce secretory IgA to limit penetration of bacteria into tissues are not to be overlooked. Researches on germ-free mice reveal that absence of microbes causes significant immune defects at structure levels like decreased Peyer's patches, lamina propria, and isolated lymphoid follicles. At cellular levels, defects like decreased intestinal $CD8^+$ T cells and $CD4^+$ T helper 17 cells and reduced B-cell production of secretory IgA have been reported [8] (Fig. 13.1).

The microbiome at a specific niche is believed to cause local as well as systemic effects on host biology [7]. Chronic low-grade intestinal inflammation seen in irritable bowel syndrome (IBS) [9] and intense intestinal autoimmunity observed in the inflammatory bowel disease (IBD) [8] are examples of local effects of the

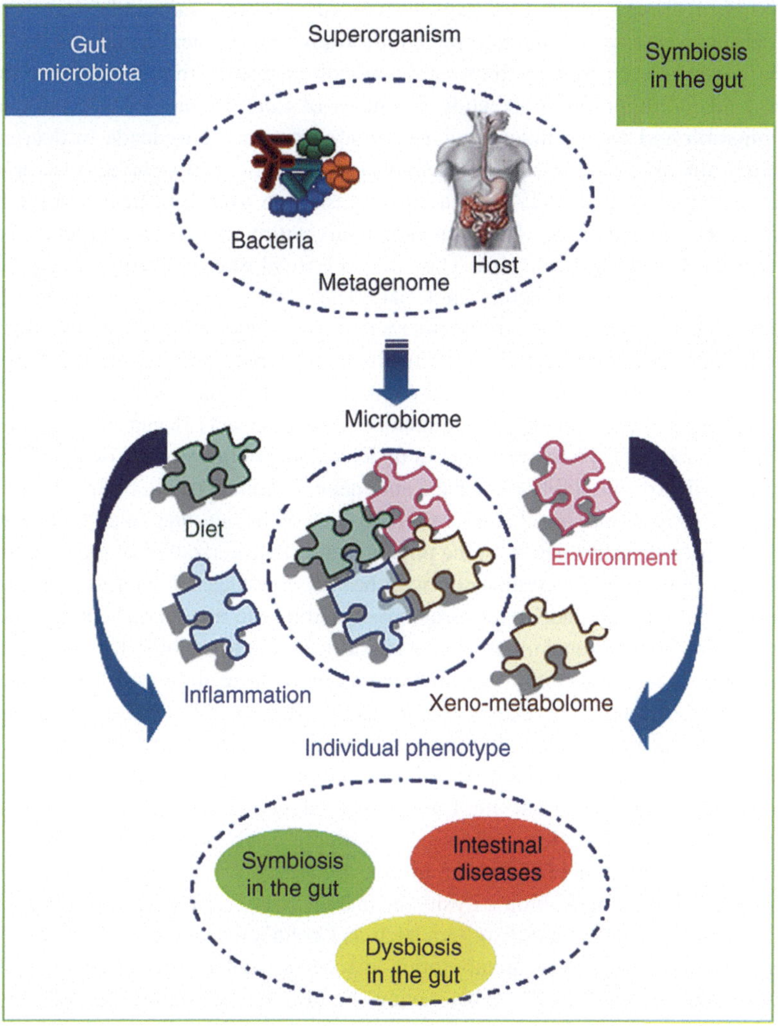

Fig. 13.1 The human gut microbiota as a superorganism. The largest microbiome is accredited as that seen in our gastrointestinal tract. It is influenced by several external factors, such as diet, inflammation stage, environment, and xeno-metabolome. Each microbiome refers an individual phenotype, able to describe symbiosis-, dysbiosis-, and disease-related gut conditions (Adapted from Putignani et al. (2014) Pediatr Res 76:2–10)

microbiome on host biology. The systemic effect of microbiome on host biology is evident from the fact that the gut microbiome contributes to the etiology of experimental disease models affecting remote organ systems. Immune cells stimulated at the intestinal site, like microbe-sensing antigen processing cells (APCs) and adaptive immune cells, are trafficked to distal tissue sites by systemic diffusion [7].

The bidirectional interaction between the gut-intestine tract and brain via the microbiome is crucial for maintaining equilibrium between health and disease. The normal microbiome promotes increased metabolic capacity, immune system maturation, and SCFA production, while dysbiosis of microbiome leads to increased inflammation and overabundance of *Enterobacteriaceae*. This leads to decreased intestinal mucus, immune cell differentiation, gut-associated lymphoid tissue and metabolic capacity, and SCFA production, causing an overall ill health in the host [4]. Recent studies indicate that microorganisms within and among the body habitats exhibit intricate relationships. They play a critical role in driving physical factors such as oxygen, moisture, and pH. They, also, play a role in the host immunological regulation microbial interactions. All these point out the fact that the microbiome plays more crucial role in health and disease than expected so far [10].

The crucial intervention by intestinal communities implicated in diseases like allergies [11], late-onset autism [12], inflammatory bowel disease [13], cancer [14], obesity, and type 2 diabetes [15] reinstates that our understanding of microbial ecology will have a direct bearing on our ability to manage and maintain human health in the future. The major move toward studying changes in composition of the intestinal microbiota in relation to diseases relies primarily on the phylogenetic characterization of the microbiota of diseased individuals in comparison with healthy candidates. Given the substantial interindividual and intraindividual variations in addition to age-related changes in the composition of the intestinal microbiota, it is a difficult task to establish the presence and relative abundance of specific microbial communities in relation with human health status. However, options are realistic to use specific changes in compositional and functional diversity of microbiota as biomarkers for health or specific diseases.

The gut microbiota is a chief contributor in maintaining normal physiology and energy production throughout life. Energy-dependent processes like body temperature regulation, reproduction, and tissue growth are partly influenced by gut microbial energy production [16] further giving evidences for the importance of gut microbiota. The gut microbiome dysbiosis is believed to have a negative impact on gastrointestinal tract-related disorders such as Crohn's disease and ulcerative colitis, systemic diseases like metabolomic disorders, and central nervous system (CNS)-related disorders [17]. This review highlights the importance of the gut microbiome in autism which is a non-immune-mediated CNS disorder in which the role of gut microbiome has been implicated in its exacerbation.

13.2 Plausible Mechanisms Involved in Gut-Brain Axis Through Gut Microbiome Intervention: Bidirectional Communication

Gut-brain axis (GBA) can be defined as a physiological framework in which the gut microbiota communicates with the CNS and vice versa through neural, endocrine, and immune pathways. The communication being bidirectional, the role of the brain on the microbial content of the gut and how the intestinal microbiota influences the brain and behavior solicit equal attention. Homeostases of several systems like gastrointestinal tract (GI) function, appetite, and weight control are a few major factors that are significantly affected by the GBA. The microbial habitat is determined by GI motility as well as epithelial functions. Since these factors can be regulated by the CNS, any changes induced by the CNS can result in alteration and perturbation of intestinal microbiota leading to several pathological situations [9].

The enteric microbiota can assert their influence on gut homeostasis by means of regulation of bowel motility, modulation of intestine pain, immune responses, and nutrient processing [18]. Mounting evidences indicate that gut microbiome does affect the development and regulation of hypothalamic-pituitary-adrenal axis (HPA) and behavior [19]. Characterization of the bidirectional interactions between the CNS, the enteric nervous system (ENS), and the GI tract has brought about results which convincingly suggest the role of the gut microbiota in these gut-brain interactions. This type of bidirectional communication which mediates the symbiotic and pathogenic relationships between the bacteria and the mammalian host is called "microbial endocrinology" [20] or "interkingdom signaling" [21]. The gut microbiota appears to influence the development of emotional behavior, stress and pain modulation systems, and brain neurotransmitter systems. Some of the recent researches explain the multiple mechanisms involved. Endocrine and neurocrine pathways facilitate gut microbiota-to-brain signaling, and the brain, in turn, alters microbial composition and behavior through the autonomic nervous system with the active involvement of immune and humoral system [22].

The enteric microbiota does assert a vital impact on "gut-brain axis" (GBA), interacting locally with intestinal cells and the ENS and directly with the CNS through neuroendocrine and metabolic pathways (Fig. 13.2). To enumerate the main assistances rendered by bacterial colonization in terms of their absence or presence would include development and maturation of both the ENS and CNS [23], altered expression and turnover of neurotransmitters in both nervous systems [24], alterations of gut sensory-motor functions leading to delayed gastric emptying and intestinal transit [25], and enlarged cecal size [26]. The neuromuscular abnormalities caused via the intervention of microbiota results in downregulation of gene expression of enzymes related to the synthesis and transport of neurotransmitters and muscular contractile proteins [27]. These aberrations have been seen to be normalized once the experimental animals are colonized with specific bacterial species.

It has been observed that germ-free (GF) mice generally show decreased anxiety and an increased stress response. The organism with enhanced levels of ACTH and cortisol demonstrates that microbiota influences stress reactivity and anxiety-like behavior. The microbiota also regulates the set point for HPA activity and modulates the serotoninergic system. GF animals are also known to suffer from memory dysfunction with an altered expression of brain-derived neurotrophic factor (BDNF) [28].

The main principal mechanisms by which the bidirectional brain-gut-microbiota axis works have been deduced. The gut microbiota asserts its role over the brain through its influence on production, expression, and turnover of neurotransmitters

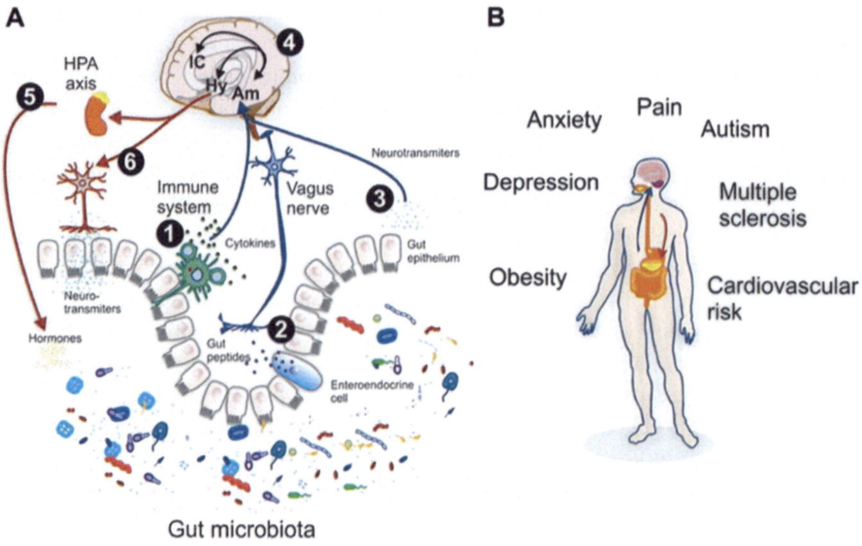

Fig. 13.2 Microbiota-gut-brain (MGB) axis. Pathways either directly or indirectly influence the bidirectional interactions between the gut microbiota and the central nervous system (CNS), involving endocrine, immune, and neural pathways (Courtesy: Augusto et al. (2013) Front Integr Neurosci 7:70)

like serotonin and gamma-aminobutyric acid (GABA) and neurotrophic factor (BDNF), protection of the intestinal barrier and tight junction integrity, modulation of enteric sensory afferents, production of various bacterial metabolites, and mucosal immune regulation, while the brain claims its influence over microbiota through their capacity to bring about alterations in mucous and biofilm production, alteration in motility, alteration of intestinal permeability, and alteration in immune function [28]. The decisive role of the gut microbiota in the production of biologically active, free catecholamines in the gut lumen of mice is established. Catecholamines are crucial in regulation of various types of body functions such as cognitive abilities, mood, and gut motility. The principal catecholamines like norepinephrine (NE) and dopamine (DA) are utilized in the CNS and peripheral nervous systems [29]. A comparative study between specific pathogen-free mice (SPF-M) and germ-free mice (GF-M) showed that substantial levels of free dopamine and norepinephrine were available in the gut lumen of SPF-M. Also the available catecholamine levels were found to be in a biologically inactive, conjugative form in GF-M. When GF-M was introduced with *Clostridium* species or SPF fecal flora, which are known to have abundant β-glucuronidase activity, drastic elevation of free catecholamines was visible indicating the role of gut microbiota and in particular the need of β-glucuronidase activity in these organisms [30]. The role of gut microbiota in regulation of sympathetic nervous system (SNS), a component in bidirectional signaling, has also been reported to take place with the help of short-chain fatty acids and ketones produced by microbiota and their promotion of sympathetic outflow via G protein-coupled receptor 41 (GPR41) [31]. The effect of short-chain fatty acids on enterochromaffin cells promotes colonic serotonin production [32] and influences memory and learning process [33]. Another plausible mechanism of microbiota-GBA interaction could be through the release of biologically active peptides from enteroendocrine cells which in turn is regulated by the nutrient availability of the microbiota [34].

When microbiota influences the brain in ways mentioned above, the GBA being a bidirectional signaling pathway, it should be an unavoidable occurrence that the brain dictates or influences microbiota composition and function through various other mechanisms. One major mechanism by which the brains assert its impact on the microbial population is exposure to stressors. A pyrosequencing study carried out to assess the effects of a single 2-hour exposure to a social stressor made a substantial alteration on colonic mucosa-associated microbial profiles of C57BL/6 mice, significantly reducing the relative proportions of the two genera and one family of highly abundant intestinal bacteria, including the genus *Lactobacillus* [35]. This can happen either directly via host-enteric microbiota signaling or indirectly via changes in the intestinal milieu. A more direct and evident influence on microbiota is facilitated by secretion of signaling molecules by neurons, immune cells, and enterochromaffin cells, under the regulation of the brain. As there are neurotransmitter receptors on bacteria, binding of neurotransmitters is capable of influencing the function of components of the microbiota that have influence on inflammatory and infection stimuli [21]. An example in this regard can be found in *Pseudomonas fluorescens* which is able to produce GABA and express GABA-binding proteins, and they also increase their necrotic-like activity on eukaryotic glial cells. This particular study also proved that GABA can regulate the lipopolysaccharide (LPS) structure and cytotoxicity in specific strains of *P. fluorescens* [36]. Another interesting dimension is the microbial biofilms in which individual groups of bacteria are found to occupy different microhabitats and metabolic niches in the human gastrointestinal tract. The prominent role of the brain in modulation of gut functions, such as motility; secretion of acids, bicarbonates, and mucus; intestinal fluid handling; and mucosal response, is important for the maintenance of the mucus layer. The individual groups of bacteria exist in different microhabitats, and metabolic niches as biofilms in these areas are also modulated by this brain function [37]. The brain affects the microbiota through various stresses induced mainly by bringing about variation in size and quality of mucus secretion [38], by delaying the recovery of the migrating motor complex pattern and inducing a transient slowing of gastric emptying [39], by increasing the frequency of cecocolonic spike-burst activity through the central release of corticotrophin-releasing factor [40]. This causes regional and global changes in gastrointestinal transit which has a profound effect on the way nutrients are delivered to the enteric microbiota. The composition and biomass of the enteric microbiota get modulated by different types of psychological stressors both in adult [41] and newborn animals [42]. When postnatal stress was induced in animal models by separating the mothers, reduction in lactobacilli was observed, disrupting the integrity of the intestinal microbiota with the appearance of stress-indicative behaviors [42]. The animals in study were also more susceptible to opportunistic infection compared to unstressed control animals.

A microbial connection has been linked to anxiety and stress, the common forms of mood disorders. Germ-free mice exhibit increased motor activity and reduced anxiety, suggesting that gut-associated pathogens can exacerbate anxiety. Decreased neurotransmitter receptors and increased tryptophan metabolism observed in the GF condition tend to suggest that the gut microbiome regulates the set point for the HPA axis [24]. When rodents were infected with the food-borne pathogen *Campylobacter jejuni*, rodents showed perceptible anxiety-like behavior. Increased c-Fos expression as a result of infection with *Campylobacter jejuni* underpins the notion that gut-associated pathogens can intensify anxiety [43]. While *Citrobacter rodentium* followed a mechanism similar to *Campylobacter jejuni* [44], *Trichuris muris* elevated anxiety via immunological and metabolic mechanisms [45]. The anxiolytic properties exhibited by the specific species of the *Lactobacillus* and *Bifidobacterium* genera are almost in contrast to the pathogenic bacteria [46].

13.3 Autism and Gut Microbiome

Autism is a non-immune-mediated CNS disorder in which the role of gut microbiome has been implicated in its exacerbation. Among many non-immune-mediated CNS disorders, autism stands out due to its prevalence and bewildering increase in epidemic levels throughout the world. As per Centers for Disease Control and Prevention data, 1 in 500 to 1 in 166 children has an autism spectrum disorder. Various physiological parameters observed in autism include autoimmune reactions, food reactions, diagnostic connection of upper GI disease, abnormal stools, autistic enterocolitis, leaky gut syndrome, excessive inflammation, aberrant glutathione levels, irregular metal, mineral levels, etc. [47], all supporting the importance of the gut microbiome. The aforementioned parameters are either influenced by microbiota or vice versa, revealing the tantalizing link between autism and gut microbiome.

Seemingly pervasive, autism spectrum disorders (ASD) are characterized by impaired communication, difficulty in social interactions, and stereotyped behavioral patterns [48]. People with autism most frequently tend to display unusual ways of learning, paying attention, and reacting to different sensations. Their imaginative skills are also known to suffer [49]. Coupling genetic predisposition with environmental factors seems to raise the risk involved in autistic children.

Considered as a psychiatric disorder, many physical symptoms common in patients with autism have largely gone unnoticed or even ignored by the medical establishment, and coinciding GI symptoms are a major example. Anecdotal reports as well as various studies indicate that gastrointestinal problems are frequently associated with autism [50] and presumed to have a correlation with autism severity [48]. Various independent studies have verified that the GI dysfunction in these children includes diarrhea, unformed stools, constipation, bloating, flatulence, etc.

[51, 52]. It was Wakefield and colleagues at the Royal Free Hospital in London who described a new variant of inflammatory bowel disease, named as autistic enterocolitis which is characterized by chronic patchy inflammation and lymphonodular hyperplasia in the terminal ileum or in the colon [53]. Intestinal permeability is another grave issue faced by children with autism, suggesting a leaky gut which can lead to neurological disability as these children are forced to absorb neurotoxic molecules across a gut membrane damaged by inflammation [54]. An abnormal level of intestinal permeability has also been documented in independently carried out studies [55, 56]. A leaky gut allows molecules to enter the bloodstream, otherwise kept at bay. Immune activation, tissue damage, and effects on the brain, including damage to brain tissue, are a few corollary problems that could emerge over time. The opioid peptides that are produced from certain diets can disrupt the normal neurotransmitter function in the brain, causing certain typical behavioral patterns observed in children with autism such as decreased socialization, decreased response to pain, abnormal language, and self-abusive or repetitive behaviors. Direct effects on the neuronal structure of the brain tissue and on the immune system by these opiate-like molecules cannot be ruled out. We need to investigate if neurotoxic and cytotoxic molecules produced by microorganisms can contribute to intensify the abovementioned situations in *autistic enterocolitis* and leaky gut and their possibility to assert greater damage by working independently or in tandem with each other.

Studies undertaken at different research labs indicate a link between the gut microbiome and ASD, although a direct causality or indirect consequence of atypical patterns of feeding and nutrition has yet to be proved. Children with ASD tend to show food selectivity with strong preferences for starches, snack, and processed food, while most tend to reject fruits, vegetables, and/or proteins [57, 58]. A chronic imbalance of gut microbiota known as intestinal dysbiosis is suspected in children with autism, and many investigators have found evidence of this imbalance in autistic patients [12]. However, most of the gut microorganisms are beneficial to host, and dozens of species of pathogenic organisms, if allowed to thrive due to the compromised immune system, can cause disease. Apart from causing local effects on gut tissue, abnormal bacteria can have an effect on the brain. The toxins that are produced by the harmful bacteria are not properly metabolized. They can build up in the brain by way of the bloodstream, resulting in confusion, delirium, and even coma. The GI inflammation and abnormal immune functions observed in children with autism may increase the abnormal levels of harmful bowel organisms, and metabolites produced by the harmful bacteria can create havoc intensifying GI inflammation, gut permeability, and abnormal immune functions. Yeasts also seem to play a negative role in vulnerable children, as yeasts are known producers of chemicals that have neurological effects and children with autism indicate elevated levels of chemicals that are found in yeasts [59].

A succession of microbial consortia studies in infants has proven that the microbiome gets enriched as associated with life events and as per the diet being introduced. A healthy microbiome is observed to be capable of assisting in the breakdown of complex plant polysaccharides, promoting digestion and overall host health [60].

Therefore, it could be predicted that deviations in the establishment and maintenance of the gut microbiome could lead to pain and discomfort. This hypothesis has been corroborated by William et al. whose study revealed that children with ASD tend to have less of *Bacteroidetes* which play a prominent role in the digestion of polysaccharides. This study could also underpin the suggestion that children with autism have anomalous carbohydrate digestion capacity and mucosal dysbiosis in the intestines. Metagenomic analysis revealed that they have a decrease in *Bacteroidetes*, an increase in the ratio of *Firmicutes* to *Bacteroidetes*, and an increase in *Betaproteobacteria* [49]. A pyrosequencing study of fecal microflora of children with autism showed significant differences in the *Actinobacteria* and *Proteobacterium* phyla in comparison with healthy controls. The same study also showed higher prevalence of *Desulfovibrio* species and *Bacteroides vulgatus* in stools of severely autistic children [12]. Basic anaerobic culturing techniques to count and isolate microorganisms, followed by polymerase chain reaction (PCR) targeting the 16S rDNA in the isolates cultivated in children with late-onset autism against neurotypical children, showed that the number and type of *Clostridium* and *Ruminococcus* species significantly differed from normal children [12]. A follow-up study by Song et al. using quantitative real-time PCR found that *Clostridium* cluster groups I and XI and *Clostridium bolteae* were significantly higher in children with autism [61]. Another culture-independent study by Parracho et al. using fluorescence in situ hybridization (FISH) reported elevated levels of *Clostridium hystolyticum* in the ASD children compared to typical children [49]. Disruption of gut microbiota might contribute for the over-colonization of neurotoxin-producing bacteria, exacerbating autistic symptoms. *Clostridium tetani* is looked upon as an organism that could induce autism [62]. *Clostridia* are known propionate producers [63], and the property of propionate to bring upon neurological effects in rats [64] has led to the hypothesis that incidence of autism is related to extensive exposure to *Clostridium* spores. A study into investigation of prevalence of four types of beneficial bacteria including *Bifidobacteria*, *Lactobacillus* spp., *E.coli*, and *Enterococcus* revealed that the children with autism had much lower levels of *Bifidobacterium*, slightly lower levels of *Enterococcus*, and much higher levels of *Lactobacillus* [65]. Given that all *Lactobacillus* strains are beneficial, their higher levels seem to be paradoxical and need to be understood. With regard to commensal bacteria which are neither beneficial nor harmful, the same study found that the autism group was more likely to have *Bacillus* spp. and less likely to have *Klebsiella oxytoca*. However, a significant piece of information procured by the same study was the finding of significantly lower levels of short-chain fatty acids observed in autistic children. This was thought to occur due to lower saccharolytic fermentation by beneficial bacteria further substantiating the suspected link between autism and gut microbiome [50].

The finding of distinctive gut microbes associated with ASD was brought about in a small pilot study using a high-throughput sequencing of the 16S rDNA gene by Kang et al. An overall less diverse gut microbiome with a lower abundance of the bacterial genera *Prevotella*, *Coprococcus*, and unclassified *Veillonellaceae* was reported in this study. The said organisms are versatile carbohydrate-degrading and/or fermenting bacteria, and the changes in the spectrum of metabolites produced from a given diet could be greatly influenced by them [66]. The possibility cannot be ruled out that the differences in microbiota composition in ASD may have negatively influenced the microbial interactions so as to result in a decreased overall diversity and reduced function. Another intriguing observation is that *Prevotella* is a genus highly enriched in populations of agrarian societies, and its depletion in ASD shall be probed if certain environmental factors like industrialization contribute to the prevalence of autism. Individuals with untreated HIV infection are reported to have uniformly high *Prevotella* [67], and a reverse pattern is observed in ASD. The rationale behind this observation can be explained as HIV patients with a suppressed adaptive immune system tend to have a higher number of *Prevotella*, whereas those with ASD have a hyperactive adaptive immune system with less *Prevotella* occurrence.

The metabolites produced by various microorganisms could be performing the same or different functions affecting the system in manifold ways, depending upon their differential abundance. It has been proven that differential abundance of bacteria-produced metabolites has the potential to directly affect neural processes. Increased urinary excretion of an abnormal phenylalanine metabolite of *Clostridia* species, namely, 3-(3-hydroxyphenyl)-3-hydroxypropionic acid (HPHPA), has been confirmed in urine samples from patients with autism. This tyrosine analogue is thought to be responsible for depletion of catecholamines, and, thus, it is believed to be a chief contributor in the exacerbation of typical autistic symptoms like stereotypical behavior, hyperactivity, and hyper-reactivity in experimental animals [59]. P-Cresol is another microbial metabolite which can initiate damage in cases of autism as they compete with neurotransmitters for enzymes and cofactors needed for sulfonation reactions in the liver [68]. A recent study using a maternal immune activation (MIA) model of ASD in mice showed a significant increase in 4-ethylphenylsulfate (4EPS), a metabolite produced by gut bacteria [69]. Bacterial tag-encoded FLX-titanium amplicon pyrosequencing carried out by De Angelis et al. showed that the highest microbial diversity presented with autism. The same study also identified higher abundance of Caloramator, *Sarcina* and *Clostridium* genera in autistic children. *Alistipes* and *Akkermansia* species were higher in autism along with almost all of the identified *Sutterellaceae* and *Enterobacteriaceae*. Concomitantly, the levels of free amino acids and volatile organic compounds of fecal samples were markedly different in autism [70]. Notably, lower levels of *Prevotella*, *Coprococcus*, and unclassified *Veillonellaceae* were observed by Kang et al. in autistic children in a study carried out using 16 s rRNA gene pyrosequencing analysis from fecal DNA samples [66].

13.3.1 Mineral Elements and Gut Microbiome

Higher average levels of several toxic metals are evident in autism severity. Lead, mercury (Hg), arsenic (As) [71], thallium, tin, and tungsten [72] are among a few of the metals which correlate with autism severity. The role of environmental pollutants such as these heavy metals in the alteration of physiological functions causing detrimental effects on health has been established. Recognized as neurotoxicants with known effects on neurodevelopment, their role in microbial dysbiosis is inferred adding to the exacerbation of autism symptoms. Although, the exact mechanism has not been deduced, the essential role of the intestinal microbiome in limiting the heavy metal body burden has been established using GF mouse studies [73]. An interesting conclusion of this particular study was that genes relevant for divalent metal transporters and oxidative pathways were expressed with significant differences depending on the microbial status of the animal along with the dose and type of metals present, suggesting the complex host-microbe interplay induced by the environmental pollutants inside the gut. The ability of metals such as Hg and As to exert toxic effects on human health has been well characterized. The volatile derivatives of these metals interact directly with host cells causing irreversible damage and aggravate the diseased state by disturbing the physiological microflora [74]. The critical role of gut microbiota in intestinal homeostasis is characterized by the fact that different types of dysbiosis cause diseases outside and inside the intestine. A study by Breton et al. proved that oral exposure to heavy metals does lead to specific changes in bacterial commensal communities. Their study showed that, relative to the control animals, test animals had notably lower numbers of *Lachnospiraceae* and higher numbers of *Lactobacillaceae* and *Erysipelotrichaceae* [73].

Chronic dietary depletion of elements like iron (Fe) [75] and zinc (Zn) [76] is found to induce significant taxonomic alterations in the gut microbial profile. Understanding the effects of Zn deficiency on the host may help to elucidate the influence of gastrointestinal microbiota on physiology from a novel perspective. The need of Zn, almost in double the amount, in conventionally raised mice against their germ-free counterparts as indicated in studies points out the role of the host microbiota in Zn homeostasis [77]. It has also been seen that optimal levels of Zn administration on various animal models had the benefits of increased concentration of short-chain fatty acids (SCFAs) [78], increased overall bacterial species richness and diversity [79], and favorable change in metabolic activity [80]. Studies have also revealed that gut microbial diversity of Zn-deficient organisms bear a resemblance to that of GI microbiota in pathological states. The mentioned study also suggested that chronic Zn deficiency may reshape the gut microbiome. The metagenome predictive analysis showed that cecal microbiome metabolism was perturbed in Zn-deficient organisms since aberrant pathways involving lipid metabolism, carbohydrate digestion, and mineral absorption were prominent [76].

With practical concern for human health, Fe is also an important trace element to study. Low Fe conditions are seen to cause a decrease in *Roseburia* spp. and an increase in *Lactobacillus* spp., along with a parallel decrease in butyrate and propionate [75]. Thus low Fe conditions may contribute to weaken the barrier effect of microbiota by strong dysbiosis and decreased production of beneficial metabolites.

13.3.2 Glutathione Metabolism and Gut Microbiota

One of the most important molecules, glutathione, is popularly called the mother of all antioxidants. This prototype antioxidant is capable of protecting cells from the harmful effects of oxidative stress and can act as a defensive agent against toxic xenobiotics [81]. Although the role of gut microbiota and dysbiosis is inferred as a cause for many pathological situations, the mechanistic insight into how the specific microbial populations lead to the progression of such disorders has not been studied extensively. It has been assumed that microbiota in the small intestine consumes glycine as well as other amino acids to support its growth and survival, curtailing the availability to the host and thus causing decreased levels of the amino acids and glutathione metabolism [82]. This could underlie the mechanism of how the gut microbiome plays a pivotal role in the exacerbation of certain metabolism-related disorders.

13.4 Conclusions

The realization is that the microbes that live inside and on us outnumber our somatic and germ cells by an estimated tenfold. This has given them the status of a supraorganism with their capacity of providing traits in human beings and has significantly increased the importance of gut microbiome in health and disease. This concept is expected to lead to a paradigm shift in the strategies involved in diagnosis, prognosis, and treatment of a few disorders involving the gut-brain axis in which autism claims a position. Without a cure so far and with limited knowledge on the etiology of the disorder, autism is a topic that needs immediate attention from the researchers. As some of the studies indicate that the gut microbiome modulates the glutathione and certain minerals like zinc, copper, and iron in the experimental organism, studies on the gut microbiome in autism can give some serendipitous insights into the etiology, diagnosis, and treatment of this condition.

References

1. Lederberg J, McCray AT (2001) Ome SweetOmics--A Genealogical Treasury of Words. The Scientist 15(7):8–8
2. Turnbaugh PJ, Gordon JI (2009) The core gut microbiome, energy balance and obesity. J Physiol 587(17):4153–4158
3. Qin J, Li R, Raes J, Arumugam M, Burgdorf KS, Manichanh C et al (2010) A human gut microbial gene catalogue established by metagenomic sequencing. Nature 464(7285):59–65
4. Glendinning L, Free A (2013) Supra-organismal interactions in the human intestine. Front Cell Infect Microbiol 4:47–47
5. Walker AW, Lawley TD (2013) Therapeutic modulation of intestinal dysbiosis. Pharmacol Res 9(1):75–86

6. Sommer F, Bäckhed F (2013) The gut microbiota—masters of host development and physiology. Nat Rev Microbiol 11(4):227–238
7. Wang Y, Kasper LH (2014) The role of microbiome in central nervous system disorders. Brain Behav Immun 38:1–12
8. Round JL, Mazmanian SK (2009) The gut microbiota shapes intestinal immune responses during health and disease. Nat Rev Immun 9(5):313–323
9. Collins SM, Bercik P (2009) The relationship between intestinal microbiota and the central nervous system in normal gastrointestinal function and disease. Gastroenterology 136(6):2003–2014
10. Faust K, Sathirapongsasuti JF, Izard J, Segata N, Gevers D, Raes J et al (2012) Microbial co-occurrence relationships in the human microbiome. PLoS Comput Biol 8(7):e1002606. https://doi.org/10.1371/journal.pcbi.1002606
11. MacDonald TT, Monteleone G (2005) Immunity inflammation and allergy in the gut. Science 307(5717):1920–1925
12. Finegold SM, Molitoris D, Song Y, Liu C, Vaisanen ML, Bolte E et al (2002) Gastrointestinal microflora studies in late-onset autism. Clin Infect Dis 35(Supplement 1):S6–S16
13. Hume G, Radford-Smith GL (2002) The pathogenesis of Crohn's disease in the 21st century. Pathology-Abingdon 34(6):561–567
14. McGarr SE, Ridlon JM, Hylemon PB (2005) Diet anaerobic bacterial metabolism and colon cancer: a review of the literature. J Clin Gastroenterol 39(2):98–109
15. Hartstra AV, Bouter KE, Bäckhed F, Nieuwdorp M (2015) Insights into the role of the microbiome in obesity and type 2 diabetes. Diabetes Care 38(1):159–165
16. Nicholson JK, Holmes E, Kinross J, Burcelin R, Gibson G, Jia W et al (2012) Host-gut microbiota metabolic interactions. Science 336(6086):1262–1267
17. Carding S, Verbeke K, Vipond DT, Corfe BM, Owen L (2015) Dysbiosis of the gut microbiota in disease. Microb Ecol Health Dis 26. https://doi.org/10.3402/mehd.v26.26191
18. Rhee SH, Pothoulakis C, Mayer EA (2009) Principles and clinical implications of the brain–gut–enteric microbiota axis. Nat Rev Gastroenterol Hepatol 6(5):306–314
19. Sudo N (2014) Microbiome HPA axis and production of endocrine hormones in the gut. Adv Exp Med Biol 817:177–194
20. Lyte M (2004) Microbial endocrinology and infectious disease in the 21st century. Trends Microbiol 12(1):14–20
21. Hughes DT, Sperandio V (2008) Inter-kingdom signalling: communication between bacteria and their hosts. Nat Rev Microbiol 6(2):111–120
22. Mayer EA, Tillisch K, Gupta A (2015) Gut/brain axis and the microbiota. J Clin Invest 125(3):926–938
23. Barbara G, Stanghellini V, Brandi G, Cremon C, Di Nardo G, De Giorgio R et al (2005) Interactions between commensal bacteria and gut sensorimotor function in health and disease. Am J Gastroenterol 100(11):2560–2568
24. Clarke G, Grenham S, Scully P, Fitzgeral P, Moloney RD, Shanahan F et al (2013) The microbiome-gut-brain axis during early life regulates the hippocampal serotonergic system in a sex-dependent manner. Mol Psychiatry 18(6):666–673
25. Abrams GD, Bishop JE (1967) Effect of the normal microbial flora on gastrointestinal motility. Proc Soc Exp Biol Med 126(1):301–304
26. Iwai H, Ishihara Y, Yamanaka J, Ito T (1973) Effects of bacterial flora on cecal size and transit rate of intestinal contents in mice. Jpn J Exp Med 43(4):297–305
27. Hooper LV, Wong MH, Thelin A, Hansson L, Falk PG, Gordon JI (2001) Molecular analysis of commensal host-microbial relationships in the intestine. Science 291(5505):881–884
28. Carabotti M, Scirocco A, Maselli MA, Severi C (2015) The gut-brain axis: interactions between enteric microbiota, central and enteric nervous systems. Ann Gastroenterol 28(2):203–209
29. Eisenhofer G, Kopin IJ, Goldstein DS (2004) Catecholamine metabolism: a contemporary view with implications for physiology and medicine. Pharmacol Rev 56(3):331–349

30. Asano Y, Hiramoto T, Nishino R, Aiba Y, Kimura T, Yoshihara K (2012) Critical role of gut microbiota in the production of biologically active, free catecholamines in the gut lumen of mice. Am J Physiol-Gastrointest Liver Physiol 303(11):G1288–G1295
31. Kimura I, Inoue D, Maeda T, Hara T, Ichimura A, Miyauchi S et al (2011) Short-chain fatty acids and ketones directly regulate sympathetic nervous system via G protein-coupled receptor 41 (GPR41). Proc Natl Acad Sci U S A 108(19):8030–8035
32. Reigstad CS, Salmonson CE, Rainey JF, Szurszewski JH, Linden DR, Sonnenburg JL et al (2015) Gut microbes promote colonic serotonin production through an effect of short-chain fatty acids on enterochromaffin cells. FASEB J 29(4):1395–1403
33. Stefanko DP, Barrett RM, Ly AR, Reolon GK, Wood MA (2009) Modulation of long-term memory for object recognition via HDAC inhibition. Proc Natl Acad Sci U S A 106(23):9447–9452
34. Uribe A, Alam M, Johansson O, Midtvedt T, Theodorsson E (1994) Microflora modulates endocrine cells in the gastrointestinal mucosa of the rat. Gastroenterology 107(5):1259–1269
35. Galley JD, Nelson MC, Yu Z, Dowd SE, Walter J, Kumar PS et al (2014) Exposure to a social stressor disrupts the community structure of the colonic mucosa-associated microbiota. BMC Microbiol 14(1):189. https://doi.org/10.1186/1471-2180-14-189
36. Guthrie GD, Nicholson-Guthrie CS (1989) gamma-Aminobutyric acid uptake by a bacterial system with neurotransmitter binding characteristics. Proc Natl Acad Sci U S A 86(19):7378–7381
37. Macfarlane S, Dillon JF (2007) Microbial biofilms in the human gastrointestinal tract. J Appl Microbiol 102(5):1187–1196
38. Rubio CA, Huang CB (1991) Quantification of the sulphomucin-producing cell population of the colonic mucosa during protracted stress in rats. In Vivo 6(1):81–84
39. Gue M, Peeters T, Depoortere I, Vantrappen G, Bueno L (1989) Stress-induced changes in gastric emptying, postprandial motility, and plasma gut hormone levels in dogs. Gastroenterology 97(5):1101–1107
40. Gue M, Junien JL, Bueno L (1991) Conditioned emotional response in rats enhances colonic motility through the central release of corticotropin-releasing factor. Gastroenterology 100(4):964–970
41. Schaedler RW, Dubos RJ (1962) The fecal flora of various strains of mice. Its bearing on their susceptibility to endotoxin. J Exp Med 115(6):1149–1160
42. Bailey MT, Coe CL (1999) Maternal separation disrupts the integrity of the intestinal microflora in infant rhesus monkeys. Dev Psychobiol 35(2):146–155
43. Gaykema RP, Goehler LE, Lyte M (2004) Brain response to cecal infection with Campylobacter jejuni: analysis with Fos immunohistochemistry. Brain Behav Immun 18(3):238–245
44. Lyte M, Li W, Opitz N, Gaykema RP, Goehler LE (2006) Induction of anxiety-like behavior in mice during the initial stages of infection with the agent of murine colonic hyperplasia Citrobacter rodentium. Physiol Behav 89(3):350–357
45. Bercik P, Verdu EF, Foster JA, Macri J, Potter M, Huang X et al (2010) Chronic gastrointestinal inflammation induces anxiety-like behavior and alters central nervous system biochemistry in mice. Gastroenterology 139(6):2102–2112
46. Ohland CL, Kish L, Bell H, Thiesen A, Hotte N, Pankiv E et al (2013) Effects of Lactobacillus helveticus on murine behavior are dependent on diet and genotype and correlate with alterations in the gut microbiome. Psychoneuroendocrinology 38(9):1738–1747
47. Jepson B, Johnson J (2007) Changing the course of autism: A scientific approach for parents and physicians. 1st edn, Sentient Publications Boulder. ISBN-10: 1591810612
48. Williams BL, Hornig M, Parekh T, Lipkin WI (2012) Application of novel PCR-based methods for detection, quantitation, and phylogenetic characterization of Sutterella species in intestinal biopsy samples from children with autism and gastrointestinal disturbances. MBio 3(1). https://doi.org/10.1128/mBio.00261-11
49. Parracho HM, Bingham MO, Gibson GR, McCartney AL (2005) Differences between the gut microflora of children with autistic spectrum disorders and that of healthy children. J Med Microbiol 54(10):987–991

50. Adams JB, Johansen LJ, Powell LD, Quig D, Rubin RA (2011) Gastrointestinal flora and gastrointestinal status in children with autism–comparisons to typical children and correlation with autism severity. BMC Gastroenterol 11(1):22. https://doi.org/10.1186/1471-230X-11-22
51. Ming X, Brimacombe M, Chaaban J, Zimmerman-Bier B, Wagner GC (2008) Autism spectrum disorders: concurrent clinical disorders. J Child Neurol 23(1):6–13
52. Valicenti-McDermott M, McVicar K, Rapin I, Wershil BK, Cohen H, Shinnar S (2006) Frequency of gastrointestinal symptoms in children with autistic spectrum disorders and association with family history of autoimmune disease. J Dev Behav Pediatr 27(2):S128–S136
53. Wakefield AJ, Murch SH, Anthony A, Linnell J, Casson DM, Malik M et al (1998) Ileal-lymphoid-nodular hyperplasia, non-specific colitis, and pervasive developmental disorder in children. Lancet 352(9123):234–235
54. Wakefield AJ, Puleston JM, Montgomery SM, Anthony A, O'leary JJ, Murch SH (2002) The concept of entero-colonic encephalopathy, autism and opioid receptor ligands. Aliment Pharmacol Ther 16(4):663–674
55. d'Eufemia P, Celli M, Finocchiaro R, Pacifico L, Viozzi L, Zaccagnini M et al (1996) Abnormal intestinal permeability in children with autism. Acta Paediatr 85(9):1076–1079
56. Horvath K, Perman JA (2002) Autistic disorder and gastrointestinal disease. Cur Opin Pediatr 14(5):583–587
57. Sharp WG, Jaquess DL, Lukens CT (2013) Multi-method assessment of feeding problems among children with autism spectrum disorders. Autism Spectr Disord 7(1):56–65
58. Field D, Garland M, Williams K (2003) Correlates of specific childhood feeding problems. J Paediatr Child Health 39(4):299–304
59. Shaw W (2010) Increased urinary excretion of a 3-(3-hydroxyphenyl)-3-hydroxypropionic acid (HPHPA), an abnormal phenylalanine metabolite of Clostridia spp. in the gastrointestinal tract, in urine samples from patients with autism and schizophrenia. Nutr Neurosci 13(3):135–143
60. Koenig JE, Spor A, Scalfone N, Fricker AD, Stombaug J, Knight R et al (2011) Succession of microbial consortia in the developing infant gut microbiome. Proc Natl Acad Sci U S A 108(Supplement 1):4578–4585
61. Song Y, Liu C, Finegold SM (2004) Real-time PCR quantitation of clostridia in feces of autistic children. Appl Environ Microbiol 70(11):6459–6465
62. Sandler RH, Finegold SM, Bolte ER, Buchanan CP, Maxwell AP, Väisänen ML et al (2000) Short-term benefit from oral vancomycin treatment of regressive-onset autism. JChild Neurol 15(7):429–435
63. Elsden SR, Hilton MG (1978) Volatile acid production from threonine, valine, leucine and isoleucine by clostridia. Arch Microbiol 117(2):165–172
64. Shultz SR, MacFabe DF, Ossenkopp KP, Scratch S, Whelan J, Taylor R et al (2008) Intracerebroventricular injection of propionic acid, an enteric bacterial metabolic end-product, impairs social behavior in the rat: implications for an animal model of autism. Neuropharmacology 54(6):901–911
65. Pulikkan J, Maji A, Dhakan DB, Saxena R, Mohan B, Anto MM et al (2018) Gut microbial dysbiosis in Indian children with autism spectrum disorders. Microb Ecol 76(4):1102–1114
66. Kang DW, Park JG, Ilhan ZE, Wallstrom G, LaBaer J, Adams JB et al (2013) Reduced incidence of Prevotella and other fermenters in intestinal microflora of autistic children. PLoS One 8(7):e68322. https://doi.org/10.1371/journal.pone.0068322
67. Dillon SM, Lee EJ, Kotter CV, Austin GL, Dong Z, Hecht DK et al (2014) An altered intestinal mucosal microbiome in HIV-1 infection is associated with mucosal and systemic immune activation and endotoxemia. Mucosal Immunol 7(4):983–994
68. Clayton TA, Baker D, Lindon JC, Everett JR, Nicholson JK (2009) Pharmacometabonomic identification of a significant host-microbiome metabolic interaction affecting human drug metabolism. Proc Natl Acad Sci 106(34):14728–14733
69. Hsiao EY, McBride SW, Hsien S, Sharon G, Hyde ER, McCue T, Patterson PH (2013) Microbiota modulate behavioral and physiological abnormalities associated with neurodevelopmental disorders. Cell 155(7):1451–1463

70. De Angelis M, Piccolo M, Vannini L, Siragusa S, De Giacomo A, Serrazzanetti DI et al (2013) Fecal microbiota and metabolome of children with autism and pervasive developmental disorder not otherwise specified. PLoS One 8(10):e76993. https://doi.org/10.1371/journal.pone.0076993
71. Dickerson AS, Rahbar MH, Bakian AV, Bilder DA, Harrington RA, Pettygrove S et al (2016) Autism spectrum disorder prevalence and associations with air concentrations of lead, mercury, and arsenic. Environ Monit Assess 188(7):407. https://doi.org/10.1007/s10661-016-5405-1
72. Adams JB, Audhya T, McDonough-Means S, Rubin RA, Quig D, Geis E et al (2013) Toxicological status of children with autism vs. neurotypical children and the association with autism severity. Biol Trace Elem Res 151(2):171–180
73. Breton J, Daniel C, Dewulf J, Pothion S, Froux N, Sauty M et al (2013) Gut microbiota limits heavy metals burden caused by chronic oral exposure. Toxicol Lett 222(2):132–138
74. Meyer J, Michalke K, Kouril T, Hensel R (2008) Volatilisation of metals and metalloids: an inherent feature of methanoarchaea? Syst Appl Microbiol 31(2):81–87
75. Dostal A, Fehlbaum S, Chassard C, Zimmermann MB, Lacroix C (2013) Low iron availability in continuous in vitro colonic fermentations induces strong dysbiosis of the child gut microbial consortium and a decrease in main metabolites. FEMS Microbiol Ecol 83(1):161–175
76. Reed S, Neuman H, Moscovich S, Glahn RP, Koren O, Tako E (2015) Chronic zinc deficiency alters chick gut microbiota composition and function. Nutrients 7(12):9768–9784
77. Smith JC, McDaniel EG, McBean LD, Doft FS, Halsted JA (1972) Effect of microorganisms upon zinc metabolism using germfree and conventional rats. J Nutr 102(6):711–719
78. Pieper R, Vahjen W, Neumann K, Van Kessel AG, Zentek J (2012) Dose-dependent effects of dietary zinc oxide on bacterial communities and metabolic profiles in the ileum of weaned pigs. J Anim Physiol Anim Nutr 96(5):825–833
79. Vahjen W, Pieper R, Zentek J (2010) Bar-coded pyrosequencing of 16S rRNA gene amplicons reveals changes in ileal porcine bacterial communities due to high dietary zinc intake. Appl Environ Microbiol 76(19):6689–6691
80. Højberg O, Canibe N, Poulsen HD, Hedemann MS, Jensen B (2005) Influence of dietary zinc oxide and copper sulfate on the gastrointestinal ecosystem in newly weaned piglets. Appl Environ Microbiol 71(5):2267–2277
81. Pompella A, Visvikis A, Paolicchi A, De Tata V, Casini AF (2003) The changing faces of glutathione, a cellular protagonist. Biochem Pharmacol 66(8):1499–1503
82. Mardinogl A, Shoaie S, Bergentall M, Ghaffari P, Zhang C, Larsson E et al (2015) The gut microbiota modulates host amino acid and glutathione metabolism in mice. Mol Syst Biol 11(10):834. https://doi.org/10.15252/msb.20156487

Chapter 14
Metabolomic Biomarkers in Mental Disorders: Bipolar Disorder and Schizophrenia

Melissa Quintero, Danijela Stanisic, Guilherme Cruz, João G. M. Pontes, Tássia Brena Barroso Carneiro Costa, and Ljubica Tasic

Abstract Psychiatric disorders are some of the most impairing human diseases. Among them, bipolar disorder and schizophrenia are the most common. Both have complicated diagnostics due to their phenotypic, biological, and genetic heterogeneity, unknown etiology, and the underlying biological pathways, and molecular mechanisms are still not completely understood. Given the multifactorial complexity of these disorders, identification and implementation of metabolic biomarkers would assist in their early detection and diagnosis and facilitate disease monitoring and treatment responses. To date, numerous studies have utilized metabolomics to better understand psychiatric disorders, and findings from these studies have begun to converge. In this chapter, we briefly describe some of the metabolomic biomarkers found in these two disorders.

Keywords Psychiatric illness · Mental disorders · Schizophrenia · Bipolar disorder · Biomarkers · Metabolomics

14.1 Introduction

Millions of people experience mental or neuropsychiatric disorders including bipolar affective disorder and schizophrenia, among others. These mental illnesses, usually observed in childhood, adolescence, and early adulthood, are characterized by

M. Quintero · D. Stanisic · G. Cruz · T. B. B. C. Costa · L. Tasic (✉)
Laboratory of Chemical Biology, Department of Organic Chemistry, Institute of Chemistry, University of Campinas (UNICAMP), Campinas, São Paulo, Brazil
e-mail: ljubica@unicamp.br

J. G. M. Pontes
Laboratory of Microbial Chemical Biology, Institute of Chemistry, University of Campinas (UNICAMP), Campinas, São Paulo, Brazil

P. C. Guest (ed.), *Reviews on Biomarker Studies in Psychiatric and Neurodegenerative Disorders*, Advances in Experimental Medicine and Biology 1118,
https://doi.org/10.1007/978-3-030-05542-4_14

a combination of abnormal thoughts, emotions, behaviors, and disturbed perception of reality, affecting not only the person with illness but also society. While not being completely understood in terms of their causes, the symptoms of mental illnesses are scientifically valid and well known. Treatments are readily available and include psychotherapy and medication and are effective for most people, but there are no specific tests that can be used as an add-on for diagnosis. As such, there are high levels of curiosity to understand if molecular biomarkers can assist in making clearer diagnostic decisions.

Biomarker research has been an extensive success in various medical fields so far, but using biomarkers to diagnose and predict treatment response for mental disorders is still a challenge [1]. Based on observations that a specific compound may be present or altered just in patients with a given mental disorder and not in healthy individuals, a valid biomarker can be found if successfully detected. However, the fundamental definition of a psychiatric disorder is based on subjective and/or behavioral criteria which are determined clinically, making it difficult to determine whether or not a person has a particular disorder [2].

The search for biomarkers for psychiatric disorders has a long history, with earlier studies investigating molecular markers, like platelet imipramine binding or cerebrospinal 5-hydroxyindoleacetic acid (5-HIAA) in people with depression [3], or behavioral markers such as smooth pursuit eye movements in people with schizophrenia [4]. Currently, global profiling approaches, such as metabolomics, are expected to be able to pinpoint discriminating molecules as clinical biomarkers, providing an overview of the metabolic status and global biochemical events associated with a particular cellular or biological system [5]. Metabolomics has the potential to scrutinize candidate markers that will improve the diagnosis of the diseases facilitating better patient prognoses and, thus, the development of novel therapeutic strategies [6]. Metabolic markers in diagnosis are thought to be one of the most interesting categories of biomarkers, given their role in physiological processes. Since a biomarker should be detected and measured in a sample obtained using noninvasive procedures, body fluids including plasma/serum, urine, saliva, and, to some extent, cerebrospinal fluid (CSF) are thought to be useful sources for biomarker monitoring [5]. Furthermore, the scientific synergy between biomarkers and metabolomics is important as metabolomics has been used in many instances to identify novel biomarkers which can lead to new and improved therapeutic strategies for many serious and life-threatening diseases. Preliminary metabolomic signatures and some biomarkers have already been described for schizophrenia and bipolar disorder [7–11], but the endophenotype specificities are still under discussion, and identification of illness-specific biomarkers capable of adding not only to the diagnosis process but for use in monitoring prognosis or clinical response to treatment is still lacking. We need to improve the understanding of the biological abnormalities in psychiatric illnesses across conventional diagnostic boundaries. This review focuses on metabolomic biomarkers for bipolar disorder and schizophrenia.

14.1.1 Metabolomic Biomarkers

The use of metabolomics in the examination for novel biomarkers in different clinical areas is based on the hypothesis that diseases cause disruption of biochemical pathways leading to a metabolic fingerprint characteristic of the site and nature of the disease [12]. The term "biomarker" or "biological marker" was first used in 1989 as a medical subject heading to mean "measurable and quantifiable biological parameters which serve as indexes for health- and physiology-related assessments such as disease risk, psychiatric disorders, environmental exposure and its effects, disease diagnosis, metabolic processes, substance abuse, epidemiologic studies etc." In 2001 that definition was standardized and further defined. One of the earliest biomarker approaches in psychiatry [13] employed chromatography to detect a urinary metabolite, 3,4-dimethoxyphenylethylamine (later identified as *p*-tyramine [14]), which formed a controversial "pink spot" on paper chromatograms among those with schizophrenia but not in healthy individuals. Since then, genomic and transcriptomic approaches have dominated biomarker discovery efforts in psychiatric disorders. However, the global study of metabolites (i.e., metabolomics) emerged later as a promising approach for identification of potential diagnostic and treatment response biomarkers for psychiatric disorders [15].

Currently, there are several types of metabolomic biomarkers, which enable response prediction or dynamical description of both disease progression and treatment effectiveness. The integration between clinical metabolomics and pharmacology may allow the discovery of more meaningful biomarkers which could enable the development of individualized treatment methods. The success of this integration depends on the explicit consideration of study designs and data analysis techniques that can effectively quantify sources of biological variability [16].

To date, the most explored bioanalytical platforms in metabolomics research are nuclear magnetic resonance (NMR) and mass spectrometry (MS) [17]. The recent rapid development of a range of analytical platforms including gas chromatography, liquid chromatography techniques like high-performance liquid chromatography (HPLC) or ultra-performance liquid chromatography (UPLC), and capillary electrophoresis, and the use of these as hyphenated analytical methods coupled to MS or NMR, created new possibilities for biomarker research. Other techniques popular in metabolomic research include magnetic resonance imaging (MRI) and high-resolution magic angle spinning spectroscopy (HR-MAS). Bioanalytical platforms enable separation, detection, characterization, and quantification of metabolites, and then we can relate this information to altered metabolic pathways. Due to the complexity of the metabolome and the diverse properties of metabolites, no single analytical platform can detect all of the metabolites in a biological sample. The combined use of modern instrumental analytical approaches has helped to increase the coverage of detected metabolites that cannot be achieved by single-analysis techniques [18–20]. Integrated platforms have been frequently used to provide sensitive and reliable detection of thousands of metabolites in a biofluid sample (Fig. 14.1).

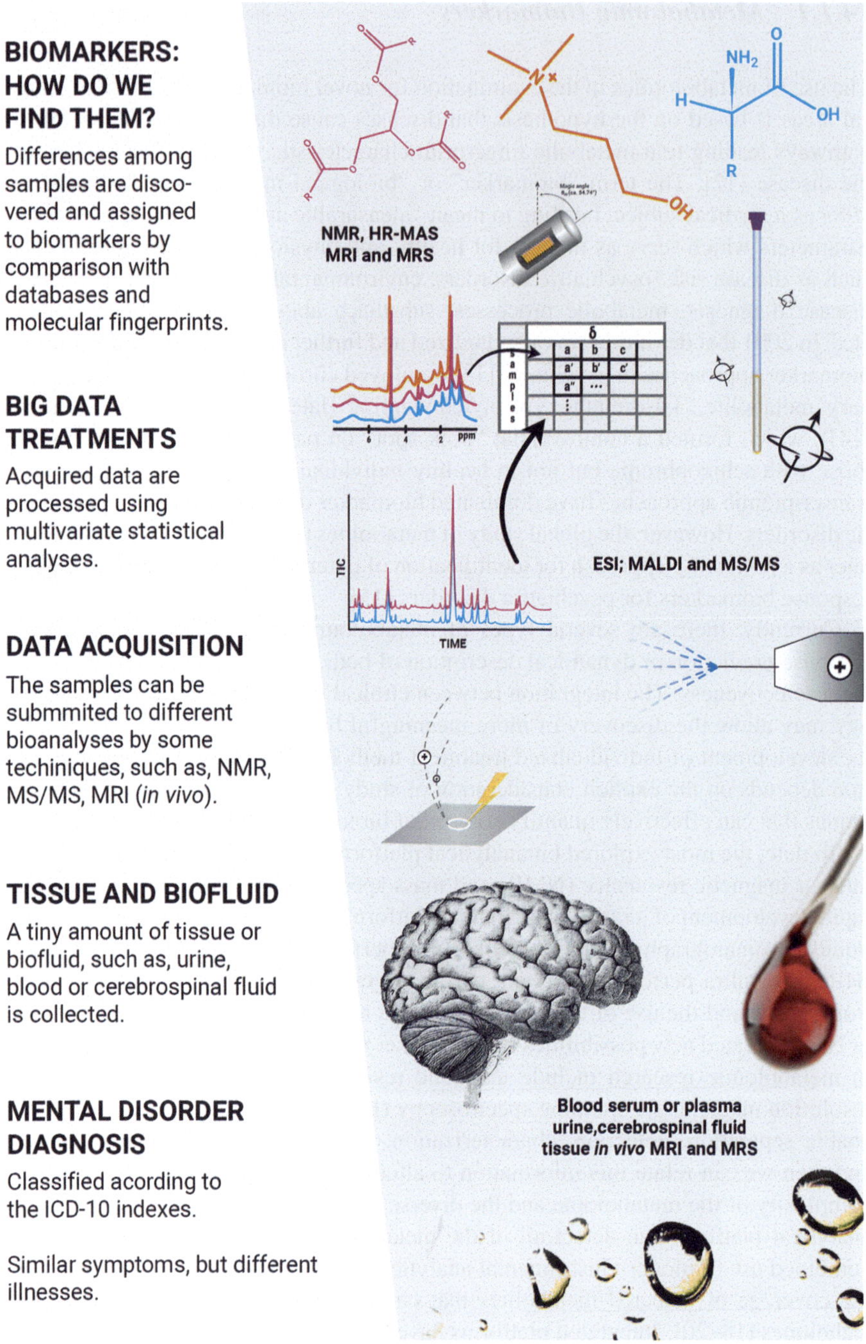

Fig. 14.1 Workflow in untargeted metabolomic studies. After clinical evaluation and sample collection from at least two groups (e.g., disease and healthy), an appropriate analytical method or combination of more than one should be used for sample assessment and the resulting large amounts of data analyzed for differences that might point to mental disorder biomarkers

Bioanalytical platforms in metabolomics start from a comparative analysis among at least two groups of samples and rely on big data processing by chemometrics. The most used of the data processing techniques are multivariate statistical analyses such as principal component analysis (PCA) to explore the data and detect outliers [21, 22], partial least squares discriminant analysis (PLS-DA) and orthogonal partial least squares discriminant analysis (OPLS-DA) to classify the samples and identify the main metabolites responsible for the discrimination [23, 24], and logistic regression to evaluate the metabolites as biomarkers [25]. The list of tools available for spectral processing and data analysis are available in Alonso et al. [19].

14.2 Biomarkers in Psychiatric Disorders

14.2.1 Metabolomic Biomarkers in Bipolar Disorder

Bipolar disorder is chronic and recurrent disorder that affects around 1% of the global population [26–28], leading to disabilities in young people (mostly between 18 and 44 years of age), such as cognitive and functional impairment, and increased mortality particularly from suicide and cardiovascular disease [29, 30]. Bipolar disorder is characterized as biphasic moods that include depression and mania (in some cases hypomania) and which occur as recurrent episodes of changes in behavior of individuals. There are several subclassifications of this disorder including bipolar I, bipolar II, cyclothymia, and other atypical forms, depending on severity and duration of depressive and manic episodes [29]. The diagnostic error still leads to erroneous treatment, increasing patient suffering, and, therefore, greater suicide risk [31].

Bipolar disorder is a major cause of comorbidity because it directly affects the productivity of people making the disease a socioeconomic problem, besides being a major cause of death, and 25% of those affected by the disease attempt suicide with 11% success [30, 32–34]. Bipolar disorder is a historically known disease with well-documented epidemiology [30]. In the early nineteenth century, it was believed that the cause of most diseases was a problem of uric acid metabolism. By exerting influence on common diseases such as gout, high blood pressure, and renal calculus, it was also attributed to dementia, schizophrenia, mania, depression, and bipolar disorder [31]. Lithium was discovered as a chemical element in the same century. In the tests performed with its salts, mainly lithium carbonate, it was found that it is capable of dissolving crystals of uric acid. Based on this information, psychiatrists started prescribing lithium carbonate (Li_2CO_3) for mental disorders. This study could be considered as one of the first metabolomic research instances. Other studies showed that Li^+ was effective against bipolar disorder [31, 35, 36].

Bipolar disorder biomarkers have been studied in blood, serum, urine, and plasma by ^{1}H NMR [10, 37–39], by gas chromatography (GC)-MS [38, 40, 41], and through in vivo brain imaging experiments [42–44]. Table 14.1 lists some of metabolites that are altered for this type of mental disorder including

Table 14.1 Specific biomarkers related to bipolar disorders

Biomarkers	Sample type	Analytical platforms
Acetate-choline [39]	Human blood serum	NMR
Acetone [10, 45]	Female urine	NMR
N-Acetyl-aspartate [46]	Brain	MRS
N-Acetyl-aspartate (NAA)/PCr-Cr [47]	White matter	MRI
N-Acetyl-aspartyl-glutamate [46]	Brain	MRS
N-Acetyl-aspartyl-glutamic acid [37]	Human blood serum	NMR
N-Acetyl-glutamic acid [48]	Serum	CE-TOFMS
N-Acetyl-phenylalanine [37]	Human blood serum	NMR
Adipic acid [10]	Urine	GC-MS
β-Alanine [10, 38, 48]	Urine	^{1}H-NMR, GC-MS, CE-TOFMS
γ-Aminobutyric acid (GABA) [42, 49]	Brain, temporal lobes	MRS, MRI
β-Amino-*iso*butyric acid [10]	Urine	GC-MS
Amygdalin [37]	Human blood serum	NMR
Arabitol [10]	Urine	GC-MS
Arginine [48]	Serum	CE-TOFMS
Azelaic acid [10, 38]	Urine	NMR, GC-MS
Choline [9, 37, 45]	Human blood serum, urine, male urine	NMR
Citrulline [50]	Plasma of drug-free patients	CE-TOFMS
Creatine [42–44, 46]	Brain, white and gray matter, cerebrospinal fluid	MRS, MRI
Formate [10, 45]	Male urine	NMR
Glutamate [39, 42–44]	Brain, human blood serum, white and gray matter, cerebrospinal fluid	MRS, MRI, NMR
Glutamine [37, 42–44]	Human blood serum, brain, white and gray matter, cerebrospinal fluid	NMR, MRS, MRI
Glycine [10]	Urine	GC-MS
Glycerol-phosphocholine + phosphocholine (GPC + PC) [47]	White matter	MRI
Glycolate [10]	Urine	NMR
5-HIAA [51]	White matter	HPLC, DTI
5-HT serotonin neurotransmitter [51]	White matter	HPLC, DTI
4-Hydroxybenzoic acid [10]	Urine	GC-MS
α-Hydroxybutyrate [9, 10, 38, 45]	Urine	NMR, GC-MS
5-Hydroxy-hexanoic acid [10]	Urine	GC-MS
D-Hydroxy-pyrimidine [10, 38, 41]	Urine	NMR, GC-MS
*Iso*butyrate [9]	Urine	NMR
*Iso*citric acid (*iso*citrate) [52]	CFS (rats)	CE-TOFMS
α-Ketoglutaric acid [37]	Human blood serum	NMR

(continued)

Table 14.1 (continued)

Biomarkers	Sample type	Analytical platforms
Kynurenine, kynurenic acid and 3-hydroxy-kynurenine [51]	White matter	HPLC, DTI
Lactate[10]	Urine	NMR
Lipids and lipoamide [37, 39]	Human blood serum	NMR
Mannitol [10]	Urine	GC-MS
Methylmalonic acid [10]	Urine	GC-MS
N-Methyl-nicotinamide [45]	Urine	NMR
Myo-inositol [37, 39, 53]	Human blood serum	NMR
Oxalacetate [45]	Female urine	NMR
Phenylalanine [10]	Urine	GC-MS
Phosphatidic acid (44:4) [54]	Human blood serum	UHPLC-MS
Phosphatidic acid (48:8 (OH)) [54]	Human blood serum	UHPLC-MS
Phosphatidylethanolamine (42:5) [54]	Human blood serum	UHPLC-MS
Phosphatidylglycerol (32:4 (OH)) [54]	Human blood serum	UHPLC-MS
Phosphatidylinositol (40:3) [54]	Human blood serum	UHPLC-MS
Phosphocreatine-creatine (PCr-Cr) [44, 47]	White matter and gray matter, cerebrospinal fluid	MRI
Propionate [10]	Urine	NMR
Pseudouridine [38]	Urine	NMR, GC-MS
Pyruvate	Serum	CE-TOFMS
Pyroglutamic acid [10]	Urine	GC-MS
Serine [48]	Serum	CE-TOFMS
Triacylglycerol (42:3) [54]	Human blood serum	UHPLC-MS
Tryptophan [51]	White matter	HPLC, DTI
5-Hydroxyindole-3-acetic acid (5-HIAA) [51]	White matter	HPLC, DTI
5-Hydroxytryptamine (5-HT) serotonin neurotransmitter [51]	White matter	HPLC, DTI

α-hydroxybutyrate, choline, *iso*butyrate, and *N*-methylnicotinamide which differentiate patients with bipolar disorders from healthy individuals, while propionate, formate, 2,3-dihydroxybutanoic acid, phenylalanine, 2,4-dihydroxypyrimidine, and β-alanine were differentiated in patients with bipolar disorder and mild cognitive disorders [9, 41, 55].

In some studies, such as ^{1}H-NMR analysis of anterior cingulate cortex of brain tissue, a higher concentration of *myo*-inositol in patients with bipolar disorder was found and when patients underwent treatment with lithium-based drugs, the concentration of *myo*-inositol was decreased considerably [53]. These data were reinforced by the metabolomic analysis of blood serum [37]. *Myo*-inositol is synthesized from special plasma membrane lipids by inositol monophosphatase type II (IMPase II), which is a candidate enzyme for Li^+ since it is inhibited by Li^+ in tests using pharmacologically relevant concentrations in vitro [56]. Myo-inositol acts as a secondary messenger and is responsible for immobilizing calcium (Ca^{2+}) stored inside the cell, which acts as a prosthetic agent of other enzymes such as hexokinase. In general, *myo*-inositol is able to initiate several metabolic paths vital to cellular functions [57]. The action of Li^+ may be linked to the competitive inhibition of inositol monophosphatase II, which would lead to a decrease in *myo*-inositol concentrations in neurons, which depend on this enzyme to synthesize it since *myo*-inositol does not cross the blood-brain barrier [58].

On the other hand, glutamate, glutamine, creatine, and the ratios among these metabolites are the most cited in the literature, and NMR, MRI, LC-MS, and HPLC techniques have been used for their detection [37, 39, 42–44]. Glutamate is one of the most abundant neurotransmitters in the brain, as a precursor of glutamine, γ-aminobutyric acid (GABA), and glutathione. It is one of the structural components of proteins and an intermediate metabolite [59–61]. The differences between metabolites indicated by Sethi et al. [37] lead to conclusion that some possible biomarkers for bipolar disorder are α-ketoisovaleric acid, α-ketoglutaric acid, *N*-acetyl-aspartyl glutamic acid, *N*-acetyl-phenylalanine, and glutamine, which were detected or altered only in bipolar disorder patients, while *N*-acetyl-alanine was detected only in controls [37]. *N*-acetyl-aspartate, *N*-acetyl-aspartyl-glutamate, creatine, and phosphocreatine are some of the metabolites suggested by Haarman et al. [46] as the levels of these were decreased in the left hippocampus after MRS quantification using LC Model. The authors hypothesized that these results are a consequence of diminished neuronal integrity in this region. *N*-acetyl-aspartate is the second most abundant substance in the central nervous system, formed in mitochondria from acetyl-CoA and aspartate. Phosphocreatine and creatine are important buffers in the brain and muscle tissues, providing energy and a constant ATP concentration, and a decrease in their concentrations might be due to cell death [46].

A lipidomic study in blood serum employing ultra-HPLC (UHPLC) coupled with high-resolution MS (HRMS) identified lipid profiles of bipolar disorder patients and healthy individuals. Sphingolipids and glycerolipids were increased, while glycerophospholipids were decreased in patient serum samples. Of the lipids with greater differential proportions in the groups, the authors concluded that phospholipid biosynthesis is the most altered pathway in bipolar disorder patients [54].

Recently, Soeiro-de-Souza et al. [43] investigated and compared healthy individuals with bipolar disorder I and II patients, who were treated with Li^+, anticonvulsants, and antipsychotics. The authors carried out brain imaging using magnetic resonance spectroscopy and quantified glutamate, glutamine, and creatine and their ratios after treatment. One of their conclusions was that bipolar disorder patients and controls had the same volume of white matter but lower volumes of anterior cingulate cortex and a higher volume of CSF. Also, the change in the concentration of glutamate was observed in treated patients. Moreover, they found a lower ratio of glutamate/creatine during medication and a constant glutamine/creatine ratio. By comparing bipolar subjects with healthy individuals, the higher glutamate/creatine and glutamine/creatine ratios were observed after omitting medications, which might be indicative that their increments were not caused by medication use [43]. The concentration of glutamine was high, and the glutamate/glutamine ratio and glutamate concentrations were lower in the patient group when compared to the healthy controls. Pairing the groups by age and gender, lower glutamate concentrations and glutamate/glutamine ratios were observed in the bipolar patients [42].

Zheng et al. [9] examined urinary samples using a NMR-based metabonomic method and found that choline, *iso*butyrate, α-hydroxybutyrate, and *N*-methylnicotinamide could be good urinary metabolite biomarkers for bipolar disorder (AUC of 0.89 for training samples and 0.86 in the sample test). They found that the increase of α-hydroxybutyrate was derived from elevated levels of α-ketobutyrate, which is consequence of a higher conversion rate of cystathionine to cysteine. The altered levels of choline in bipolar patients' urine could be due to the abnormalities in the acetylcholine neurotransmission precursor mechanism and abnormal phospholipid metabolism [9, 62]. The higher concentration of *iso*butyrate in urine is a result of a lower blood concentration of valine, which can influence an increase in concentration of catecholamines and serotonin. The *N*-methylnicotina mide concentration was decreased in bipolar subjects being an end product of nicotinamide processing in the tryptophan-NAD^+ pathway. Tryptophan is a precursor of kynurenine and serotonin metabolism, and the decrease in metabolites of nicotinic acid leads to an increase in kynurenine metabolism [9]. Five metabolites were specified as potential biomarkers (pseudouridine, β-alanine, α-hydroxybutyrate, 2,4-dihydroxypyrimidine, and azelaic acid) in bipolar disorder by a combined ^{1}H-NMR/GC-MS approach [38]. Previous studies from the same group suggested that 2,4-dihydroxypyrimidine, *N*-methylnicotinamide, choline, *iso*butyrate, and α-hydroxybutyrate are as possible metabolites for diagnosis of bipolar disorder using urine samples examined by GC-MS analysis [9, 38, 41].

Another study that connects kynurenic acid and metabolites from kynurenine metabolism to bipolar disorder is from Poletti et al. [51]. Kynurenine and kynurenic acid were determined from bipolar disorder patient plasma using HPLC and diffusion tensor imaging (DTI) in an evaluation of white matter microstructure. There is a hypothesis that serotonergic turnover in bipolar disorder could influence the white matter of brain microstructure as that changes in myelination and integrity of myelin can influence signal speed and communication with different brain areas [63–65]. Through application of DTI, it was possible to confirm the hypothesis in a homogeneous sample of patients with bipolar disorder and to link kynurenine acid and 5-hydroxyacetic acid with changes in white matter microstructure. The result demonstrated an inefficient turnover of serotonin, but this did not affect tryptophan, despite the lower concentration of the latter in blood samples [51].

14.2.2 Metabolomic Biomarkers in Schizophrenia

Schizophrenia is a chronic psychiatric disorder with a heterogeneous genetic and neurobiological background that influences early brain development and is expressed as a combination of psychotic symptoms such as hallucinations and delusions, along with organizational, motivational, and cognitive dysfunctions [66, 67].

It starts in late adolescence or early adulthood and affects approximately 0.5–1% of the world population with high heritability [68].

Abnormalities in neurotransmission have provided the basis for theories on the pathophysiology of schizophrenia. Most of these theories center on either an excess or a deficiency of neurotransmitters such as dopamine, serotonin, and glutamate. Other theories implicate aspartate, glycine, and γ-aminobutyric acid (GABA) as part of the neurochemical imbalance in schizophrenia [69]. Another theory for the symptoms of schizophrenia involves the activity of glutamate, the major excitatory neurotransmitter in the brain. This theory has arisen in response to the finding that phencyclidine and ketamine, both noncompetitive NMDA/glutamate antagonists, could induce schizophrenia-like symptoms [70]. However, there is not a consensus on whether there is an increase [71, 72] or a decrease [73, 74] of these metabolite concentrations. Metabolites, like tele-methylhistamine (t-MH), were detected by GC-MS in CSF samples of schizophrenia patients. Other studies showed that t-MH is associated with an increase of histaminergic activity and desensitization of neural tissues due to decreases of histamine H1 and H2 receptors and an increase of histamine concentrations [75].

Several metabolomic studies have also pointed to some metabolic abnormalities (Table 14.2). He et al. [8] demonstrated differences in amino acid and lipid metabolism in medicated and non-medicated schizophrenia patients when compared to the control group. Subsequent analyses of these potentially relevant metabolites as well as analysis of known schizophrenia risk genes identified the glutamine and arginine signaling pathways as possible risk factors. Another study [71] raised the possibility that there are at least two different schizophrenia-related risk pathways, and these are involved in glucoregulation and proline metabolism. In addition to glutamine and arginine metabolism, amino acids altered in plasma or CSF of schizophrenia patients have been linked to nitrogen compound biosynthetic processes. The finding of changes in certain lipids, fatty acids, and amino acids has implicated phospholipid synthesis [85]. The use of metabolic profiles in CSF from drug-naïve patients compared with matched controls found elevated glucose concentrations in patients, whereas the serum glucose concentration showed no differences [7]. Lipid analysis in plasma from patients with schizophrenia compared with that of healthy individuals revealed significantly lower concentrations of lipids [87]. Potential metabolite markers consisted of several fatty acids and ketone bodies, and the presumed upregulated fatty acid catabolism may result from an insufficiency of glucose supply in the brains of patients with schizophrenia [72]. Also, glycine and serine are amino acids that have frequently been reported as markers of schizophrenia due to their significant changes in the metabolic profile among patients and healthy individuals [84, 88, 89].

Noninvasive analyses of schizophrenia patient expired breaths indicated a higher level of alkanes when compared to healthy individuals. The level of ethane and pentane, which cross the pulmonary alveolar membrane, is result of dead cells due to peroxidation of cell membrane lipids. Other alkanes such as butane were not detected. Furthermore, pentane is not a specific biomarker for schizophrenia since it

Table 14.2 Specific biomarkers related to schizophrenia

Biomarkers	Sample type	Analytical platforms
Acetoacetate [72]	Urine	GC-TOFMS and ^{1}H-NMR
Acetone [72]	Urine	GC-TOFMS and ^{1}H-NMR
N-Acetyl-aspartate [76]	Serum	GC-MS
Cis-aconitic acid [72]	Urine	GC-TOFMS
Allantoin [77]	Serum	GC-MS
2-Aminoadipic acid [72]	Urine	GC-TOFMS
α-Aminobutyrate (AABA) [72]	Serum Urine	GC-TOFMS
γ-Aminobutyric acid (GABA) [77]	Blood	LC-ESI/MS/MS
Arachidonic acid [78]	Serum	HPLC-MS HPLC fluorescence
Arginine [8]	Plasma	MS
Aspartic acid [40, 72, 76]	Serum Peripheral blood mononuclear cells (PBMC)	GC-MS GC-TOFMS
Benzoic acid [40]	PBMC	GC-MS
Betaine [77]	Blood and plasma	CE-TOFMS
1,3-Bisphosphoglycerate [76]	Serum	GC-MS
Carbon disulfide [79]	Breath	GC-MS
Catechol [72]	Urine	GC-TOFMS
Cholesterol [76, 80]	Serum	GC-MS
Citrate [7, 72, 76]	Serum CSF	GC-MS GC-TOFMS and ^{1}H-NMR
Citric acid [55]	PBMC	GC-MS
Creatine [77]	Blood and plasma	CE-TOFMS
Creatinine [40]	PBMC	GC-MS
Cystine [72]	Serum	GC-TOFMS
6-Deoxy-mannofuranose [80]	Serum	GC-MS
Dihydroxyacetone phosphate [55]	PBMC	GC-MS
2,3-Dihydroxybutanoic acid [72]	Urine	GC-TOFMS
3,4-Dihydroxyphenylacetic acid (DOPAC) [77]	Blood	LC-ESI/MS/MS
3,4-Dimethoxyphenethylamine (3,4-DMPEA) [81]	Urine	MS
Dopamine (DA) [40, 77]	PBMC Blood	GC-MS LC-ESI/MS/MS
Eicosanoic acid [80]	Serum	GC-MS
Eicosenoic acid [72]	Serum	GC-TOFMS
Erythrose [76]	Serum	GC-MS
Ethane [82]	Breath	GC-MS
2-Ethyl-3-hydroxypropionic acid [72]	Urine	GC-TOFMS

(continued)

Table 14.2 (continued)

Biomarkers	Sample type	Analytical platforms
Fructose [55]	PBMC	GC-MS
Fructose 6-phosphate [55]	PBMC	GC-MS
Fumaric acid [40]	PBMC	GC-MS
Galactose oxime [80]	Serum	GC-MS
Glucose [7, 55, 72, 76]	PBMC Urine CSF Serum	GC-MS, GC-TOFMS and ^{1}H-NMR
Glucose 6-phosphate [55]	PBMC	GC-MS
Gluconic acid [77]	Blood and plasma	CE-TOFMS
Glucuronic acid [76]	Serum	GC-MS
Glutamate [72, 77, 78, 83]	Blood, plasma, urine	CE-TOFMS; LC-ESI/MS/MS GC-TOFMS HPLC-MS
Glutamine [8]	Plasma CSF	MS
γ-Glutamylcysteine [78]	Serum	HPLC-MS HPLC fluorescence
Glutathione [78]	Serum	HPLC-MS HPLC fluorescence
Glyceraldehyde-3-phosphate [55]	PBMC	GC-MS
Glycerate [72]	Serum	GC-TOFMS
Glycerate 3-phosphate [55]	PBMC	GC-MS
Glycerol [40, 76]	PBMC Serum	GC-MS
Glycerol 3-phosphate [55]	PBMC	GC-MS
Glycine [76, 84]	Plasma Serum	GC-MS
Glycocyamine [72]	Urine	GC-TOFMS
Heptadecanoic acid [80]	Serum	GC-MS
Hexadecanoic acid (palmitic acid) [72, 76]	Serum	GC-MS GC-TOFMS
Histidine [8]	Plasma	MS
Homoserine [40]	PBMC	GC-MS
Homovanillic acid (HVA) [77]	Blood	LC-ESI/MS/MS
Hydroxyacetic acid [72]	Urine	GC-TOFMS
3-Hydroxyadipic acid [72]	Urine	GC-TOFMS
2-Hydroxybutyrate [72]	Serum Urine	GC-TOFMS
3-Hydroxybutyrate [72, 78]	Serum Urine	GC-TOFMS and ^{1}H-NMR; HPLC-MS
2-Hydroxyethyl palmitate [40]	PBMC	GC-MS

(continued)

Table 14.2 (continued)

Biomarkers	Sample type	Analytical platforms
5-Hydroxyindole-3-acetic acid (5-HIAA) [77]	Blood	LC-ESI/MS/MS
Hydroxylamine [40]	PBMC	GC-MS
3-Hydroxysebacic acid [72]	Urine	GC-TOFMS
5-Hydroxytryptamine (5-HT) [77, 78]	Blood Serum	LC-ESI/MS/MS HPLC-MS HPLC fluorescence
Inositol [40]	PBMC	GC-MS
Isoleucine [72]	Urine	GC-TOFMS
α-Ketoglutarate [76]	Serum	GC-MS
L-Kynurenine [78]	Serum	HPLC-MS HPLC fluorescence
Lactate [7, 55, 72, 76, 78]	PBMC Serum CSF	GC-MS; GC-TOFMS; ^{1}H-NMR; HPLC-MS
Lactobionic acid [76]	Serum	GC-MS
Linoleic acid [72, 76, 78]	Serum	HPLC-MS HPLC fluorescence GC-MS GC-TOFMS
Malate [72]	Serum	GC-TOFMS
Maltose [40]	PBMC	GC-MS
3-Methoxy-4-hydroxyphenylglycol (MHPG) [77]	Blood	LC-ESI/MS/MS
Tele-methylhistamine [85, 86]	CSF	GC-MS
Methyl phosphate [40]	PBMC	GC-MS
Myo-inositol [72, 76]	Serum	GC-MS GC-TOFMS
Norepinephrine (NE) [77]	Blood	LC-ESI/MS/MS
Octadecanoic acid (stearic acid) [72, 76]	Serum	GC-MS GC-TOFMS
(9*Z*)-Octadec-9-enoic acid (oleic acid) [72, 80]	Serum	GC-MS GC-TOFMS
Octanoic acid [40]	PBMC	GC-MS
Ornithine [8]	Plasma	MS
2-Oxoglutarate [72]	Serum	GC-TOFMS
1-Oxoproline [80]	Serum	GC-MS
PC ae C38:6 [8]	Plasma	MS
Pentadecanoic acid [80]	Serum	GC-MS
Pentane [79, 82]	Breath	GC-MS
4-Pentenoic acid [72]	Urine	GC-TOFMS
Phenylalanine [72]	Serum	GC-TOFMS
Pipecolinic acid [72]	Urine	GC-TOFMS

(continued)

Table 14.2 (continued)

Biomarkers	Sample type	Analytical platforms
2-Piperidinecarboxylic acid [80]	Serum	GC-MS
Pyroglutamic acid (5-oxoproline) [40, 72]	PBMC Urine Serum	GC-MS GC-TOFMS
Pyruvate [55, 72]	PBMC Serum	GC-MS GC-TOFMS
Ribose 5-phosphate [55]	PBMC	GC-MS
Serine [72, 78, 84]	Plasma, serum	GC-MS, GC-TOFMS, HPLC-MS
Sorbitol [40]	PBMC	GC-MS
Suberic acid [72]	Urine	GC-TOFMS
Succinic acid [55]	PBMC	GC-MS
Tetradecanoic acid [72]	Serum	GC-TOFMS
Threonic acid [72]	Urine	GC-TOFMS
Threonine [78]	Serum	HPLC-MS HPLC fluorescence
α-Tocopherol [40]	PBMC	GC-MS
γ-Tocopherol [40, 76]	Serum PBMC	GC-MS
-Tryptophan [76, 78]	Serum	HPLC-MS HPLC fluorescence GC-MS
L-Tyrosine [78]	Serum	HPLC-MS HPLC fluorescence
Uric acid [76]	Serum	GC-MS
Valine [40, 72]	PBMC Urine	GC-MS GC-TOFMS
Vanillylmandelic acid (VMA) [77]	Blood	LC-ESI/MS/MS

has been reported to occur in other diseases such as heart attack, rheumatoid arthritis, and nutritional deficiency [79, 82].

Another biomarker that has been studied is trimethylglycine (also known as betaine) since it participates in homocysteine metabolism. More specifically, it participates in the conversion of homocysteine to methionine which is mediated by betaine-homocysteine methyltransferase in human tissues. Koike et al. [83] observed a decrease of betaine levels in schizophrenia patients and consequently an increase of the homocysteine concentration in the brain, which can lead to a high oxidative stress and a neural damage. In addition to betaine, hydroxylamine, pyroglutamic acid, γ-tocopherol, and α-tocopherol have been listed as markers of the increase in oxidative stress in schizophrenia patients [83]. The higher concentrations of hydroxylamine in patients compared to healthy individuals suggest a higher accumulation of reactive oxygen species (ROS) in patients with schizophrenia, while lower tocopherol concentrations indicate a decrease in antioxidative defense [40].

As already stated, biomarker candidates found for BD and schizophrenia are numerous, but these have still not been tested in geographically different environments and using greater number of samples as means of validation. In order to apply these in clinical research, there must be coordinated efforts around the globe to verify which of the cited biomarkers are applicable and universal. It is also expected that this will help to expand research in modern and multidisciplinary psychiatry and design biomarker-based user friendly tests that can be performed as simple clinical trials with easy to read and interpretable data.

14.3 Samples, Methodology, and Techniques: Concerns

To guarantee the quality of the data and the biological interpretations, it is important to properly select the sample type, the method of sample preparation, and the preprocessing procedures. Another critical issue is the selection of an appropriate control group [90, 91]. When immediate analysis is not feasible, to prevent changes in the composition of the samples, it is necessary to store them at −80 °C, but cycles of freezing and thawing should be avoided [90, 92–94].

The cellular components of the blood (erythrocytes, leukocytes, and platelets) can be excluded using centrifugation, obtaining the plasma or serum. The supernatant portions of these are derived with or without the addition of anticoagulants, respectively. Plasma is composed of a variety of substances such as proteins, peptides, and electrolytes. The precipitation of proteins can be carried out with trichloroacetic acid (TCA) and cold methanol to reduce their potential interference in the analysis [90].

Plasma, serum, and CSF are naturally buffered. However, the pH in urine samples can vary and should be controlled by addition of a sodium phosphate buffer (Na_2HPO_4/NaH_2PO_4) [90–93].

The sodium salt of 3-trimethylsilylpropionic acid (TSP) should not be used as a NMR standard reference in samples containing high concentrations of proteins. An alternative is the use of formate or internal standards, such as the lactate signal (3H, d, 1.33 ppm, $^3J = 7$ Hz) in plasma or serum samples [94, 95].

Analysis of urine samples presents several advantages compared to blood or CSF, since urine collection is easy and noninvasive and enables more extensive sampling frequency. Furthermore, the sample preparation is less complex due to the absence, or low level, of proteins and peptides [90, 92, 96]. However, it is important to define the time of urine collection since the concentration of metabolites can vary during the day due to dietary, lifestyle, and pharmaceutical interference [90, 92].

The collection of CSF is invasive, and, therefore, routine sampling and longitudinal studies are not desirable. One of the main concerns of using CSF is the potential of blood contamination during the sampling process. There are some known protocols to decrease blood contamination in CSF [90, 93].

Numerous methods of sample preparation prior to MS-based analysis have been reported. These are diverse extraction protocols that will lead to the observation of different fractions in the metabolite profile [90, 94]. However, the signals are usually filtered using specific pulse sequences in NMR-based analysis instead of applying sample preparation techniques, such as protein precipitation or ultrafiltration [94–97].

For further information, there are many described procedures for collection and preparation of serum [91, 94, 95, 97], plasma [91, 94], urine [91, 92, 94], and CSF [98] samples. Barbosa et al. described the detailed procedures involved in serum lipidomics [97].

14.3.1 Softwares and Databases

NMR spectral processing can be performed using Topspin or MestreNova software. The Chenomx commercial package provides a tool that can be used from the point of spectral processing to metabolite identification. For data statistical analysis, there are free online platforms including PRIMe [99] and MetaboAnalyst [100] and free or commercial software, such as R, MVAPACK [101], MATLAB, Statistica, and the Unscrambler. The open-access databases Human Metabolome Database (HMDB) [102], Biological Magnetic Resonance Data Bank (BMRB) [103], and Madison-Qingdao Metabolomics Consortium Database (MMCD) [104] can be used to assist in the spectral assignments of metabolites. A list and discussion about the database and software for NMR metabolomics are available in the literature [105].

Current software available for processing mass spectra includes MassHunter (Agilent^R), Thermo™ Xcalibur™ (LC-MS), MestreNova (NMR and MS), and others. Some specific software for both GC-MS and LC-MS preprocessing includes MetAlign™ [106], MZmine 2 (LC-MS) [107], XCMS [108], and SpectConnect (GC-MS) [109].

The statistical analysis of mass spectra data can be performed using the Agilent MassHunter Profinder (data extraction) and Mass Profiler Professional (MPP-Agilent^R). Data mining can also be performed, and chemometric analysis can be used for analysis of volatile organic compounds in GC-MS analysis. Finally, ANOVA analysis and unsupervised statistical methods can also be applied such as PCA and HCA [110].

Some software is also available that can be used for image data processing generated from imaging mass spectrometry (IMS) such as BioMap (Novartis) or flexImaging (Bruker Daltonics), and the FlexAnalysis system (Bruker Daltonics) can be used for quantification.

14.3.2 NMR × MS

Standard NMR pulse sequences frequently used in metabolomic studies are the water suppression pulse sequence Watergate [111, 112], T_2-edited CPMG pulse sequence for filtering out signals of larger molecules [113, 114], and diffusion-ordered spectroscopy (DOSY)-edited pulse sequence to detect assemblies of lipids [94, 95, 97].

The development of high-resolution ^{1}H magic angle spinning (MAS) NMR spectroscopy has extended the capability of metabolomic studies since this makes it possible to analyze intact tissues without a pre-treatment step. The experiments are the same ones as those used in liquid state NMR metabolomics [12, 94, 96, 115].

Since each technique has its strengths and limitations, the combination of NMR and MS techniques for metabolic fingerprinting and profiling is growing and showing the capacity to improve results by covering a more comprehensive range of metabolites [94, 116, 117]. This can be done using hyphenated techniques such as HPLC-NMR-MS, in which parallel NMR and MS analyses are performed using a splitting of the eluting HPLC fraction. Another alternative is to aliquot the sample for analysis by NMR and MS separately. Some authors have already applied this approach in bipolar disorder and schizophrenia metabolomic studies.

Generally, when it is necessary to perform analyses in which the samples need to be analyzed over a period of time, NMR spectroscopy is chosen because of the greater reproducibility of the resulting spectra, while GC-MS or LC-MS are commonly used when it is necessary to achieve a higher sensitivity. However, this requires high purity of samples and sometimes laborious preparation of the metabolic components, such as derivatization of the samples [118]. Furthermore, the testing of the range of molecules analyzed according to their functional groups will be conditioned to the extraction solvent that is used in preparation of the samples.

In addition to the possibility of analyzing biofluids and the preparation care already mentioned previously, tissue analysis by IMS has been explored as in the work of Matsumoto et al. [119]. IMS is a technique that has allowed the analysis of complex surfaces of samples (in 2D or 3D) and can result in the display of hundreds to thousands of compounds per sample surface [120, 121].

14.4 Conclusion

Bipolar disorder and schizophrenia are chronic mental disorders that affect the population and can notably interfere with the normal life of a person. Modern medicine still struggles with the accurate diagnosis and treatment of these diseases. Recent and promising metabolomic discoveries in mental disorders will help in the selection of appropriate drug treatments, improve human health significantly, and ameliorate the quotidian effects on individuals. Acetone, *N*-acetyl-aspartate, GABA, creatine, glutamine, glutamate, glycine, 5-hydroxyindole-3-acetic acid (5-HIAA), myoinositol, α-ketoglutarate, L-kynurenine, lactate, phenylalanine, pyruvate, and tryptophan are the most frequently occurring biomarkers cited in the literature as being altered in bipolar disorder and schizophrenia patients [10, 37, 39, 40, 42, 43, 45–47, 49, 51, 53, 55, 72, 76–78]. Both disorders have some similar symptoms and share some common metabolites, but the question is whether these metabolites originate from the same source and metabolic pathway. It is anticipated that further studies will help to determine this.

References

1. Venigalla H, Mekala HM, Hassan M, Zain H (2017) An update on biomarkers in psychiatric disorders—are we aware, do we use in our clinical practice? Ment Health Fam Med 13:471–479
2. Boksa P (2013) A way forward for research on biomarkers for psychiatric disorders. J Psychiatry Neurosci 38(2):75–77
3. Gjerris A, Sørensen AS, Rafaelsen OJ, Werdelin L, Alling C, Linnoila M (1987) 5-HT and 5-HIAA in cerebrospinal fluid in depression. J Affect Disord 12(1):13–22
4. Ross RG, Olincy A, Harris JG, Sullivan B, Radant A (2000) Smooth pursuit eye movements in schizophrenia and attentional dysfunction: adults with schizophrenia, ADHD, and a normal comparison group. Biol Psychiatry 48(3):197–203
5. Yeung PK (2018) Metabolomics and biomarkers for drug discovery. Metabolites 8(1)
6. Lista S, Giegling I, Rujescu D (2014) Schizophrenia: blood-serum-plasma metabolomics. In: Martins-de-Souza D (ed) Proteomics and metabolomics in psychiatry. Karger Publishers, Basel, p 27–44. ISBN-10: 3318025992
7. Holmes E, Tsang TM, Huang JT-J, Leweke FM, Koethe D, Gerth CW et al (2006) Metabolic profiling of CSF: evidence that early intervention may impact on disease progression and outcome in schizophrenia. PLoS Med 3(8):e327. https://doi.org/10.1371/journal.pmed.0030327
8. He Y, Yu Z, Giegling I, Xie L, Hartmann AM, Prehn C, Adamski J et al (2012) Schizophrenia shows a unique metabolomics signature in plasma. Transl Psychiatry 2:e149. https://doi.org/10.1038/tp.2012.76
9. Zheng P, Wei Y-D, Yao G-E, Ren G-P, Guo J, Zhou C-J et al (2013) Novel urinary biomarkers for diagnosing bipolar disorder. Metabolomics 9(4):800–808
10. Chen J-J, Zhou C-J, Liu Z, Fu Y-Y, Zheng P, Yang D-Y et al (2015) Divergent urinary metabolic phenotypes between major depressive disorder and bipolar disorder identified by a combined GC-MS and NMR spectroscopic metabonomic approach. J Proteome Res 14(8):3382–3389
11. Sethi S, Brietzke E (2015) Omics-based biomarkers: application of metabolomics in neuropsychiatric disorders. Int J Neuropsychopharmacol 19(3):yv096. https://doi.org/10.1093/ijnp/pyv096
12. Lindon JC, Holmes E, Nicholson JK (2003) Peer reviewed: so what's the deal with metabonomics? Anal Chem 75(17):384A–391A
13. Friedhoff AJ, van Winkle E (1962) The characteristics of an amine found in urine of schizophrenic patients. J Nerv Ment Dis 135:550–555
14. Boulton AA, Pollitt RJ, Majer JR (1967) Identity of a urinary "pink spot" in schizophrenia and Parkinson's disease. Nature 215:132–134
15. Money TT, Bousman CA (2013) Metabolomics of psychotic disorders. Metabolomics 3:117. https://doi.org/10.4172/2153-0769.1000117
16. Kohler I, Hankemeier T, van der Graaf PH, Knibbe CAJ, van Hasselt JGC (2017) Integrating clinical metabolomics-based biomarker discovery and clinical pharmacology to enable precision medicine. Eur J Pharm Sci 109S:S15–S21
17. Fuhrer T, Zamboni N (2015) High-throughput discovery metabolomics. Curr Opin Biotechnol 31:73–78
18. Zhang A, Sun H, Wang P, Han Y, Wang X (2012) Modern analytical techniques in metabolomics analysis. Analyst 137(2):293–300
19. Alonso A, Marsal S, Julià A (2015) Analytical methods in untargeted metabolomics: state of the art in 2015. Front Bioeng Biotechnol 3:23. https://doi.org/10.3389/fbioe.2015.00023
20. Riekeberg E, Powers R (2017) New frontiers in metabolomics: from measurement to insight. F1000Res 6:1148. https://doi.org/10.12688/f1000research.11495.1
21. Wold S, Esbensen K, Geladi P (1987) Principal component analysis. Chemometrics Intellig Lab Syst 2(1):37–52
22. Bro R, Smilde AK (2014) Principal component analysis. Anal Methods 6(9):2812–2831

23. Bylesjö M, Rantalainen M, Cloarec O, Nicholson JK, Holmes E, Trygg J (2006) OPLS discriminant analysis: combining the strengths of PLS-DA and SIMCA classification. J Chemom 20(8-10):341–351
24. Berrueta LA, Alonso-Salces RM, Héberger K (2007) Supervised pattern recognition in food analysis. J Chromatogr A 1158(1-2):196–214
25. Peng C-YJ, Lee KL, Ingersoll GM (2002) An introduction to logistic regression analysis and reporting. J Educ Res 96(1):3–14
26. Merikangas KR, Jin R, He J-P, Kessler RC, Lee S, Sampson NA et al (2011) Prevalence and correlates of bipolar spectrum disorder in the world mental health survey initiative. Arch Gen Psychiatry 68(3):241–251
27. Kessler RC, Merikangas KR, Wang PS (2007) Prevalence, comorbidity, and service utilization for mood disorders in the United States at the beginning of the twenty-first century. Annu Rev Clin Psychol 3:137–158
28. Sagar R, Pattanayak RD (2017) Potential biomarkers for bipolar disorder: where do we stand? Indian J Med Res 145(1):7–16
29. Vieta E, Berk M, Schulze TG, Carvalho AF, Suppes T, Calabrese JR et al (2018) Bipolar disorders. Nat Rev Dis Primers 4:18008. https://doi.org/10.1038/nrdp.2018.8
30. Hilty DM, Leamon MH, Lim RF, Kelly RH, Hales RE (2006) A review of bipolar disorder in adults. Psychiatry 3(9):43–55
31. Shorter E (2009) The history of lithium therapy. Bipolar Disord 11(Suppl2):4–9
32. Coppen A, Healy D (1996) Biological psychiatry in Britain. In: Healy D (ed) The psychopharmacologists: interviews by David Healey. Chapman and Hall, London, pp 265–286
33. Marwaha S, Sal N, Bebbington P (2014) Adult psychiatric morbidity survey—bipolar disorder. In: NHS digital. https://digital.nhs.uk/data-and-information/publications/statistical/adult-psychiatric-morbidity-survey/adult-psychiatric-morbidity-survey-survey-of-mental-health-and-wellbeing-england-2014. Accessed 27 Sep 2018
34. Murray CJL, Lopez AD (1996) The global burden of disease. Harvard University Press, Cambridge, MA. ISBN:0-9655466-0-8
35. Cade JFJ (1949) Lithium salts in the treatment of psychotic excitement. Med J Aust 2(10):349–352
36. Malhi GS, Tanious M, Das P, Coulston CM, Berk M (2013) Potential mechanisms of action of lithium in bipolar disorder. CNS Drugs 27(2):135–153
37. Sethi S, Pedrini M, Rizzo LB, Zeni-Graiff M, Mas CD, Cassinelli AC et al (2017) 1H-NMR T2-edited, and 2D-NMR in bipolar disorder metabolic profiling. Int J Bipolar Disord 5(1):23. https://doi.org/10.1186/s40345-017-0088-2
38. Chen J-J, Liu Z, Fan S-H, Yang D-Y, Zheng P, Shao W-H et al (2014) Combined application of NMR- and GC-MS-based metabonomics yields a superior urinary biomarker panel for bipolar disorder. Sci Rep 4:5855. https://doi.org/10.1038/srep05855
39. Sussulini A, Prando A, Maretto DA, Poppi RJ, Tasic L, Banzato CEM et al (2009) Metabolic profiling of human blood serum from treated patients with bipolar disorder employing 1H NMR spectroscopy and chemometrics. Anal Chem 81(23):9755–9763
40. Liu M-L, Zheng P, Liu Z, Xu Y, Mu J, Guo J et al (2014) GC-MS based metabolomics identification of possible novel biomarkers for schizophrenia in peripheral blood mononuclear cells. Mol BioSyst 10(9):2398–2406
41. Xu X-J, Zheng P, Ren G-P, Liu M-L, Mu J, Guo J et al (2014) 2,4-Dihydroxypyrimidine is a potential urinary metabolite biomarker for diagnosing bipolar disorder. Mol Biosyst 10(4):813–819
42. Soeiro-de-Souza MG, Henning A, Machado-Vieira R, Moreno RA, Pastorello BF, da Costa LC et al (2015) Anterior cingulate Glutamate-Glutamine cycle metabolites are altered in euthymic bipolar I disorder. Eur Neuropsychopharmacol 25(12):2221–2229
43. Soeiro-de-Souza MG, Otaduy MCG, Machado-Vieira R, Moreno RA, Nery FG, Leite C et al (2018) Anterior cingulate cortex glutamatergic metabolites and mood stabilizers in euthymic

bipolar I disorder patients: a proton magnetic resonance spectroscopy study. Biol Psychiatry Cogn Neurosci Neuroimaging 3(12):985-991
44. Atagün Mİ, Şıkoğlu EM, Can SS, Uğurlu GK, Kaymak SU, Çayköylü A et al (2018) Neurochemical differences between bipolar disorder type I and II in superior temporal cortices: a proton magnetic resonance spectroscopy study. J Affect Disord 235:15–19
45. Chen J-J, Huang H, Zhao L-B, Zhou D-Z, Yang Y-T, Zheng P et al (2014) Sex-specific urinary biomarkers for diagnosing bipolar disorder. PLoS One 9(12):e115221. https://doi.org/10.1371/journal.pone.0115221
46. Haarman BCM, Riemersma-Van der Lek RF (2016) Volume, metabolites and neuroinflammation of the hippocampus in bipolar disorder. Bipolar Disord 56:21–33
47. Tannous J, Cao B, Stanley J, Amaral-Silva H, Soares J (2018) Metabolite abnormalities in the anterior white matter of patients with pediatric bipolar disorder. Biol Psychiatry 83(9, Supplement):S263. https://doi.org/10.1016/j.biopsych.2018.02.678
48. Yoshimi N, Futamura T, Kakumoto K, Salehi AM, Sellgren CM, Holmén-Larsson J et al (2016) Blood metabolomics analysis identifies abnormalities in the citric acid cycle, urea cycle, and amino acid metabolism in bipolar disorder. BBA Clin 5:151–158
49. Atagün Mİ, Şıkoğlu EM, Soykan Ç, Serdar Süleyman C, Ulusoy-Kaymak S, Çayköylü A et al (2017) Perisylvian GABA levels in schizophrenia and bipolar disorder. Neurosci Lett 637:70–74
50. Kageyama Y, Kasahara T, Morishita H, Mataga N, Deguchi Y, Tani M et al (2017) Search for plasma biomarkers in drug-free patients with bipolar disorder and schizophrenia using metabolome analysis. Psychiatry Clin Neurosci 71(2):115–123
51. Poletti S, Myint AM, Schüetze G, Bollettini I, Mazza E, Grillitsch D et al (2018) Kynurenine pathway and white matter microstructure in bipolar disorder. Eur Arch Psychiatry Clin Neurosci 268(2):157–168
52. Yoshimi N, Futamura T, Bergen SE, Iwayama Y, Ishima T, Sellgren C et al (2016) Cerebrospinal fluid metabolomics identifies a key role of isocitrate dehydrogenase in bipolar disorder: evidence in support of mitochondrial dysfunction hypothesis. Mol Psychiatry 21(11):1504–1510
53. Davanzo P, Thomas MA, Yue K, Oshiro T, Belin T, Strober M et al (2001) Decreased anterior cingulate myo-inositol/creatine spectroscopy resonance with lithium treatment in children with bipolar disorder. Neuropsychopharmacology 24(4):359–369
54. Ribeiro HC, Klassen A, Pedrini M, Carvalho MS, Rizzo LB, Noto MN et al (2017) A preliminary study of bipolar disorder type I by mass spectrometry-based serum lipidomics. Psychiatry Res 258:268–273
55. Liu M-L, Zhang X-T, Du X-Y, Fang Z, Liu Z, Xu Y et al (2015) Severe disturbance of glucose metabolism in peripheral blood mononuclear cells of schizophrenia patients: a targeted metabolomic study. J Transl Med 13:226. https://doi.org/10.1186/s12967-015-0540-y
56. Atack JR, Rapoport SI, Varley CL (1993) Characterization of inositol monophosphatase in human cerebrospinal fluid. Brain Res 613(2):305–308
57. Berridge MJ (1993) Inositol trisphosphate and calcium signalling. Nature 361(6410): 315–325
58. Manji HK, Potter WZ, Lenox RH (1995) Signal transduction pathways. Molecular targets for lithium's actions. Arch Gen Psychiatry 52(7):531–543
59. Yüksel C, Öngür D (2010) Magnetic resonance spectroscopy studies of glutamate-related abnormalities in mood disorders. Biol Psychiatry 68(9):785–794
60. Bak LK, Schousboe A, Waagepetersen HS (2006) The glutamate/GABA-glutamine cycle: aspects of transport, neurotransmitter homeostasis and ammonia transfer. J Neurochem 98(3):641–653
61. Erecińska M, Silver IA (1990) Metabolism and role of glutamate in mammalian brain. Prog Neurobiol 35(4):245–296
62. Frey BN, Stanley JA, Nery FG, Monkul ES, Nicoletti MA, Chen H-H et al (2007) Abnormal cellular energy and phospholipid metabolism in the left dorsolateral prefrontal cortex of

medication-free individuals with bipolar disorder: an in vivo 1H MRS study. Bipolar Disord 9(Suppl 1):119–127

63. Oertel-Knöchel V, Reinke B, Alves G, Jurcoane A, Wenzler S, Prvulovic D et al (2014) Frontal white matter alterations are associated with executive cognitive function in euthymic bipolar patients. J Affect Disord 155:223–233
64. Lu LH, Zhou XJ, Keedy SK, Reilly JL, Sweeney JA (2011) White matter microstructure in untreated first episode bipolar disorder with psychosis: comparison with schizophrenia. Bipolar Disord 13(7-8):604–613
65. Benedetti F, Yeh P-H, Bellani M, Radaelli D, Nicoletti MA, Poletti S et al (2011) Disruption of white matter integrity in bipolar depression as a possible structural marker of illness. Biol Psychiatry 69(4):309–317
66. Patel KR, Cherian J, Gohil K, Atkinson D (2014) Schizophrenia: overview and treatment options. PT 39(9):638–645
67. Owen MJ, Sawa A, Mortensen PB (2016) Schizophrenia. Lancet 388(10039):86–97
68. Kahn RS, Sommer IE, Murray RM, Meyer-Lindenberg A, Weinberger DR, Cannon TD et al (2015) Schizophrenia. Nat Rev Dis Primers 1:15067. https://doi.org/10.1038/nrdp.2015.67
69. Lavretsky H (2008) History of schizophrenia as a psychiatric disorder. In: Mueser KT, Jeste DV (eds) Clinical handbook of schizophrenia. Guilford Publications, New York, pp 3–13. ISBN:9781609182373
70. Jentsch JD, Roth RH (1999) The neuropsychopharmacology of phencyclidine: from NMDA receptor hypofunction to the dopamine hypothesis of schizophrenia. Neuropsychopharmacology 20:201–225
71. Orešič M, Tang J, Seppänen-Laakso T, Mattila I, Saarni SE, Saarni SI et al (2011) Metabolome in schizophrenia and other psychotic disorders: a general population-based study. Genome Med 3(3):19. https://doi.org/10.1186/gm233
72. Yang J, Chen T, Sun L, Zhao Z, Qi X, Zhou K et al (2013) Potential metabolite markers of schizophrenia. Mol Psychiatry 18(1):67–78
73. Hashimoto K, Fukushima T, Shimizu E, Komatsu N, Watanabe H, Shinoda N et al (2003) Decreased serum levels of D-serine in patients with schizophrenia: evidence in support of the N-methyl-D-aspartate receptor hypofunction hypothesis of schizophrenia. Arch Gen Psychiatry 60(6):572–576
74. Palomino A, González-Pinto A, Aldama A, González-Gómez C, Mosquera F, González-García G et al (2007) Decreased levels of plasma glutamate in patients with first-episode schizophrenia and bipolar disorder. Schizophr Res 95(1-3):174–178
75. Prell GD, Green JP, Kaufmann CA, Khandelwal JK, Morrishow AM, Kirch DG et al (1995) Histamine metabolites in cerebrospinal fluid of patients with chronic schizophrenia: their relationships to levels of other aminergic transmitters and ratings of symptoms. Schizophr Res 14(2):93–104
76. Xuan J, Pan G, Qiu Y, Yang L, Su M, Liu Y et al (2011) Metabolomic profiling to identify potential serum biomarkers for schizophrenia and risperidone action. J Proteome Res 10(12):5433–5443
77. Cai H-L, Zhu R-H, Li H-D (2010) Determination of dansylated monoamine and amino acid neurotransmitters and their metabolites in human plasma by liquid chromatography-electrospray ionization tandem mass spectrometry. Anal Biochem 396(1):103–111
78. Fukushima T, Iizuka H, Yokota A, Suzuki T, Ohno C, Kono Y et al (2014) Quantitative analyses of schizophrenia-associated metabolites in serum: serum D-lactate levels are negatively correlated with gamma-glutamylcysteine in medicated schizophrenia patients. PLoS One 9(7):e101652. https://doi.org/10.1371/journal.pone.0101652
79. Phillips M, Sabas M, Greenberg J (1993) Increased pentane and carbon disulfide in the breath of patients with schizophrenia. J Clin Pathol 46(9):861–864
80. Al Awam K, Haußleiter IS, Dudley E, Donev R, Brüne M, Juckel G et al (2015) Multiplatform metabolome and proteome profiling identifies serum metabolite and protein signatures as prospective biomarkers for schizophrenia. J Neural Transm 122(Suppl 1):S111–S122

81. Creveling CR, Daly JW (1967) Identification of 3,4-dimethoxyphenethylamine from schizophrenic urine by mass spectrometry. Nature 216(5111):190–191
82. Ross BM, Shah S, Peet M (2011) Increased breath ethane and pentane concentrations in currently unmedicated patients with schizophrenia. OJPsych 1(01):1–7
83. Koike S, Bundo M, Iwamoto K, Suga M, Kuwabara H, Ohashi Y et al (2014) A snapshot of plasma metabolites in first-episode schizophrenia: a capillary electrophoresis time-of-flight mass spectrometry study. Transl Psychiatry 4:e379. https://doi.org/10.1038/tp.2014.19
84. Baruah S, Waziri R, Hegwood TS, Mallis LM (1991) Plasma serine in schizophrenics and controls measured by gas chromatography-mass spectrometry. Psychiatry Res 37(3):261–270
85. Petrovchich I, Sosinsky A, Konde A, Archibald A, Henderson D, Maletic-Savatic M et al (2016) Metabolomics in schizophrenia and major depressive disorder. Front Biol 11(3):222–231
86. Ito C (2004) The role of the central histaminergic system on schizophrenia. Drug News Perspect 17(6):383–387
87. Kaddurah-Daouk R, McEvoy J, Baillie RA, Lee D, Yao JK, Doraiswamy PM et al (2007) Metabolomic mapping of atypical antipsychotic effects in schizophrenia. Mol Psychiatry 12(10):934–945
88. Kantrowitz JT, Malhotra AK, Cornblatt B, Silipo G, Balla A, Suckow RF et al (2010) High dose D-serine in the treatment of schizophrenia. Schizophr Res 121(1-3):125–130
89. Nunes EA, MacKenzie EM, Rossolatos D, Perez-Parada J, Baker GB, Dursun SM (2012) D-serine and schizophrenia: an update. Expert Rev Neurother 12(7):801–812
90. Nunes de Paiva MJ, Menezes HC, de Lourdes CZ (2014) Sampling and analysis of metabolomes in biological fluids. Analyst 139(15):3683–3694
91. Li N, Song YP, Tang H, Wang Y (2016) Recent developments in sample preparation and data pre-treatment in metabonomics research. Arch Biochem Biophys 589:4–9
92. Pontes JGM, Brasil AJM, Cruz GCF, de Souza RN, Tasic L (2017) NMR-based metabolomics strategies: plants, animals and humans. Anal Methods 9:1078–1096
93. Smolinska A, Blanchet L, Buydens LMC, Wijmenga SS (2012) NMR and pattern recognition methods in metabolomics: from data acquisition to biomarker discovery: a review. Anal Chim Acta 750:82–97
94. Beckonert O, Keun HC, Ebbels TMD, Bundy J, Holmes E, Lindon JC et al (2007) Metabolic profiling, metabolomic and metabonomic procedures for NMR spectroscopy of urine, plasma, serum and tissue extracts. Nat Protoc 2(11):2692–2703
95. Pontes JGM, Brasil AJM, Cruz GCF, de Souza RN, Tasic L (2017) ^{1}H NMR metabolomic profiling of human and animal blood serum samples. Methods Mol Biol 1546:275–282
96. Lenz EM, Wilson ID (2007) Analytical strategies in metabonomics. J Proteome Res 6(2):443–458
97. Barbosa BS, Martins LG, Costa TBBC, Cruz G, Tasic L (2018) Qualitative and quantitative NMR approaches in blood serum lipidomics. Methods Mol Biol 1735:365–379
98. Stoop MP, Coulier L, Rosenling T, Shi S, Smolinska AM, Buydens L et al (2010) Quantitative proteomics and metabolomics analysis of normal human cerebrospinal fluid samples. Mol Cell Proteomics 9(9):2063–2075
99. Sakurai T, Yamada Y, Sawada Y, Matsuda F, Akiyama K, Shinozaki K et al (2013) PRIMe update: innovative content for plant metabolomics and integration of gene expression and metabolite accumulation. Plant Cell Physiol 54(2):e5–e5. https://doi.org/10.1093/pcp/pcs184
100. Chong J, Soufan O, Li C, Caraus I, Li S, Bourque G et al (2018) MetaboAnalyst 4.0: towards more transparent and integrative metabolomics analysis. Nucleic Acids Res 46(W1):W486–W494
101. Worley B, Powers R (2014) MVAPACK: a complete data handling package for NMR metabolomics. ACS Chem Biol 9(5):1138–1144
102. Wishart DS, Feunang YD, Marcu A, Guo AC, Liang K, Vázquez-Fresno R et al (2018) HMDB 4.0: the human metabolome database for 2018. Nucleic Acids Res 46:D608–D617
103. Ulrich EL, Akutsu H, Doreleijers JF, Harano Y, Ioannidis YE, Lin J, Livny M, Mading S, Maziuk D, Miller Z, Nakatani E, Schulte CF, Tolmie DE, Kent Wenger R, Yao H, Markley JL (2007) BioMagResBank. Nucleic Acids Res 36(Database issue):D402–D408

104. Cui Q, Lewis IA, Hegeman AD, Anderson ME, Li J, Schulte CF et al (2008) Metabolite identification via the Madison Metabolomics Consortium Database. Nat Biotechnol 26:162–164
105. Ellinger JJ, Chylla RA, Ulrich EL, Markley JL (2012) Databases and software for NMR-based metabolomics. Curr Metabolomics 1:28–40
106. Lommen A (2009) MetAlign: interface-driven, versatile metabolomics tool for hyphenated full-scan mass spectrometry data preprocessing. Anal Chem 81(8):3079–3086
107. Pluskal T, Castillo S, Villar-Briones A, Orešič M (2010) MZmine 2: modular framework for processing, visualizing, and analyzing mass spectrometry-based molecular profile data. BMC Bioinformatics 11:395. https://doi.org/10.1186/1471-2105-11-395
108. Smith CA, Want EJ, O'Maille G, Abagyan R, Siuzdak G (2006) XCMS: processing mass spectrometry data for metabolite profiling using nonlinear peak alignment, matching and identification. Anal Chem 78(3):779–787
109. Styczynski MP, Moxley JF, Tong LV, Walther JL, Jensen KL, Stephanopoulos GN (2007) Systematic identification of conserved metabolites in GC/MS data for metabolomics and biomarker discovery. Anal Chem 79(3):966–973
110. Chen W, Ma C, Miao A, Pang S, Qi D, Wang W (2017) Chemometric methods for the analysis of graftage-related black tea aroma variation by solid phase mirco-extraction and gas chromatography-mass spectrometry. Agilent Technologies. https://www.agilent.com/cs/library/applications/5991-8330EN.pdf
111. Piotto M, Saudek V, Sklenář V (1992) Gradient-tailored excitation for single-quantum NMR spectroscopy of aqueous solutions. J Biomol NMR 2(6):661–665
112. Liu M, Mao X, Ye C, Huang H, Nicholson JK, Lindon JC (1998) Improved WATERGATE pulse sequences for solvent suppression in NMR spectroscopy. J Magn Reson 132(1):125–129
113. Carr HY, Purcell EM (1954) Effects of diffusion on free precession in nuclear magnetic resonance experiments. Phys Rev 94:630–638
114. Meiboom S, Gill D (1958) Modified spin-echo method for measuring nuclear relaxation times. Rev Sci Instrum 29:688–691
115. Sethi S, Hayashi MAF, Barbosa BS, Pontes JGM, Tasic L, Brietzke E (2017) Lipidomics, biomarkers, and schizophrenia: a current perspective. Adv Exp Med Biol 965:265–290
116. Nagana Gowda GA, Raftery D (2017) Recent advances in NMR-based metabolomics. Anal Chem 89(1):490–510
117. Marshall DD, Lei S, Worley B, Huang Y, Garcia-Garcia A, Franco R et al (2015) Combining DI-ESI–MS and NMR datasets for metabolic profiling. Metabolomics 11(2):391–402
118. López-López Á, López-Gonzálvez Á, Barker-Tejeda TC, Barbas C (2018) A review of validated biomarkers obtained through metabolomics. Expert Rev Mol Diagn 18(6):557–575
119. Matsumoto J, Sugiura Y, Yuki D, Hayasaka T, Goto-Inoue N, Zaima N et al (2011) Abnormal phospholipids distribution in the prefrontal cortex from a patient with schizophrenia revealed by matrix-assisted laser desorption/ionization imaging mass spectrometry. Anal Bioanal Chem 400(7):1933–1943
120. Amstalden Van Hove ER, Smith DF, Heeren RMA (2010) A concise review of mass spectrometry imaging. J Chromatogr A 1217:3946–3954
121. Bodzon-Kulakowska A, Sude P (2016) Imaging mass spectrometry: instrumentation, applications, and combination with other visualization techniques. Mass Espectrom Rev 35(1):147–169

Chapter 15
Early Detection and Treatment of Patients with Alzheimer's Disease: Future Perspectives

Francesca L. Guest

Abstract Alzheimer's disease affects approximately 6% of people over the age of 65 years. It is characterized as chronic degeneration of cortical neurons, with loss of memory, cognition and executive functions. As the disease progresses, it is accompanied by accumulation of amyloid plaques and neurofibrillary tangles in key areas of the brain, leading to a loss of neurogenesis and synaptic plasticity in the hippocampus, along with changes in the levels of essential neurotransmitters such as acetylcholine and glutamate. Individuals with concomitant diseases such as depression, diabetes and cardiovascular disorders have a higher risk of developing Alzheimer's disease, and those who have a healthier diet and partake in regular exercise and intellectual stimulation have a lower risk of developing the disorder. This chapter describes the advances made in early diagnosis of Alzheimer's disease as this could help to improve outcomes for the patients by facilitating earlier treatment.

Keywords Alzheimer's disease · Biomarkers · Imaging · Proteomics · Metabolomics · Lab-on-a-chip · Smartphone monitoring

15.1 Introduction

Alzheimer's disease is the most prevalent form of dementia in the aged population. It affects approximately 0.6% of the world population and occurs in 6% of people over the age of 65 years [1]. As the average life expectancy continues to increase, this percentage is expected to increase to 1.2% of the world's population by the year 2050 [2, 3]. Alzheimer's disease is characterized as a chronic degeneration of cortical neurons, resulting in decline of memory and cognition, along with loss of executive function and the manifestation of behavioural abnormalities [4]. As the disease progresses, amyloid and neurofibrillary proteins accumulate in localized areas of the brain, forming plaques and tangles which can disrupt neuronal signalling and ultimately lead to the loss of neurons and brain tissue [5]. This results in loss of

F. L. Guest (✉)
Taunton and Somerset NHS Trust, Musgrove Park Hospital, Taunton, Somerset, UK

P. C. Guest (ed.), *Reviews on Biomarker Studies in Psychiatric and Neurodegenerative Disorders*, Advances in Experimental Medicine and Biology 1118,
https://doi.org/10.1007/978-3-030-05542-4_15

neurogenesis and synaptic plasticity in the hippocampus, the main region of the brain involved in regulation of cognition and memory [6, 7]. These effects are associated with alterations in the levels of neurotransmitters such as acetylcholine and glutamate, which are essential in both brain and bodily function [8, 9]. Ultimately, these combined deficits are fatal, with median survival times of 4.2 and 5.7 years after diagnosis for males and females, respectively [10].

Long-term memories do not appear to be affected in the early stages of Alzheimer's disease although the associated memory loss can impair daily life at an increasing level as the illness advances. Alzheimer's disease is categorized into three clinical stages associated with specific symptoms (Table 15.1) [11]. Similar to many diseases, the development of Alzheimer's disease has been linked to advancing age and shows a gender preference, with twice as many females over the age of 65 developing the disease compared with males [12]. The higher prevalence seen in females may be due to changes incurred after menopause and driven by deficits of the hormone oestrogen. In addition, specific genes have been identified as risk factors although relevant studies have not been conclusive [13–15]. However,

Table 15.1 Clinical stages of Alzheimer's disease

Stage 1 Early-stage symptoms (lasts 2–5 years)	Difficulties in recalling the right word or name
	Repetition in conversations
	Difficulties in performing tasks
	Not remembering details of newly learned information
	Losing or misplacing objects
	Problems with planning, organizational skills and problem-solving
	Getting lost in a familiar place
	Forgetting important dates
	Feelings of moodiness, depression or withdrawal
Stage 2 Middle-stage symptoms (lasts 2–10 years)	Forgetting events regarding the past
	Feeling withdrawn, moody or angry in some situations
	Unable to recall addresses or telephone numbers
	Confused about locations, times or dates
	Difficulties controlling bladder/bowels
	Changes in sleep patterns
	Sometimes wandering and becoming lost
	Changes in behaviour
	Becoming increasingly dependent
Stage 3 Late-stage symptoms (lasts 1–3 years)	Requiring full-time care
	Loss of awareness of recent experiences and environment
	Increasingly disorientated
	Loss of physical skills like walking and swallowing
	Increasing difficulties of communication
	Increasingly vulnerable to infection
	Experiencing delusions/hallucinations
	Increasingly aggressive/violent

individuals with concomitant diseases such as depression, anxiety, diabetes, obesity and cardiovascular disorders are known to have a higher risk of developing Alzheimer's disease [16]. In contrast, people who have led a healthier life with regard to a nourishing diet, regular exercise, intellectual stimulation and low alcohol intake show a lower risk of developing the disorder [17].

15.2 Physical Signs of Alzheimer's Disease

At the molecular level, the symptoms of Alzheimer's disease are associated with neuritic plaques and neurofibrillary tangles, related to accumulation of amyloid-beta peptide (Aβ) and to hyperphosphorylation of the microtubule-associated tau protein in neurons (Fig. 15.1) [18]. The Aβ peptide is generated by proteolytic processing of the larger amyloid precursor protein. Abnormal processing of this precursor largely results in two different versions of the peptide, consisting of either 40 amino acids ($A\beta_{1-40}$) or a carboxy-terminally extended form ($A\beta_{1-42}$) containing two additional amino acids. The $A\beta_{1-42}$ peptide is "sticky" in nature and forms toxic insoluble plaques in the brain [19]. In turn, this leads to perturbed cytoskeletal changes, neuronal dysfunction and cell death. Associated with the clinical stages outlined above, stage 1 continues until the accumulation of Aβ in the brain leads to a triggering of the amyloid cascade. In stage 2, the early pathological features may be present, showing physical changes to the brain. In the final stage, the clinical dementia is associated with accumulation of plaques and neurofibrillary tangles in key brain areas [4].

15.3 The Benefits of Early Detection

As is the case with most disorders, early diagnosis of Alzheimer's disease could help improve outcomes for the patients. For example, early detection could lead to earlier access to treatment, advice and support, which could slow the progression of the disease [20]. Diagnosis is carried out through an interview of the individual by a doctor or specialist. The interview normally considers parameters such as memory, higher thinking and other abilities and is administered largely verbally with written components and instructions to motor responses or simple tasks, such as the Mini Mental State Examination (MMSE), Montreal Cognitive Assessment (MoCA), The Addenbrooke's Cognitive Examination-Revised (ACE-R), Cambridge Cognitive Examination (CAMCOG) and the Repeatable Battery for the Assessment of Neuropsychological Status (RBANS). There are also a number of tests designed to measure subtle cognitive changes which have been applied to aid in patient assessment. For example, the Face Name Associative Memory Exam (FNAME) assesses memory retrieval [21]. In this test, the patients are shown pictures of unfamiliar faces with associated names, and then they are asked to recall as many of these as they can by assigning names to faces. The Short-Term Memory Binding

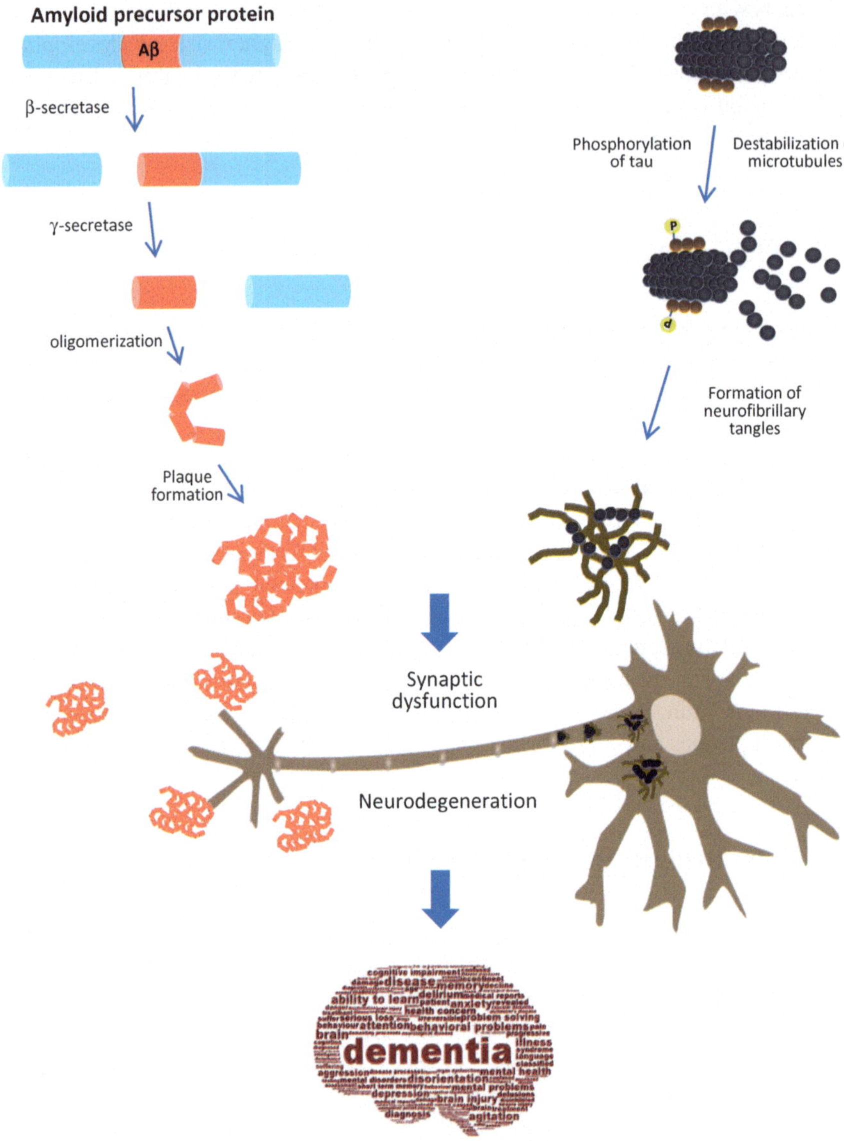

Fig. 15.1 Alzheimer's disease is characterized by extracellular deposits of amyloid plaques and intraneuronal neurofibrillary tangles, leading to neurodegeneration and dementia

test examines a patient's ability to recognize differences between two arrays of coloured geometric shapes [22]. In the spatial pattern separation task, an individual is shown an image containing a localized shape (such as a small circle) and then asked to pinpoint the location of this circle on another image containing multiple similar shapes [23].

In some cases, patients may be given a magnetic resonance imaging (MRI) or computerized tomography (CT) scan of the brain, to determine whether or not atrophy of the hippocampus or other brain areas has occurred [20]. The use of radiographic brain imaging may also rule out other possible causes of Alzheimer's disease-like symptoms such as a stroke or development of a brain tumour. Focused testing based on brain imaging or detection of biomarkers in the cerebrospinal fluid or the blood may help to identify those individuals at risk of developing the disease even at the earliest pre-symptomatic stages [24–27]. As previously stated, earlier detection leads to intervention earlier in the pathogenic process, which is thought to produce a greater clinical benefit [28, 29].

15.4 Treatment Approaches

Several treatment approaches currently approved aim to improve the lives of people suffering with Alzheimer's disease. These methods include receiving support, medications, nutritional alterations and activities. The main Food and Drug Administration (FDA)-approved medications in use for Alzheimer's disease consist of three main strategic classes: (1) inhibition of cholinesterase, (2) stimulating acetylcholine synthesis and (3) antagonism of N-methyl-D-aspartate (NMDA) receptors (Table 15.2) [30–35]. Although it was the first drug to be marketed for treatment of Alzheimer's disease, tacrine has been withdrawn from clinical use due to side effects [36]. However, similar drugs may re-enter the market with the synthesis of tacrine-like derivatives which show a reduced side effect profile [36, 37]. Donepezil, galantamine or rivastigmine is commonly prescribed to treat patients in the early of middle stages of Alzheimer's disease and may help to decrease memory problems and improve concentration and focus. Memantine may be prescribed for those patients in the middle to late stages of the disease to improve mental abilities and daily living by helping to counteract neuronal damage due to pathological activation of NMDA receptors via excess glutamatergic neurotransmission [35, 38]. In addition, antidepressants such as fluoxetine, citalopram or sertraline fluoxetine may be prescribed for Alzheimer's disease patients suffering from depression- or anxiety-related

Table 15.2 FDA-approved drugs for treatment of Alzheimer's disease

Drug	Class	Mechanism of action
Donepezil	Cholinesterase inhibitor	Prevents acetylcholine breakdown in the brain
Galantamine	Cholinesterase inhibitor/ acetylcholine agonist	Prevents acetylcholine breakdown and stimulates acetylcholine production
Tacrine	Cholinesterase inhibitor/ acetylcholine agonist	Prevents acetylcholine breakdown and stimulates acetylcholine production
Rivastigmine	Cholinesterase inhibitor	Prevents acetylcholine and butyrylcholine breakdown in the brain
Memantine	NMDA receptor antagonist	Blocks toxic effects of excessive glutamate and regulates glutamate synthesis

symptoms [39]. Likewise, the monoamine oxidase inhibitor mirtazapine dosed at night can aid sleep, which is often disturbed in Alzheimer's disease patients; medications such as trazodone can be prescribed for agitation, and lorazepam can be used in late-stage patients with aggression.

There are also behavioural methods that patients can adopt to manage memory difficulties such as setting up daily routines and reminders and using smartphone apps linked with an online calendar [40, 41]. It is also important that Alzheimer's disease patients maintain their daily activities as much as possible, considering the preponderance of evidence showing that both mental and physical exercise can lead to improvement in some deficits [42, 43].

15.5 New Treatments for Alzheimer's Disease

Although all of the FDA-approved drugs for treatment of Alzheimer's disease directly target neurotransmitter pathways, a number of alternate therapeutic avenues are currently being tested in clinical trials. Several compounds have been tested that inhibit the β-secretase and γ-secretase enzymes, which are involved in proteolytic generation of $A\beta_{1-42}$ [44]. However, the development of compounds targeting the γ-secretase has been discontinued due to off-target effects [45, 46]. Some of the inhibitors of β-secretase are already FDA-approved for diabetes, including the peroxisome proliferator-activated receptor-gamma agonist pioglitazone [47, 48]. Administration of this drug to patients with mild to moderate Alzheimer's disease was shown to reduce Aβ levels along with increased cerebral blood flow and better cognitive test results [49–51]. In addition, vaccines aimed at reducing the Aβ load are undergoing clinical testing, although it appears that these may only be effective in the early stages of the disease [9, 52]. Along the same lines, a vaccine called AADvac1, which inhibits build-up of tau bundles in neurons, is currently undergoing testing in early clinical investigations. Thus far, the drug has successfully passed through safety testing in a randomized, double-blind, placebo-controlled, phase 1 trial [53–55]. Further studies are needed to determine if this vaccine has clinical efficacy.

15.5.1 Statins

Several investigations have demonstrated that high cholesterol levels are linked to amyloid deposition [56]. Because of this, a number of clinical studies have tested the effects of cholesterol-lowering drugs (such as statins) on various endpoint measures in Alzheimer's disease patients [57–59]. These compounds may be helpful through their ability to decrease inflammation and increase blood flow. There is some evidence that cholesterol-lowering drugs may aid cognition, lower Aβ-induced neurotoxicity and prevent dementia; however studies in this area have, as yet, been inconclusive. As with all new treatment approaches, further clinical trials are required to determine if these compounds are efficacious.

15.5.2 *Anti-inflammatory Agents*

There is compelling evidence that inflammation and oxidative reduction processes are disturbed in individuals with Alzheimer's disease. The hallmarks of Alzheimer's disease are the neurofibrillary tangles and Aβ plaques, although carbonyl and oxidative stress are also present [60]. Furthermore, pro-inflammatory activation of astroglia and microglia has also been found in Alzheimer's disease patients [61]. Therefore, a number of anti-inflammatory drugs have been tested in attempts to improve symptoms [62]. When tested, compounds such as cyclooxygenase inhibitors and glucocorticoids demonstrated little efficacy with some adverse effects [63]. Another anti-inflammatory agent called etanercept, which targets tumour necrosis factor, showed some efficacy with improvement of cognitive symptoms in a case study [64] and is now undergoing testing in clinical trials [65]. In a similar manner, the anti-inflammatory/antioxidant compound "curcumin" has shown neuroprotective functions such as preventing tau aggregation, inhibiting the formation of new Aβ aggregates and disaggregating existing ones [66, 67]. Therefore, clinical trials investigating the effects of this compound in Alzheimer's disease patients should be performed.

15.5.3 *Caffeine*

Several studies have been conducted which indicate that higher coffee consumption is associated with a lower risk of developing Alzheimer disease [68]. A more recent study showed that there was "J-shaped" association between coffee intake and cognitive disorders, with the lowest risk occurring at a consumption level of 1–2 cups of coffee per day [69]. The mechanism of action of caffeine occurs through antagonism of various subclasses of adenosine receptors, although xanthine metabolites, such as theobromine and theophylline, may also contribute to the beneficial effects on brain function [70]. However, further randomized trials and cohort studies are warranted to confirm these findings, and additional research should be conducted to identify the caffeine metabolites responsible for any protective effects on the brain.

15.5.4 *Diet and Physical Exercise*

Increasing evidence indicates that healthy eating and increased physical exercise may be helpful in the prevention of Alzheimer's disease [71–73]. A clinical study found that an administration of a medicinal food called CerefolinNAC® over 18 months significantly reduced cortical and hippocampal atrophy rates in patients with Alzheimer's disease as well as those with cognitive impairment due to vascular dementia [74]. Another study showed that low-dose administration of a triglyceride-based medicinal food called Axona® might be effective in those with mild Alzheimer's disease [75]. In addition, the traditional Mediterranean diet has shown promise for reducing the risk of developing senile dementia and for decreasing symptoms after diagnosis. A recent 3-year brain imaging study found lower amyloid deposition and neurodegeneration in middle-aged adults with higher Mediterranean diet adherence ($n = 34$) compared to those who followed this diet less strictly ($n = 36$) [76]. Nevertheless, further clinical studies are needed including higher numbers of individuals as well as Alzheimer's patients at different disease stages. Other potential non-pharmaceutical treatments include physical exercise, which is already known to have beneficial effects in variety of diseases [72, 73]. It is already known that physical activity can reduce effects commonly seen as hallmark characteristics of Alzheimer's disease. For example, studies have reported that exercise can reduce oxidative stress and inflammation [77] and improve insulin signalling in many tissues of the body [78]. Exercise has also been found to increase neurogenesis in the hippocampus [79]. A number of trials are now underway to test the effects of aerobic exercise on cognition, cerebrospinal fluid biomarker levels and MRI endpoints in elderly patients with mild cognitive impairment [80, 81]. It should be noted that some studies have already reported negative outcomes on endpoint measurements such as the levels of cerebrospinal fluid biomarkers [82] and clinically important cognitive outcomes [83]. Therefore, further studies are needed before solid conclusions can be made.

15.6 Imaging Biomarkers for Diagnosis and Monitoring of Alzheimer's Disease

In Alzheimer's disease, some pathophysiological abnormalities precede overt clinical symptoms by several years, and many of these changes can be detected by biomarkers [84]. Currently, six biomarkers have been incorporated as clinical diagnostic criteria and have also been incorporated into some clinical studies. These are based on either Aβ plaque deposition or tau hyperphosphorylation. Cerebrospinal fluid levels of $A\beta_{1-42}$, total tau and phosphorylated tau (Thr_{181}/Thr_{231}) have already been used for diagnosis and risk assessment in prodromal phases for mild cognitive impairment and dementia [85]. Magnetic resonance imaging (MRI) allows the visualization of structural and functional abnormalities through different stages of disease pathophysiology [86]. Volumetric MRI has been the most widely used approach for visualization of hippocampal atrophy, and newer techniques such as diffusion-tensor imaging (DTI) and functional MRI can provide additional information on structural and functional neuronal connectivity [87–89].

15.6.1 Aβ Deposition

The use of positron emission tomography (PET) in combination with various molecular imaging agents can detect several aspects of Alzheimer's disease pathophysiology, such as Aβ plaque deposition and accumulation of tau tangles, along with perturbations of neurotransmitter receptors, metabolism and neuroinflammation [90]. The most frequently used PET tracer for detecting Aβ plaque deposition in vivo is ^{11}C-Pittsburgh compound-B (PiB) [91]. Due to the short half-life of ^{11}C (20 min), other compounds containing an isotope with a longer half-life, such as ^{18}F (110 min), have also been developed [92]. In mild cognitive impairment patients, a positive amyloid PET scan is a useful biomarker indicating that these individuals have a high risk of developing Alzheimer's disease [93]. Along these same lines, two of the ^{18}F compounds, florbetapir and flutemetamol, have been used for detection of both preclinical and overt Alzheimer's disease [94–96]. Another PET study used a combination of ^{18}F-fluoro-D-glucose (FDG) and ^{18}F-florbetaben (FBB) and found that Aβ deposition was correlated with cortical dysfunction [97]. Together, these findings indicate that PET amyloid imaging is a useful biomarker for identifying pre-symptomatic individuals at high risk of conversion to Alzheimer's disease.

15.6.2 Inflammation

Inflammation in the brain may be an early marker of Alzheimer's disease due to microglial activation following the formation of amyloid plaques [98, 99]. PET imaging using ^{11}C-L-deuterium-deprenyl (DED), which detects astrocytic monoamine oxidase B, found increased binding in patients with mild cognitive impairment, indicating that astrocytosis may occur early in Alzheimer's disease pathophysiology [100, 101]. Taken together, these findings indicate that microglia may serve as a novel drug target for Alzheimer's disease.

15.6.3 Neurofibrillary Tangles

The misfolding and aggregation of hyperphosphorylated tau into neurofibrillary tangles is known to be a key component of Alzheimer's disease pathophysiology [102]. A number of radiotracers tau PET imaging have been developed, including ^{11}C-PBB3 (2-((1E,3E)-4-(6-(^{11}C-methylamino)pyridin-3-yl)buta-1,3-dienyl) benzo[d]thiazol-6-ol), ^{18}F-THK-5105 (6-[(3-^{18}F-fluoro-2-hydroxy)propoxy]-2-(4-dimethylaminophenyl)quinoline), ^{18}F-THK-5117 (2-(4-methylaminophenyl)-6-[(3-^{18}F]-fluoro-2-hydroxy)propoxy]quinoline) and ^{18}F-T808 (2-(4-(2-^{18}F]-fluoroethyl) piperidin-1-yl)benzo[4,5]imidazo[1,2-α]pyrimidine) [103]. A study using ^{11}C-PBB3 found that high levels of tau tangles were associated with ageing, low-level education, cognitive performance and high Aβ plaque burden [104]. This was consistent with another study using ^{18}F-THK-5117 as the PET tracer [105]. This latter study found that the progression of neurofibrillary pathology was associated with cognitive deterioration in Alzheimer's disease patients.

15.6.4 Metabolism

The glucose analogue 2-deoxy-2-^{18}F-fluoro-D-glucose (^{18}F-FDG) has been used to assess cerebral glucose metabolism as correlate of neuronal function in specific brain regions [106]. In Alzheimer's disease patients, the metabolic signature normally shows hypometabolism in the parietotemporal association, medial temporal, posterior cingulate and frontal cortices [107–110]. Another study showed progressive reductions in PET glucose metabolism occurred years before the appearance of clinical symptoms in those who later developed Alzheimer's disease [111]. In addition, the losses in the hippocampus occurred before those in cortical regions.

15.7 Biomarkers in Body Fluids

15.7.1 Cerebrospinal Fluid

Since, the cerebrospinal fluid is in contact with the central nervous system, molecular biomarkers in this body fluid can be used to monitor brain function. For example, cerebrospinal fluid measurements of the levels of $A\beta_{1-42}$, $A\beta_{1-40}$, tau and hyperphosphorylated tau can serve as specific biomarkers for risk of developing Alzheimer's disease, as well as for diagnosis and monitoring of disease progression [84, 85]. In general, higher $A\beta_{1-42}$ levels tend to be associated with greater impairments in cognition, and studies have found that just measuring this peptide alone can discriminate Alzheimer's disease patients from controls and from individuals with mild cognitive impairment, with a sensitivity of approximately 90% [112]. The ratio of $A\beta_{1-42}/A\beta_{1-40}$ appears to have even stronger diagnostic accuracy for Alzheimer's disease and concordance with Aβ PET analyses, compared to measuring levels of $A\beta_{1-42}$ alone [113–116].

It is known that mild cognitive impairment patients progress to dementia at a higher rate compared to healthy age-matched controls [117], and this is associated with cerebrospinal fluid levels of $A\beta_{1-42}$, total tau, and phosphorylated tau_{181} [112, 118]. However, these biomarkers lack sufficient discriminatory power to distinguish patients who will progress to Alzheimer's disease compared to those who develop other neurodegenerative conditions such as Lewy body dementia, frontotemporal dementia, vascular dementia and Creutzfeldt-Jakob disease [119–121]. As a potential biomarker of synaptic degeneration, a number of studies have analysed cerebrospinal fluid levels of neurogranin. These showed that high levels of this protein were predictive of prodromal Alzheimer's disease in mild cognitive impairment patients and were also predictive of hippocampal losses measured by MRI and FDG PET [122]. Other synaptic proteins such as synaptosomal-associated protein 25 and synaptotagmin-1 have also been analysed in cerebrospinal fluid and found to be altered in prodromal dementia and Alzheimer's disease subjects [123, 124]. These findings will require validation using larger cohorts which include patients with various neurodegenerative conditions. However, it should be stressed that cerebrospinal fluid biomarkers have limited clinical applications, given their invasive nature. Thus, more recent efforts have focussed on developing biomarker tests for Alzheimer's disease in blood plasma and serum, which are less invasively obtained [125].

15.7.2 Blood, Serum and Plasma

Due to the mixed pathology in dementia patients, it is critical that any identified biomarkers can distinguish Alzheimer's disease from other related disorders, including frontotemporal dementia, Parkinson's disease, amyotrophic lateral sclerosis and psychiatric disorders. Furthermore, there is an urgent need to standardize biomarker discovery approaches across different laboratories to increase the potential of translating the most successful biomarker-based tests into clinical use [122]. Plasma levels of $A\beta_{1-42}$ and $A\beta_{1-40}$ may be correlated with Alzheimer's disease although some conflicting information has been reported on this [126, 127]. In addition, plasma levels of these peptides show poor correlation with the corresponding cerebrospinal fluid levels [128]. Another study found that although the $A\beta_{1-42}/A\beta_{1-40}$ ratio did show a weak correlation between plasma and cerebrospinal fluid, the ratio was significantly lower in patients with dementia due to Alzheimer's disease compared to those with dementia due to other causes [129]. In contrast the plasma levels of phosphorylated tau_{181} show a stronger correlation with brain Aβ deposition and neurofibrillary tangles as determined by PET [130]. In addition, a recent prospective study of 5309 elderly females found that high serum levels of two tau fragments (tau-A and tau-C) were associated with a lower risk of dementia and Alzheimer's disease [131].

One multiplex immunoassay study of serum identified a panel comprised of 30 biomarkers which distinguished Alzheimer's disease patients from cognitively normal controls with a sensitivity of 88% and a specificity of 82% [132]. Another multiplex immunoassay profiling study of plasma from cognitively healthy, mild cognitive impairment and Alzheimer's disease individuals reported the identification of ten analytes that were correlated with disease severity and progression [133]. In these two studies, only one protein, apolipoprotein CIII, was found in common. This highlights a potential problem in reproducibility of such studies across different laboratories.

A number of other factors with known roles in brain functions have also been identified in serum or plasma in association with either brain pathologies or risk of developing Alzheimer's disease, including thyroid hormone [134], neurofilament light chain [135], acylated ghrelin [136], sphingolipids [26], vitamin D [137] and microRNAs [138–140]. MicroRNAs are short noncoding RNAs that are involved in regulation of posttranscriptional gene expression throughout the body, including the brain [141]. It is therefore logical to conclude that they are involved in regulation of brain functions such as cognition and learning. A number of other studies have shown that circulating cytokines may also be implicated in Alzheimer's disease progression [142, 143]. Multiplex assays for cytokines have been crucial in screening for peripheral inflammatory processes that may be involved in Alzheimer's disease through investigations of factors such as interleukin (IL)-1, IL-6, IL-7, IL-8, serum amyloid A, clusterin, CCL15, CXCL9 and tumour necrosis factor receptor 1 [144–148]. As mentioned above, it is important that standardization across laboratories occurs to maximize cross comparison of findings. It is also imperative that all biomarker candidates are carefully scrutinized and validated prior to translation into clinical studies.

15.8 Future Perspectives

15.8.1 Advances in Imaging

New ways of detecting Alzheimer's disease risk years before full manifestation of the disease are constantly emerging. These include advances in PET and MRI for visualizing plaque deposition and neurofibrillary tangles, which are known to precede the first cognitive deficits. Recently, evidence has documented pathological processes in the retinas of patients suffering from mild cognitive impairment, Alzheimer's disease and other neurodegenerative disorders [149]. For example, optic nerve and retinal ganglion cell degeneration, along with abnormal electroretinography responses, has been identified in patients with mild cognitive impairment and Alzheimer's disease [150]. Such changes can be detected by optical coherence tomography, which provides high-resolution two-dimensional cross-sectional imaging and three-dimensional volumetric measurements of the retina in vivo (Fig. 15.2). A meta-analysis published in 2015 found that optical coherence tomography can be used to detect losses in the peripapillary retinal nerve fibre layer in mild cognitive impairment and Alzheimer's disease patients [151]. Furthermore, a proof-of-principle study demonstrated that Aβ plaques can also be detected by optical coherence tomography retinal imaging [152]. Taken together, this may offer a non-invasive, high-resolution imaging means of detecting the disease years before symptoms are apparent and neurodegeneration occurs, theoretically allowing for more effective intervention.

15.8.2 Lab-on-a-Chip

Given the high prevalence and burden of neurodegenerative disorders such as Alzheimer's disease, there is considerable space in the market for novel and more effective diagnostic and treatment approaches. To meet these needs, antibody-based biomarker tests have now been developed on handheld devices which are approximately the size of a credit card [153, 154]. Such devices have been termed lab-on-a-chip platforms and offer the possibility of inexpensive, on-site, user-friendly

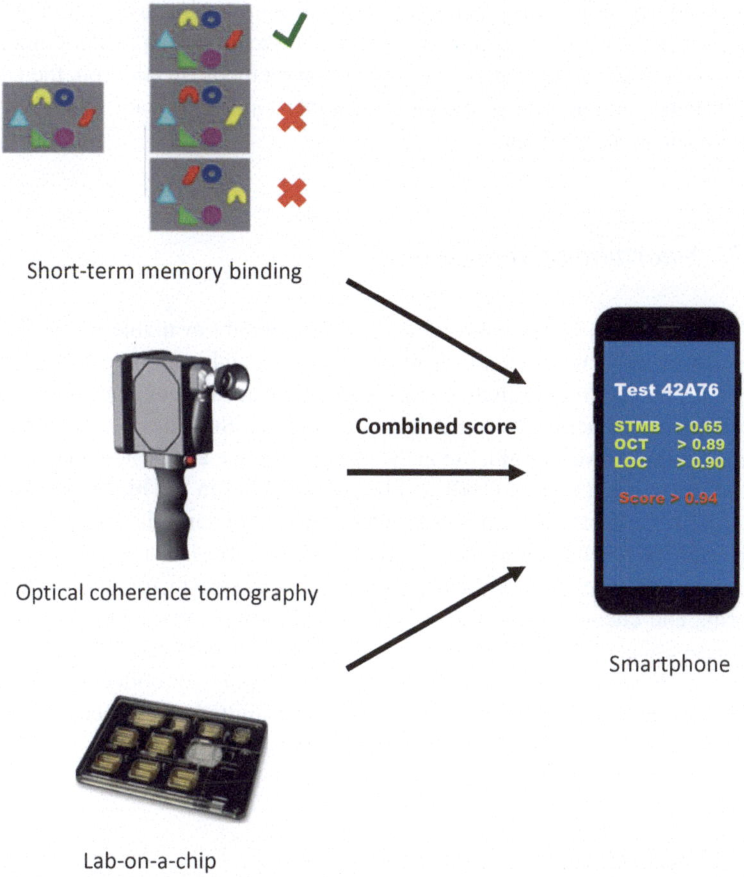

Fig. 15.2 New technologies which could allow early detection of Alzheimer's disease years before overt symptoms occur. Signals from multiple technologies could be amalgamated as a single biomarker score. The example shows a combination of cognition testing (STMB), visualization of amyloid deposits on the retina (OCT) and monitoring the presence of predictive biomarkers in the blood (LOC). Identification of high-risk individuals at an early disease stage would enable earlier treatment for better patient outcomes. The same approach could also be used to monitor treatment efficacy

analyses. Furthermore, results can be delivered in less than 20 min. The lab-on-a-chip system is modular which allows development of various assay formats. Immunoassay-based microcards have so far been developed for detection of protein markers such as C-reactive protein [155] and small molecules like vitamin D [156], using either electrochemical or optical read-outs. These tests have several advantages over existing methods, including their user friendliness, no required expertise, rapid throughput and relatively low cost. Testing involves addition of a blood sample to the card, and this card is inserted into a small reader to yield a biomarker score in the time span of a visit to the doctor's office (Fig. 15.2). Combined with additional information from other tests, this will help healthcare professionals to make informed decisions regarding therapeutic intervention at an earlier point in the disease course than can be achieved using current methods. In addition, the lab-on-a-chip system contains a port, enabling transmission of the data to a computer, smartphone or wearable device.

15.8.3 Smartphone Applications

While various biomarker-monitoring devices are already available on the market, the benefits of these will increase from their use in combination with smartphone apps [157]. This facilitates timely results without the need for expensive equipment and promotes awareness and responsibility in patients. Such devices are already in use for monitoring levels of glucose in the blood and urine and protein in urine [158, 159] and coagulation factors [160] and hormones [161] in the blood. Smartphone apps can also be useful in setting reminders for Alzheimer's disease patients and show benefits as a behavioural modification tool [162, 163]. The nearly ubiquitous nature of smartphones makes them a useful delivery system for interventions by monitoring and encouraging positive behavioural changes that could help to mitigate disease effects [164]. Furthermore, the readouts of biomarker tests have been linked to smartphone-based readouts, using camera optics for collection and transmission of the data [165–167], and these can be transmitted to a database for analysis with the results sent back to the user in real time.

15.9 Conclusions

A number of therapeutic strategies have been tested for treatment of Alzheimer's disease, although most of these have shown little or no effect on disease progression in advanced patients. However, it is anticipated that treatment of patients earlier in the disease course may slow disease progression. This will require validated biomarker tests for early detection and accurate diagnosis during the preclinical stages of the disease. As Alzheimer's disease arises from heterogeneous aetiologies, more accurate and sensitive tests should be developed that include a combination of imaging, blood tests and clinical cognitive assessment. For example, biomarkers from various sources, such as optical coherence tomography, lab-on-a-chip testing and clinical assessment, can be combined from each patient using a complex algorithm with smartphone readouts to achieve the highest predictive values [2]. In 2011, O'Bryant and co-workers developed an algorithm composed of a blood profile of 30 proteins and demographic data that could distinguish patients from controls with greater accuracy than could be achieved using the protein signature alone [132]. Further similar studies are warranted, as earlier detection can lead to earlier and therefore more effective treatment options to help improve the lives of individuals affected by this debilitating disease.

References

1. Kalaria RN, Maestre GE, Arizaga R, Friedland RP, Galasko D, Hall K et al (2008) Alzheimer's disease and vascular dementia in developing countries: prevalence, management, and risk factors. Lancet Neurol 7:812–826
2. Prince M, Bryce R, Albanese E, Wimo A, Ribeiro W, Ferri CP (2013) The global prevalence of dementia: a systematic review and metaanalysis. Alzheimers Dement 9(1):63–75 e62. https://doi.org/10.1016/j.jalz.2012.11.007
3. https://www.alz.co.uk/research/worldalzheimerreport2016sheet.pdf
4. Perneczky R (ed) (2018) Biomarkers for preclinical Alzheimer's disease (Neuromethods), 1st edn. Humana Press, Clifton. ISBN-10: 1493976737
5. Kazim SF, Iqbal K (2016) Neurotrophic factor small-molecule mimetics mediated neuroregeneration and synaptic repair: emerging therapeutic modality for Alzheimer's disease. Mol Neurodegener 11(1):50. https://doi.org/10.1186/s13024-016-0119-y
6. Mu Y, Gage FH (2011) Adult hippocampal neurogenesis and its role in Alzheimer's disease. Mol Neurodegener 6:85. https://doi.org/10.1186/1750-1326-6-85
7. Battaglia F, Wang HY, Ghilardi MF, Gashi E, Quartarone A, Friedman E et al (2007) Cortical plasticity in Alzheimer's disease in humans and rodents. Biol Psychiatry 62:1405–1412
8. Davies P, Maloney AJ (1976) Selective loss of central cholinergic neurons in Alzheimer's disease. Lancet 2(8000):1403
9. Hung SY, Fu WM (2017) Drug candidates in clinical trials for Alzheimer's disease. J Biomed Sci 24(1):47. https://doi.org/10.1186/s12929-017-0355-7
10. Larson EB, Shadlen MF, Wang L, McCormick WC, Bowen JD, Teri L et al (2004) Survival after initial diagnosis of Alzheimer disease. Ann Intern Med 140(7):501–509
11. Guest PC (2017) Biomarkers and mental illness: it's not all in the mind, 1st edn. Copernicus, Göttingen. ISBN-10: 3319460870
12. Barron AM, Pike CJ (2012) Sex hormones, aging, and Alzheimer's disease. Front Biosci (Elite Ed) 4:976–997
13. McGeer PL, McGeer EG (2001) Polymorphisms in inflammatory genes and the risk of Alzheimer disease. Arch Neurol 58(11):1790–1792
14. van Exel E, Eikelenboom P, Comijs H, Frölich M, Smit JH, Stek ML et al (2009) Vascular factors and markers of inflammation in offspring with a parental history of late-onset Alzheimer disease. Arch Gen Psychiatry 66(11):1263–1270
15. Moustafa AA, Hassan M, Hewedi DH, Hewedi I, Garami JK, Al Ashwal H et al (2018) Genetic underpinnings in Alzheimer's disease—a review. Rev Neurosci 29(1):21–38
16. Ford E, Greenslade N, Paudyal P, Bremner S, Smith HE, Banerjee S et al (2018) Predicting dementia from primary care records: a systematic review and meta-analysis. PLoS One 13(3):e0194735. https://doi.org/10.1371/journal.pone.0194735
17. Friedman DB, Becofsky K, Anderson LA, Bryant LL, Hunter RH, Ivey SL et al (2015) Public perceptions about risk and protective factors for cognitive health and impairment: a review of the literature. Int Psychogeriatr 27(8):1263–1275
18. De-Paula VJ, Radanovic M, Diniz BS, Forlenza OV (2012) Alzheimer's disease. Subcell Biochem 65:329–352
19. Huse JT, Doms RW (2000) Closing in on the amyloid cascade: recent insights into the cell biology of Alzheimer's disease. Mol Neurobiol 22(1–3):81–98
20. Rentz DM, Parra Rodriguez MA, Amariglio R, Stern Y, Sperling R, Ferris S (2013) Promising developments in neuropsychological approaches for the detection of preclinical Alzheimer's disease: a selective review. Alzheimers Res Ther 5:58. https://doi.org/10.1186/alzrt222
21. Rentz DM, Amariglio RE, Becker JA, Frey M, Olson LE, Frishe K et al (2011) Face-name associative memory performance is related to amyloid burden in normal elderly. Neuropsychologia 49(9):2776–2783
22. Cecchini MA, Yassuda MS, Bahia VS, de Souza LC, Guimarães HC, Caramelli P et al (2017) Recalling feature bindings differentiates Alzheimer's disease from frontotemporal dementia. J Neurol 264(10):2162–2169

23. Holden HM, Hoebel C, Loftis K, Gilbert PE (2012) Spatial pattern separation in cognitively normal young and older adults. Hippocampus 22(9):1826–1832
24. Counts SE, Ikonomovic MD, Mercado N, Vega IE, Mufson EJ (2017) Biomarkers for the early detection and progression of Alzheimer's disease. Neurotherapeutics 14(1):35–53
25. Villemagne VL, Rowe CC, Barnham KJ, Cherny R, Woodward M, Bozinosvski S et al (2017) A randomized, exploratory molecular imaging study targeting amyloid β with a novel 8-OH quinoline in Alzheimer's disease: the PBT2-204 IMAGINE study. Alzheimers Dement (N Y) 3(4):622–635
26. Varma VR, Oommen AM, Varma S, Casanova R, An Y, Andrews RM et al (2018) Brain and blood metabolite signatures of pathology and progression in Alzheimer disease: a targeted metabolomics study. PLoS Med 15(1):e1002482. https://doi.org/10.1371/journal.pmed.1002482
27. Nabers A, Perna L, Lange J, Mons U, Schartner J, Güldenhaupt J et al (2018) Amyloid blood biomarker detects Alzheimer's disease. EMBO Mol Med. 10(5). https://doi.org/10.15252/emmm.201708763
28. Herman L, Atri A, Salloway S (2017) Alzheimer's disease in primary care: the significance of early detection, diagnosis, and intervention. Am J Med 130(6):756. https://doi.org/10.1016/j.amjmed.2017.04.001
29. Wattmo C, Wallin ÅK (2017) Early- versus late-onset Alzheimer's disease in clinical practice: cognitive and global outcomes over 3 years. Alzheimers Res Ther 9(1):70. https://doi.org/10.1186/s13195-017-0294-2
30. http://www.who.int/medicines/areas/priority_medicines/BP6_11Alzheimer.pdf
31. Manning FC (1994) Tacrine therapy for the dementia of Alzheimer's disease. Am Fam Physician 50(4):819–826
32. Gauthier S, Panisset M, Nalbantoglu J, Poirier J (1997) Alzheimer's disease: current knowledge, management and research. CMAJ 157(8):1047–1052
33. Nordberg A, Svensson AL (1998) Cholinesterase inhibitors in the treatment of Alzheimer's disease: a comparison of tolerability and pharmacology. Drug Saf 19(6):465–480
34. Grutzendler J, Morris JC (2001) Cholinesterase inhibitors for Alzheimer's disease. Drugs 61(1):41–52
35. Doraiswamy PM (2002) Non-cholinergic strategies for treating and preventing Alzheimer's disease. CNS Drugs 16(12):811–824
36. Mezeiova E, Korabecny J, Sepsova V, Hrabinova M, Jost P, Muckova L et al (2017) Development of 2-methoxyhuprine as novel lead for Alzheimer's disease therapy. Molecules 22(8). https://doi.org/10.3390/molecules22081265
37. Oset-Gasque MJ, Marco-Contelles J (2017) New tacrines as anti-Alzheimer's disease agents. The (Benzo)Chromeno-PyranoTacrines. Curr Top Med Chem 17(31):3349–3360
38. Doggrell S (2003) Is memantine a breakthrough in the treatment of moderate-to-severe Alzheimer's disease? Expert Opin Pharmacother 4(10):1857–1860
39. Modrego PJ (2010) Depression in Alzheimer's disease. Pathophysiology, diagnosis, and treatment. J Alzheimers Dis 21(4):1077–1087
40. El Haj M, Gallouj K, Antoine P (2017) Google calendar enhances prospective memory in Alzheimer's disease: a case report. J Alzheimers Dis 57(1):285–291
41. Brown EL, Ruggiano N, Li J, Clarke PJ, Kay ES, Hristidis V (2017) Smartphone-based health technologies for dementia care: opportunities, challenges, and current practices. J Appl Gerontol 1:733464817723088. https://doi.org/10.1177/0733464817723088
42. Brown BM, Peiffer JJ, Sohrabi HR, Mondal A, Gupta VB, Rainey-Smith SR et al (2012) Intense physical activity is associated with cognitive performance in the elderly. Transl Psychiatry 2:e191. https://doi.org/10.1038/tp.2012.118
43. Huntley JD, Gould RL, Liu K, Smith M, Howard RJ (2015) Do cognitive interventions improve general cognition in dementia? A meta-analysis and meta-regression. BMJ Open 5(4):e005247. https://doi.org/10.1136/bmjopen-2014-005247
44. Jhee S, Shiovitz T, Crawford AW, Cutler NR (2001) Beta-amyloid therapies in Alzheimer's disease. Expert Opin Investig Drugs 10(4):593–605

45. Pollack SJ, Lewis H (2005) Secretase inhibitors for Alzheimer's disease: challenges of a promiscuous protease. Curr Opin Investig Drugs 6(1):35–47
46. Chen GF, Xu TH, Yan Y, Zhou YR, Jiang Y, Melcher K (2017) Amyloid beta: structure, biology and structure-based therapeutic development. Acta Pharmacol Sin 38(9):1205–1235
47. Nisha CM, Kumar A, Nair P, Gupta N, Silakari C, Tripathi T et al (2016) Molecular docking and in silico ADMET study reveals acylguanidine 7a as a potential inhibitor of β-secretase. Adv Bioinforma 2016:9258578. https://doi.org/10.1155/2016/9258578
48. Galimberti D, Scarpini E (2017) Pioglitazone for the treatment of Alzheimer's disease. Expert Opin Investig Drugs 26(1):97–101
49. Correia SC, Santos RX, Perry G, Zhu X, Moreira PI, Smith MA (2011) Insulin-resistant brain state: the culprit in sporadic Alzheimer's disease? Ageing Res Rev 10(2):264–273
50. Sato T, Hanyu H, Hirao K, Kanetaka H, Sakurai H, Iwamoto T (2011) Efficacy of PPAR-γ agonist pioglitazone in mild Alzheimer disease. Neurobiol Aging 32(9):1626–1633
51. Cheng H, Shang Y, Jiang L, Shi TL, Wang L (2016) The peroxisome proliferators activated receptor-gamma agonists as therapeutics for the treatment of Alzheimer's disease and mild-to-moderate Alzheimer's disease: a meta-analysis. Int J Neurosci 126(4):299–307
52. Wisniewski T, Drummond E (2016) Developing therapeutic vaccines against Alzheimer's disease. Expert Rev Vaccines 15(3):401–415
53. Grüninger F (2015) Invited review: drug development for tauopathies. Neuropathol Appl Neurobiol 41(1):81–96
54. Godyń J, Jończyk J, Panek D, Malawska B (2016) Therapeutic strategies for Alzheimer's disease in clinical trials. Pharmacol Rep 68(1):127–138
55. Novak P, Schmidt R, Kontsekova E, Zilka N, Kovacech B, Skrabana R et al (2017) Safety and immunogenicity of the tau vaccine AADvac1 in patients with Alzheimer's disease: a randomised, double-blind, placebo-controlled, phase 1 trial. Lancet Neurol 16(2):123–134
56. Wang C, Shou Y, Pan J, Du Y, Liu C, Wang H (2018) The relationship between cholesterol level and Alzheimer's disease-associated APP proteolysis/Aβ metabolism. Nutr Neurosci 11:1–11. https://doi.org/10.1080/1028415X.2017.1416942
57. Schultz BG, Patten DK, Berlau DJ (2018) The role of statins in both cognitive impairment and protection against dementia: a tale of two mechanisms. Transl Neurodegener 7:5. https://doi.org/10.1186/s40035-018-0110-3
58. Li HH, Lin CL, Huang CN (2018) Neuroprotective effects of statins against amyloid β-induced neurotoxicity. Neural Regen Res 13(2):198–206
59. Chu CS, Tseng PT, Stubbs B, Chen TY, Tang CH, Li DJ et al (2018) Use of statins and the risk of dementia and mild cognitive impairment: a systematic review and meta-analysis. Sci Rep 8(1):5804. https://doi.org/10.1038/s41598-018-24248-8
60. Münch G, Schinzel R, Loske C, Wong A, Durany N, Li JJ et al (1998) Alzheimer's disease—synergistic effects of glucose deficit, oxidative stress and advanced glycation endproducts. J Neural Transm 105:439–461
61. Wong A, Luth HJ, Deuther-Conrad W, Dukic-Stefanovic S, Gasic-Milenkovic J, Arendt T et al (2001) Advanced glycation endproducts co-localize with inducible nitric oxide synthase in Alzheimer's disease. Brain Res 920:32–40
62. Venigalla M, Sonego S, Gyengesi E, Sharman MJ, Münch G (2016) Novel promising therapeutics against chronic neuroinflammation and neurodegeneration in Alzheimer's disease. Neurochem Int 95:63–74
63. Figueiredo-Pereira ME, Corwin C, Babich J (2016) Prostaglandin J2: a potential target for halting inflammation-induced neurodegeneration. Ann N Y Acad Sci 1363:125–137
64. Camargo CHF, Justus FF, Retzlaff G, Blood MRY, Schafranski MD (2015) Action of anti-TNF-α drugs on the progression of Alzheimer's disease: a case report. Dement Neuropsychol 9(2):196–200

65. Butchart J, Brook L, Hopkins V, Teeling J, Püntener U, Culliford D et al (2015) Etanercept in Alzheimer disease: a randomized, placebo-controlled, double-blind, phase 2 trial. Neurology 84(21):2161–2168
66. Mazzanti G, Di Giacomo S (2016) Curcumin and resveratrol in the management of cognitive disorders: what is the clinical evidence?. Molecules 21(9). https://doi.org/10.3390/molecules21091243
67. Barbara R, Belletti D, Pederzoli F, Masoni M, Keller J, Ballestrazzi A et al (2017) Novel Curcumin loaded nanoparticles engineered for Blood-Brain Barrier crossing and able to disrupt Abeta aggregates. Int J Pharm 526(1–2):413–424
68. Liu QP, Wu YF, Cheng HY, Xia T, Ding H, Wang H et al (2016) Habitual coffee consumption and risk of cognitive decline/dementia: a systematic review and meta-analysis of prospective cohort studies. Nutrition 32:628–636
69. Wu L, Sun D, He Y (2017) Coffee intake and the incident risk of cognitive disorders: a dose-response meta-analysis of nine prospective cohort studies. Clin Nutr 36(3):730–736
70. Camandola S, Plick N, Mattson MP (2018) Impact of coffee and cacao purine metabolites on neuroplasticity and neurodegenerative disease. Neurochem Res. https://doi.org/10.1007/s11064-018-2492-0
71. Thaipisuttikul P, Galvin JE (2012) Use of medical foods and nutritional approaches in the treatment of Alzheimer's disease. Clin Pract (Lond) 9:199–209
72. Farina N, Rusted J, Tabet N (2014) The effect of exercise interventions on cognitive outcome in Alzheimer's disease: a systematic review. Int Psychogeriatr 26:9–18
73. Bertram S, Brixius K, Brinkmann C (2016) Exercise for the diabetic brain: how physical training may help prevent dementia and Alzheimer's disease in T2DM patients. Endocrine 53(2):350–363
74. Shankle WR, Hara J, Barrentine LW, Curole MV (2016) CerefolinNAC therapy of hyperhomocysteinemia delays cortical and white matter atrophy in Alzheimer's disease and cerebrovascular disease. J Alzheimers Dis 54(3):1073–1084
75. Ohnuma T, Toda A, Kimoto A, Takebayashi Y, Higashiyama R, Tagata Y et al (2016) Benefits of use, and tolerance of, medium-chain triglyceride medical food in the management of Japanese patients with Alzheimer's disease: a prospective, open-label pilot study. Clin Interv Aging 11:29–36
76. Berti V, Walters M, Sterling J, Quinn CG, Logue M, Andrews R et al (2018) Mediterranean diet and 3-year Alzheimer brain biomarker changes in middle-aged adults. Neurology 90(20):e1789–e1798. https://doi.org/10.1212/WNL.0000000000005527
77. Pedersen BK (2017) Anti-inflammatory effects of exercise: role in diabetes and cardiovascular disease. Eur J Clin Investig 47(8):600–611
78. Zanuso S, Sacchetti M, Sundberg CJ, Orlando G, Benvenuti P, Balducci S (2017) Exercise in type 2 diabetes: genetic, metabolic and neuromuscular adaptations. A review of the evidence. Br J Sports Med 51(21):1533–1538
79. Rendeiro C, Rhodes JS (2018) A new perspective of the hippocampus in the origin of exercise-brain interactions. Brain Struct Funct 223(6):2527–2545. https://doi.org/10.1007/s00429-018-1665-6
80. Devenney KE, Sanders ML, Lawlor B, Olde Rikkert MGM, Schneider S, NeuroExercise Study Group (2017) The effects of an extensive exercise programme on the progression of Mild Cognitive Impairment (MCI): study protocol for a randomised controlled trial. BMC Geriatr 17(1):75. https://doi.org/10.1186/s12877-017-0457-9
81. Karssemeijer EG, Bossers WJ, Aaronson JA, Kessels RP, Olde Rikkert MG (2017) The effect of an interactive cycling training on cognitive functioning in older adults with mild dementia: study protocol for a randomized controlled trial. BMC Geriatr 17(1):73. https://doi.org/10.1186/s12877-017-0464-x
82. Jensen CS, Portelius E, Høgh P, Wermuth L, Blennow K, Zetterberg H et al (2017) Effect of physical exercise on markers of neuronal dysfunction in cerebrospinal fluid in patients with Alzheimer's disease. Alzheimers Dement (N Y) 3(2):284–290

83. Barreto PS, Demougeot L, Vellas B, Rolland Y (2017) Exercise training for preventing dementia, mild cognitive impairment, and clinically meaningful cognitive decline: a systematic review and meta-analysis. J Gerontol A Biol Sci Med Sci 73(11):1504–1511. https://doi.org/10.1093/gerona/glx234
84. Cavedo E, Lista S, Khachaturian Z, Aisen P, Amouyel P, Herholz K et al (2014) The road ahead to cure Alzheimer's disease: development of biological markers and neuroimaging methods for prevention trials across all stages and target populations. J Prev Alzheimers Dis 1(3):181–202
85. Baldeiras I, Santana I, Leitão MJ, Gens H, Pascoal R, Tábuas-Pereira M et al (2018) Addition of the Aβ42/40 ratio to the cerebrospinal fluid biomarker profile increases the predictive value for underlying Alzheimer's disease dementia in mild cognitive impairment. Alzheimers Res Ther 10(1):33. https://doi.org/10.1186/s13195-018-0362-2
86. Tang X, Cai F, Ding DX, Zhang LL, Cai XY, Fang Q (2018) Magnetic resonance imaging relaxation time in Alzheimer's disease. Brain Res Bull 140:176–189. https://doi.org/10.1016/j.brainresbull.2018.05.004
87. Matsuda H (2016) MRI morphometry in Alzheimer's disease. Ageing Res Rev 30:17–24
88. Tae WS, Ham BJ, Pyun SB, Kang SH, Kim BJ (2018) Current clinical applications of diffusion-tensor imaging in neurological disorders. J Clin Neurol 14(2):129–140
89. Hua J, Liu P, Kim T, Donahue M, Rane S, Chen JJ et al (2018) MRI techniques to measure arterial and venous cerebral blood volume. Neuroimage. pii: S1053-8119(18)30115-0. https://doi.org/10.1016/j.neuroimage.2018.02.027
90. Schilling LP, Zimmer ER, Shin M, Leuzy A, Pascoal TA, Benedet AL et al (2016) Imaging Alzheimer's disease pathophysiology with PET. Dement Neuropsychol 10(2):79–90
91. Klunk WE, Engler H, Nordberg A, Wang Y, Blomqvist G, Holt DP et al (2004) Imaging brain amyloid in Alzheimer's disease with Pittsburgh compound-B. Ann Neurol 55(3):306–319
92. Vandenberghe R, Van Laere K, Ivanoiu A, Salmon E, Bastin C, Triau E et al (2010) 18F-flutemetamol amyloid imaging in Alzheimer disease and mild cognitive impairment: a phase 2 trial. Ann Neurol 68(3):319–329
93. Okello A, Koivunen J, Edison P, Archer HA, Turkheimer FE, Någren K et al (2009) Conversion of amyloid positive and negative MCI to AD over 3 years: an 11C-PIB PET study. Neurology 73(10):754–760
94. Rosenberg PB, Wong DF, Edell SL, Ross JS, Joshi AD, Brašić JR et al (2013) Cognition and amyloid load in Alzheimer disease imaged with florbetapir F 18(AV-45) positron emission tomography. Am J Geriatr Psychiatry 21(3):272–278
95. Fischer FU, Wolf D, Scheurich A, Fellgiebel A (2015) Alzheimer's disease neuroimaging I. Altered whole-brain white matter networks in preclinical Alzheimer's disease. Neuroimage Clin 8:660–666
96. Thal DR, Beach TG, Zanette M, Heurling K, Chakrabarty A, Ismail A et al (2015) [(18) F]flutemetamol amyloid positron emission tomography in preclinical and symptomatic Alzheimer's disease: specific detection of advanced phases of amyloid-beta pathology. Alzheimers Dement 11(8):975–985
97. Chiaravalloti A, Castellano AE, Ricci M, Barbagallo G, Sannino P, Ursini F et al (2018) Coupled imaging with [^{18}F]FBB and [^{18}F]FDG in AD subjects show a selective association between amyloid burden and cortical dysfunction in the brain. Mol Imaging Biol 20(4):659–666
98. Akiyama H, Barger S, Barnum S, Bradt B, Bauer J, Cole GM et al (2000) Inflammation and Alzheimer's disease. Neurol Aging 21(3):383–421
99. Cagnin A, Brooks DJ, Kennedy AM, Gunn RN, Myers R, Turkheimer FE et al (2001) In-vivo measurement of activated microglia in dementia. Lancet 358(9280):461–457
100. Carter SF, Scholl M, Almkvist O, Wall A, Engler H, Långström B et al (2012) Evidence for astrocytosis in prodromal Alzheimer disease provided by 11C-deuterium-L-deprenyl: a multitracer PET paradigm combining 11C-Pittsburgh compound B and 18F-FDG. J Nuclear Med 53(1):37–46

101. Rodriguez-Vieitez E, Nordberg A (2018) Imaging neuroinflammation: quantification of astrocytosis in a multitracer PET approach. Methods Mol Biol 1750:231–251
102. Sebastián-Serrano Á, de Diego-García L, Díaz-Hernández M (2018) The neurotoxic role of extracellular Tau protein. Int J Mol Sci 19(4). https://doi.org/10.3390/ijms19040998
103. Zimmer ER, Leuzy A, Gauthier S, Rosa-Neto P (2014) Developments in Tau PET imaging. Can J Neurol Sci 41(5):547–553
104. Shimada H, Kitamura S, Shinotoh H, Endo H, Niwa F, Hirano S et al (2016) Association between Aβ and tau accumulations and their influence on clinical features in aging and Alzheimer's disease spectrum brains: a [¹¹C]PBB3-PET study. Alzheimers Dement (Amst) 6:11–20
105. Ishiki A, Okamura N, Furukawa K, Furumoto S, Harada R, Tomita N et al (2015) Longitudinal assessment of Tau pathology in patients with Alzheimer's disease using [18F]THK-5117 positron emission tomography. PLoS One 10(10):e0140311. https://doi.org/10.1371/journal.pone.0140311
106. Schaffer C, Sarad N, DeCrumpe A, Goswami D, Herrmann S, Morales J et al (2015) Biomarkers in the diagnosis and prognosis of Alzheimer's disease. J Lab Autom 20(5):589–600
107. Minoshima S, Giordani B, Berent S, Frey KA, Foster NL, Kuhl DE (1997) Metabolic reduction in the posterior cingulate cortex in very early Alzheimer's disease. Ann Neurol 42(1):85–94
108. Silverman DH, Small GW, Chang CY, Lu CS, Kung De Aburto MA et al (2001) Positron emission tomography in evaluation of dementia: regional brain metabolism and longterm outcome. JAMA 286(17):2120–2127
109. Mosconi L (2005) Brain glucose metabolism in the early and specific diagnosis of Alzheimer's disease. FDG-PET studies in MCI and AD. Eur J Nucl Med Mol Imaging 32(4):486–510
110. Scholl M, Damian A, Engler H (2014) Fluorodeoxyglucose PET in neurology and psychiatry. PET Clin 9(4):371–390
111. Mosconi L, Mistur R, Switalski R, Tsui WH, Glodzik L, Li Y et al (2009) FDG-PET changes in brain glucose metabolism from normal cognition to pathologically verified Alzheimer's disease. Eur J Nucl Med Mol Imaging 36(5):811–822
112. Hansson O, Zetterberg H, Buchhave P, Londos E, Blennow K, Minthon L (2006) Association between CSF biomarkers and incipient Alzheimer's disease in patients with mild cognitive impairment: a follow-up study. Lancet Neurol 5:228–234
113. Hansson O, Zetterberg H, Buchhave P, Andreasson U, Londos E, Minthon L et al (2007) Prediction of Alzheimer's disease using the CSF Abeta42/Abeta40 ratio in patients with mild cognitive impairment. Dement Geriatr Cogn Disord 23:316–320
114. Wiltfang J, Esselmann H, Bibl M, Hull M, Hampel H, Kessler H et al (2007) Amyloid beta peptide ratio 42/40 but not A beta 42 correlates with phospho-Tau in patients with low- and high-CSF A beta 40 load. J Neurochem 101:1053–1059
115. Lewczuk P, Matzen A, Blennow K, Parnetti L, Molinuevo JL, Eusebi P et al (2017) Cerebrospinal fluid Abeta42/40 corresponds better than Abeta42 to amyloid PET in Alzheimer's disease. J Alzheimers Dis 55:813–822
116. Dumurgier J, Schraen S, Gabelle A, Vercruysse O, Bombois S, Laplanche JL et al (2015) Cerebrospinal fluid amyloid-β 42/40 ratio in clinical setting of memory centers: a multicentric study. Alzheimers Res Ther 7(1):30. https://doi.org/10.1186/s13195-015-0114-5
117. De Roeck EE, Engelborghs S, Dierckx E (2016) Next generation brain health depends on early Alzheimer disease diagnosis: from a timely diagnosis to future population screening. J Am Med Dir Assoc 17(5):452–453
118. Vos SJB, Verhey F, Frolich L, Kornhuber J, Wiltfang J, Maier W et al (2015) Prevalence and prognosis of Alzheimer's disease at the mild cognitive impairment stage. Brain 138:1327–1338
119. Bjerke M, Zetterberg H, Edman Å, Blennow K, Wallin A, Andreasson U (2011) Cerebrospinal fluid matrix metalloproteinases and tissue inhibitor of metalloproteinases in combination with subcortical and cortical biomarkers in vascular dementia and Alzheimer's disease. J Alzheimers Dis 27(3):665–676

120. Slaets S, Le Bastard N, Martin JJ, Sleegers K, Van Broeckhoven C, De Deyn PP et al (2013) Cerebrospinal fluid Aβ1-40 improves differential dementia diagnosis in patients with intermediate P-tau181P levels. J Alzheimers Dis 36(4):759–767
121. Llorens F, Schmitz M, Ferrer I, Zerr I (2016) CSF biomarkers in neurodegenerative and vascular dementias. Prog Neurobiol 138-140:36–53
122. Blennow K, Zetterberg H (2018) The past and the future of Alzheimer's disease fluid biomarkers. J Alzheimers Dis 62(3):1125–1140
123. Brinkmalm A, Brinkmalm G, Honer WG, Frolich L, Hausner L, Minthon L et al (2014) SNAP-25 is a promising novel cerebrospinal fluid biomarker for synapse degeneration in Alzheimer's disease. Mol Neurodegener 9:53. https://doi.org/10.1186/1750-1326-9-53
124. Ohrfelt A, Brinkmalm A, Dumurgier J, Brinkmalm G, Hansson O, Zetterberg H et al (2016) The pre-synaptic vesicle protein synaptotagmin is a novel biomarker for Alzheimer's disease. Alzheimers Res Ther 8:41. https://doi.org/10.1186/s13195-016-0208-8
125. Blennow K (2017) A review of fluid biomarkers for Alzheimer's disease: moving from CSF to blood. Neurol Ther 6(Suppl 1):15–24
126. Soares HD, Chen Y, Sabbagh M, Roher A, Schrijvers E, Breteler M (2009) Identifying early markers of Alzheimer's disease using quantitative multiplex proteomic immunoassay panels. Ann N Y Acad Sci 1180:56–67
127. Perneczky R, Guo LH (2016) Plasma proteomics biomarkers in Alzheimer's disease: latest advances and challenges. Methods Mol Biol 1303:521–529
128. Le Bastard N, Aerts L, Leurs J, Blomme W, De Deyn PP, Engelborghs S (2009) No correlation between time-linked plasma and CSF Abeta levels. Neurochem Int 55(8):820–825
129. Vogelgsang J, Shahpasand-Kroner H, Vogelgsang R, Streit F, Vukovich R, Wiltfang J (2018) Multiplex immunoassay measurement of amyloid-β₄₂ to amyloid-β₄₀ ratio in plasma discriminates between dementia due to Alzheimer's disease and dementia not due to Alzheimer's disease. Exp Brain Res 236(5):1241–1250
130. Mielke MM, Hagen CE, Xu J, Chai X, Vemuri P, Lowe VJ et al (2018) Plasma phospho-tau181 increases with Alzheimer's disease clinical severity and is associated with tau- and amyloid-positron emission tomography. Alzheimers Dement. pii: S1552-5260(18)30067-0. https://doi.org/10.1016/j.jalz.2018.02.013
131. Neergaard JS, Dragsbæk K, Christiansen C, Karsdal MA, Brix S, Henriksen K (2018) Two novel blood-based biomarker candidates measuring degradation of tau are associated with dementia: a prospective study. PLoS One 13(4):e0194802. https://doi.org/10.1371/journal.pone.0194802
132. O'Bryant SE, Xiao G, Barber R, Reisch J, Hall J, Cullum CM et al (2011) A blood-based algorithm for the detection of Alzheimer's disease. Dement Geriatr Cogn Disord 32(1):55–62
133. Hye A, Riddoch-Contreras J, Baird AL, Ashton NJ, Bazenet C, Leung R et al (2014) Plasma proteins predict conversion to dementia from prodromal disease. Alzheimers Dement 10(6):799–807
134. Choi HJ, Byun MS, Yi D, Sohn BK, Lee JH, Lee JY et al (2017) Associations of thyroid hormone serum levels with in-vivo Alzheimer's disease pathologies. Alzheimers Res Ther 9(1):64. https://doi.org/10.1186/s13195-017-0291-5
135. Weston PSJ, Poole T, Ryan NS, Nair A, Liang Y, Macpherson K et al (2017) Serum neurofilament light in familial Alzheimer disease: a marker of early neurodegeneration. Neurology 89(21):2167–2175
136. Cao X, Zhu M, He Y, Chu W, Du Y, Du H (2018) Increased serum acylated ghrelin levels in patients with mild cognitive impairment. J Alzheimers Dis 61(2):545–552
137. Ouma S, Suenaga M, Bölükbaşı Hatip FF, Hatip-Al-Khatib I, Tsuboi Y et al (2018) Serum vitamin D in patients with mild cognitive impairment and Alzheimer's disease. Brain Behav 8(3):e00936. https://doi.org/10.1002/brb3.936
138. Guo R, Fan G, Zhang J, Wu C, Du Y, Ye H et al (2017) A 9-microRNA signature in serum serves as a noninvasive biomarker in early diagnosis of Alzheimer's disease. J Alzheimers Dis 60(4):1365–1377

139. Wei H, Xu Y, Xu W, Zhou Q, Chen Q, Yang M et al (2018) Serum exosomal miR-223 serves as a potential diagnostic and prognostic biomarker for dementia. Neuroscience 379:167–176
140. Yang TT, Liu CG, Gao SC, Zhang Y, Wang PC (2018) The serum exosome derived MicroRNA-135a, -193b, and -384 were potential Alzheimer's disease biomarkers. Biomed Environ Sci 31(2):87–89
141. Denk J, Oberhauser F, Kornhuber J, Wiltfang J, Fassbender K, Schroeter ML et al (2018) Specific serum and CSF microRNA profiles distinguish sporadic behavioural variant of frontotemporal dementia compared with Alzheimer patients and cognitively healthy controls. PLoS One 13(5):e0197329. https://doi.org/10.1371/journal.pone.0197329
142. Wu Y, Xu J, Xu J, Cheng J, Jiao D, Zhou C et al (2017) Lower serum levels of miR-29c-3p and miR-19b-3p as biomarkers for Alzheimer's disease. Tohoku J Exp Med 242(2):129–136
143. Magalhães TNC, Weiler M, Teixeira CVL, Hayata T, Moraes AS, Boldrini VO et al (2017) Systemic inflammation and multimodal biomarkers in amnestic mild cognitive impairment and Alzheimer's disease. Mol Neurobiol 55(7):5689–5697. https://doi.org/10.1007/s12035-017-0795-9
144. Lai KSP, Liu CS, Rau A, Lanctôt KL, Köhler CA, Pakosh M et al (2017) Peripheral inflammatory markers in Alzheimer's disease: a systematic review and meta-analysis of 175 studies. J Neurol Neurosurg Psychiatry 88(10):876–882
145. Chen A, Oakley AE, Monteiro M, Tuomela K, Allan LM, Mukaetova-Ladinska EB et al (2016) Multiplex analyte assays to characterize different dementias: brain inflammatory cytokines in poststroke and other dementias. Neurobiol Aging 38:56–67. https://doi.org/10.1016/j.neurobiolaging.2015.10.021
146. Choi C, Jeong JH, Jang JS, Choi K, Lee J, Kwon J et al (2008) Multiplex analysis of cytokines in the serum and cerebrospinal fluid of patients with Alzheimer's disease by color-coded bead technology. J Clin Neurol 4(2):84–88. https://doi.org/10.3988/jcn.2008.4.2.84
147. Delaby C, Gabelle A, Blum D, Schraen-Maschke S, Moulinier A, Boulanghien J et al (2015) Central nervous system and peripheral inflammatory processes in Alzheimer's disease: biomarker profiling approach. Front Neurol 6:181. https://doi.org/10.3389/fneur.2015.00181
148. Hochstrasser T, Marksteiner J, Defrancesco M, Deisenhammer EA, Kemmler G, Humpel C (2011) Two blood monocytic biomarkers (CCL15 and p21) combined with the mini-mental state examination discriminate Alzheimer's disease patients from healthy subjects. Dement Geriatr Cogn Dis Extra 1(1):297–309
149. Doustar J, Torbati T, Black KL, Koronyo Y, Koronyo-Hamaoui M (2017) Optical coherence tomography in Alzheimer's disease and other neurodegenerative diseases. Front Neurol 8:701. https://doi.org/10.3389/fneur.2017.00701
150. Hart NJ, Koronyo Y, Black KL, Koronyo-Hamaoui M (2016) Ocular indicators of Alzheimer's: exploring disease in the retina. Acta Neuropathol 132:767–787
151. Coppola G, Di Renzo A, Ziccardi L, Martelli F, Fadda A, Manni G et al (2015) Optical coherence tomography in Alzheimer's disease: a meta-analysis. PLoS One 10:e0134750. https://doi.org/10.1371/journal.pone.0134750
152. Schumacher S, Nestler J, Otto T, Wegener M, Ehrentreich-Förster E, Michel D et al (2012) Highly-integrated lab-on-chip system for point-of-care multiparameter analysis. Lab Chip 12:464–473
153. Peter H, Wienke J, Bier FF (2017) Lab-on-a-Chip multiplex assays. Methods Mol Biol 1546:283–294
154. Peter H, Wienke J, Guest PC, Bistolas N, Bier FF (2017) Lab-on-a-Chip proteomic assays for psychiatric disorders. Adv Exp Med Biol 974:339–349
155. Peter H, Bistolas N, Schumacher S, Laurisch C, Guest PC, Höller U et al (2018) Lab-on-a-Chip device for rapid measurement of vitamin D levels. Methods Mol Biol 1735:477–486
156. Yetisen AK, Martinez-Hurtado JL, da Cruz VF, Simsekler MC, Akram MS, Lowe CR (2014) The regulation of mobile medical applications. Lab Chip 14:833–840
157. Martinez-Hurtado JL, Yetisen AK, Yun SH (2017) Multiplex smartphone diagnostics. Methods Mol Biol 1546:295–302

158. Matías-García PR, Martinez-Hurtado JL (2018) Kidney smartphone diagnostics. Methods Mol Biol 1735:487–498
159. Vegt J, Guest PC (2018) A user-friendly app for blood coagulation disorders. Methods Mol Biol 1735:499–504
160. Matías-García PR, Martinez-Hurtado JL, Beckley A, Schmidmayr M, Seifert-Klauss V (2018) Hormonal smartphone diagnostics. Methods Mol Biol 1735:505–515
161. Hartin PJ, Nugent CD, McClean SI, Cleland I, Tschanz JT, Clark CJ et al (2016) The empowering role of mobile apps in behavior change interventions: the gray matters randomized controlled trial. JMIR Mhealth Uhealth 4(3):e93. https://doi.org/10.2196/mhealth.4878
162. Bonn SE, Alexandrou C, Hjörleifsdottir Steiner K, Wiklander K, Östenson CG, Löf M (2018) App-technology to increase physical activity among patients with diabetes type 2—the DiaCert-study, a randomized controlled trial. BMC Public Health 18(1):119. https://doi.org/10.1186/s12889-018-5026-4
163. Krishna S, Boren SA, Balas EA (2009) Healthcare via cell phones: a systematic review. Telemed J E Health 15:231–240
164. Guo T, Patnaik R, Kuhlmann K, Rai AJ, Sia SK (2015) Smartphone dongle for simultaneous measurement of hemoglobin concentration and detection of HIV antibodies. Lab Chip 15:3514–3520
165. Barbosa AI, Gehlot P, Sidapra K, Edwards AD, Reis NM (2015) Portable smartphone quantitation of prostate specific antigen (PSA) in a fluoropolymer microfluidic device. Biosens Bioelectron 70:5–14
166. Liao SC, Peng J, Mauk MG, Awasthi S, Song J, Friedman H (2016) Smart cup: a minimally-instrumented, smartphone-based point-of-care molecular diagnostic device. Sens Actuators B Chem 229:232–238
167. Yeo SJ, Choi K, Cuc BT, Hong NN, Bao DT, Ngoc NM et al (2016) Smartphone-based fluorescent diagnostic system for highly pathogenic H5N1 viruses. Theranostics 6:231–242

Index

P. C. Guest (ed.), *Reviews on Biomarker Studies in Psychiatric and Neurodegenerative Disorders*, Advances in Experimental Medicine and Biology 1118,
https://doi.org/10.1007/978-3-030-05542-4

B

N

The manufacturer's authorised representative in the EU is Springer Nature Customer Service Centre GmbH, Europaplatz 3, 69115 Heidelberg, Germany. If you have any concerns regarding our products, please contact ProductSafety@springernature.com

Printed and bound by CPI Group (UK) Ltd, Croydon, CR0 4YY

30/09/2026

02225440-0010